P. M. Ajayan, L. S. Schadler, P. V. Braun

Nanocomposite Science and Technology

Related Titles from Wiley-VCH

Caruso, F.

Colloids and Colloid Assemblies

2003, ISBN 3-527-30660-9

Decher, G., Schlenoff, J.B.

Multilayer Thin Films Sequential Assembly of Nanocomposite Materials

2003, ISBN 3-527-30440-1

Gómez-Romero, P., Sanchez, C.

Functional Hybrid Materials

2003, ISBN 3-527-30484-3

Komiyama, M., Takeuchi, T., Mukawa, T., Asanuma, H.

Molecular Imprinting From Fundamentals to Applications

2003, ISBN 3-527-30569-6

Krenkel, W.

High Temperature Ceramic Matrix Composites

2001, ISBN 3-527-30320-0

Köhler, M., Fritzsche, W.

Nanotechnology

2004, ISBN 3-527-30750-8

P. M. Ajayan, L. S. Schadler, P. V. Braun

Nanocomposite Science and Technology

WILEY-VCH GmbH & Co. KGaA

Pulickel M. Ajayan
Dept. of Materials Science and Engineering
Rensselaer Polytechnic Institute
Troy, NY 12180-3590
USA

Linda S. Schadler
Dept. of Materials Science and Engineering
Rensselaer Polytechnic Institute
Troy, NY 12180-3590
USA

Paul V. Braun
Dept. of Materials Science and Engineering
University of Illinois at Urbana-Champaign
Urbana, IL 61801
USA

1st Reprint 2004

Library of Congress Card No.:
Applied for.

British Library Cataloguing-in-Publication Data:
A catalogue record for this book is available from the British Library.

Die Deutsche Bibliothek – CIP Cataloguing-in-Publication-Data
Bibliographic information published by Die Deutsche Bibliothek
Die Deutsche Bibliothek lists this publication in the Deutsche Nationalbibliografie; detailed bibliographic data is available in the Internet at http://dnb.ddb.de

Printed in the Federal Republic of Germany

Printed on acid-free paper

Composition Mitterweger & Partner, Plankstadt
Printing Strauss Offsetdruck GmbH, Mörlenbach
Bookbinding Litges & Dopf Buchbinderei GmbH, Heppenheim
Cover Design: Gunter Schulz, Fußgönheim

ISBN 3-527-30359-6

Contents

Preface

The field of nanocomposites involves the study of multiphase material where at least one of the constituent phases has one dimension less than 100 nm. The promise of nanocomposites lies in their multifunctionality, the possibility of realizing unique combinations of properties unachievable with traditional materials. The challenges in reaching this promise are tremendous. They include control over the distribution in size and dispersion of the nanosize constituents, tailoring and understanding the role of interfaces between structurally or chemically dissimilar phases on bulk properties. Large scale and controlled processing of many nanomaterials has yet to be achieved. Our mentor as we make progress down this road is mother Nature and her quintessential nanocomposite structures, for example, bone.

We realize that a book on a subject of such wide scope is a challenging endeavour. The recent explosion of research in this area introduces another practical limitation. What is written here should be read from the perspective of a dynamic and emerging field of science and technology. Rather than covering the entire spectrum of nanocomposite science and technology, we have picked three areas that provide the basic concepts and generic examples that define the overall nature of the field. In the first chapter we discuss nanocomposites based on inorganic materials and their applications. In the second chapter polymer based nanoparticle filled composites are detailed with an emphasis on interface engineering to obtain nanocomposites with optimum performance. The third chapter is about naturally occurring systems of nanocomposites and current steps towards naturally inspired synthetic nanocomposites. Finally a short chapter contributed by our colleagues highlights the possibility of using theoretical models and simulations for understanding nanocomposite properties. We hope our readers will find the book of value to further their research interests in this fascinating and fast evolving area of nanocomposites.

Troy, July 2003 *P. M. Ajayan, L. S. Schadler and P. V. Braun*

1 Bulk Metal and Ceramics Nanocomposites

Pulickel Ajayan

1.1 Introduction

The field of nanocomposite materials has had the attention, imagination, and close scrutiny of scientists and engineers in recent years. This scrutiny results from the simple premise that using building blocks with dimensions in the nanosize range makes it possible to design and create new materials with unprecedented flexibility and improvements in their physical properties. This ability to tailor composites by using nanosize building blocks of heterogeneous chemical species has been demonstrated in several interdisciplinary fields. The most convincing examples of such designs are naturally occurring structures such as bone, which is a hierarchical nanocomposite built from ceramic tablets and organic binders. Because the constituents of a nanocomposite have different structures and compositions and hence properties, they serve various functions. Thus, the materials built from them can be multifunctional. Taking some clues from nature and based on the demands that emerging technologies put on building new materials that can satisfy several functions at the same time for many applications, scientists have been devising synthetic strategies for producing nanocomposites. These strategies have clear advantages over those used to produce homogeneous large-grained materials. Behind the push for nanocomposites is the fact that they offer useful new properties compared to conventional materials.

The concept of enhancing properties and improving characteristics of materials through the creation of multiple-phase nanocomposites is not recent. The idea has been practiced ever since civilization started and humanity began producing more efficient materials for functional purposes. In addition to the large variety of nanocomposites found in nature and in living beings (such as bone), which is the focus of chapter 3 of this book, an excellent example of the use of synthetic nanocomposites in antiquity is the recent discovery of the constitution of Mayan paintings developed in the Mesoamericas. State-of-the-art characterization of these painting samples reveals that the structure of the paints consisted of a matrix of clay mixed with organic colorant (indigo) molecules. They also contained inclusions of metal nanoparticles encapsulated in an amorphous silicate substrate, with oxide nanoparticles on the substrate

[1]. The nanoparticles were formed during heat treatment from impurities (Fe, Mn, Cr) present in the raw materials such as clays, but their content and size influenced the optical properties of the final paint. The combination of intercalated clay forming a superlattice in conjunction with metallic and oxide nanoparticles supported on the amorphous substrate made this paint one of the earliest synthetic materials resembling modern functional nanocomposites.

Nanocomposites can be considered solid structures with nanometer-scale dimensional repeat distances between the different phases that constitute the structure. These materials typically consist of an inorganic (host) solid containing an organic component or vice versa. Or they can consist of two or more inorganic/organic phases in some combinatorial form with the constraint that at least one of the phases or features be in the nanosize. Extreme examples of nanocomposites can be porous media, colloids, gels, and copolymers. In this book, however, we focus on the core concept of nanocomposite materials, i.e., a combination of nano-dimensional phases with distinct differences in structure, chemistry, and properties. One could think of the nanostructured phases present in nanocomposites as zero-dimensional (e.g., embedded clusters), 1D (one-dimensional; e.g., nanotubes), 2D (nanoscale coatings), and 3D (embedded networks). In general, nanocomposite materials can demonstrate different mechanical, electrical, optical, electrochemical, catalytic, and structural properties than those of each individual component. The multifunctional behavior for any specific property of the material is often more than the sum of the individual components.

Both simple and complex approaches to creating nanocomposite structures exist. A practical dual-phase nanocomposite system, such as supported catalysts used in heterogeneous catalysis (metal nanoparticles placed on ceramic supports), can be prepared simply by evaporation of metal onto chosen substrates or dispersal through solvent chemistry. On the other hand, material such as bone, which has a complex hierarchical structure with coexisting ceramic and polymeric phases, is difficult to duplicate entirely by existing synthesis techniques. The methods used in the preparation of nanocomposites range from chemical means to vapor phase deposition.

Apart from the properties of individual components in a nanocomposite, *interfaces* play an important role in enhancing or limiting the overall properties of the system. Due to the high surface area of nanostructures, nanocomposites present many interfaces between the constituent intermixed phases. Special properties of nanocomposite materials often arise from interaction of its phases at the interfaces. An excellent example of this phenomenon is the mechanical behavior of nanotube-filled polymer composites. Although adding nanotubes could conceivably improve the strength of polymers (due to the superior mechanical properties of the nanotubes), a noninteracting interface serves only to create weak regions in the composite, resulting in no enhancement of its mechanical properties (detailed in chapter 2). In contrast to nanocomposite materials, the interfaces in conventional composites constitute a much smaller volume fraction of the bulk material.

In the following sections of this chapter, we describe some examples of metal/ceramic nanocomposite systems that have become subjects of intense study in recent years. The various physical properties that can be tailored in these systems for specific applications is also considered, along with different approaches to synthesizing these nanocomposites.

1.2
Ceramic/Metal Nanocomposites

Many efforts are under way to develop high-performance ceramics that have promise for engineering applications such as highly efficient gas turbines, aerospace materials, automobiles, etc. Even the best processed ceramic materials used in applications pose many unsolved problems; among them, relatively low fracture toughness and strength, degradation of mechanical properties at high temperatures, and poor resistance to creep, fatigue, and thermal shock. Attempts to solve these problems have involved incorporating second phases such as particulates, platelets, whiskers, and fibers in the micron-size range at the matrix grain boundaries. However, results have been generally disappointing when micron-size fillers are used to achieve these goals. Recently the concept of nanocomposites has been considered, which is based on passive control of the microstructures by incorporating nanometer-size second-phase dispersions into ceramic matrices [2]. The dispersions can be characterized as either

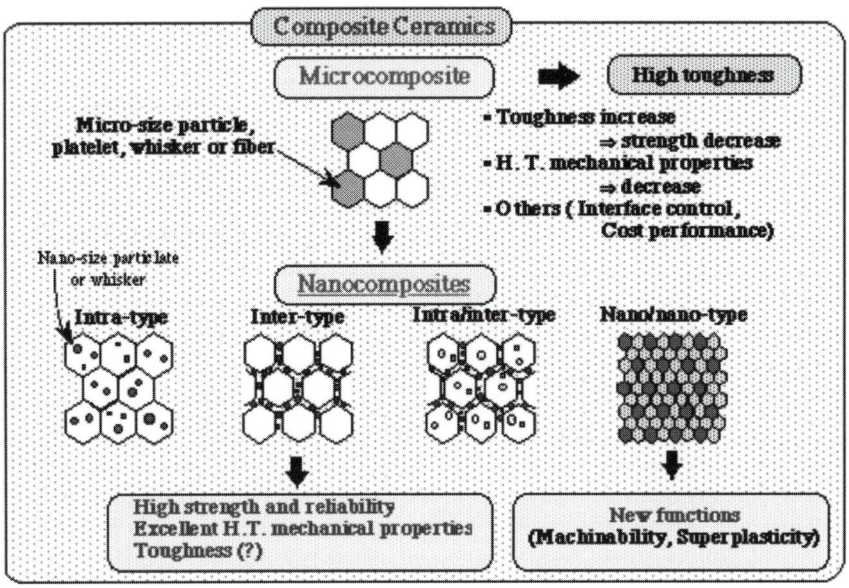

Fig. 1.1 New concept of ceramic metal nanocomposites with inter- and intra-granular designs: properties of ceramic materials can be improved by nanocomposite technology. This technique is based on passive control of the microstructures by incorporating nanometer-sized second dispersions into ceramic materials. This is a completely new method to fabricate materials with excellent mechanical properties (such as high strength and toughness), due to the desirable microstructure of ceramics (Source:[228] Reprinted with permission)

intragranular or intergranular (Figure 1.1). These materials can be produced by incorporating a very small amount of additive into a ceramic matrix. The additive segregates at the grain boundary with a gradient concentration or precipitates as molecular or cluster sized particles within the grains or at the grain boundaries. Optimized processing can lead to excellent structural control at the molecular level in most nanocomposite materials. Intragranular dispersions aim to generate and fix dislocations during the processing, annealing, cooling, and/or the in-situ control of size and shape of matrix grains. This role of dispersoids, especially on the nano scale, is important in oxide ceramics, some of which become ductile at high temperatures. The intergranular nanodispersoids must play important roles in control of the grain boundary structure of oxide (Al_2O_3, MgO) and nonoxide (Si_3N_4, SiC) ceramics, which improves their high-temperature mechanical properties [3–6]. The design concept of nanocomposites can be applied to ceramic/metal, metal/ceramic, and polymer/ceramic composite systems.

Dispersing metallic second-phase particles into ceramics improves their mechanical properties (e.g., fracture toughness). A wide variety of properties, including magnetic, electric, and optical properties, can also be, tailored in the composites due to the size effect of nanosized metal dispersions, as described later in the chapter. Conventional powder metallurgical methods and solution chemical processes like sol–gel and coprecipitation methods have been used to prepare composite powders for ceramic/metal nanocomposites such as Al_2O_3/W, Mo, Ni, Cu, Co, Fe; ZrO_2/Ni, Mo; MgO/Fe, Co, Ni; and so on. The powders are sintered in a reductive atmosphere to give homogeneous dispersions of metallic particles within the ceramic matrices. Fracture strength, toughness, and/or hardness are enhanced due to microstructural refinement by the nanodispersions and their plasticity. For transition metal particle dispersed oxide ceramic composites, ferromagnetism is a value-added supplement to the excellent mechanical properties of the composites [7,8]. In addition, good magnetic response to applied stress was found in these ceramic/ferromagnetic-metal nanocomposites, allowing the possibility of remote sensing of initiation of fractures or deformations in ceramic materials.

Nanocomposite technology is also applicable to functional ceramics such as ferroelectric, piezoelectric, varistor, and ion-conducting materials. Incorporating a small amount of ceramic or metallic nanoparticles into $BaTiO_3$, ZnO, or cubic ZrO_2 can significantly improve their mechanical strength, hardness, and toughness, which are very important in creating highly reliable electric devices operating in severe environmental conditions [9]. In addition, dispersing conducting metallic nanoparticles or nanowires can enhance the electrical properties, as described later. Dispersion of soft materials into a hard ceramic generally decreases its mechanical properties (e.g., hardness). However, in nanocomposites, soft materials added to several kinds of ceramics can improve their mechanical properties. For example, adding hexagonal boron nitride to silicon nitride ceramic can enhance its fracture strength not only at room temperature but also at very high temperatures up to 1500 °C. In addition, some of these nanocomposite materials exhibit superior thermal shock resistance and machinability because of the characteristic plasticity of one of the phases and the interface regions between that phase and the hard ceramic matrices.

Advanced bulk ceramic materials that can withstand high temperatures (>1500 °C) without degradation or oxidation are needed for applications such as structural parts of motor engines, gas turbines, catalytic heat exchangers, and combustion systems. Such hard, high-temperature stable, oxidation-resistant ceramic composites and coatings are also in demand for aircraft and spacecraft applications. Silicon nitride (Si_3N_4) and silicon carbide/silicon nitride (SiC/Si_3N_4) composites perform best in adverse high-temperature oxidizing conditions. Commercial Si_3N_4 can be used up to ~1200 °C, but the composites can withstand much higher temperatures. Such Nanocomposites are optimally produced from amorphous silicon carbonitride (obtained by the pyrolysis of compacted polyhydridomethylsilazane $[CH_3SiH\text{-}NH]_m[(CH_3)_2Si\text{-}NH]_n$ at about 1000 °C), which produces crystallites of microcrystals of Si_3N_4 and nanocrystals of SiC [10] (Figure 1.2). The oxidation resistance, determined by TGA analysis, arises from the formation of a thin (few microns) silicon oxide layer.

Processing is key to the fabrication of nanocomposites with optimized properties. Some examples of commonly used processes for creating nanocomposites are discussed below.

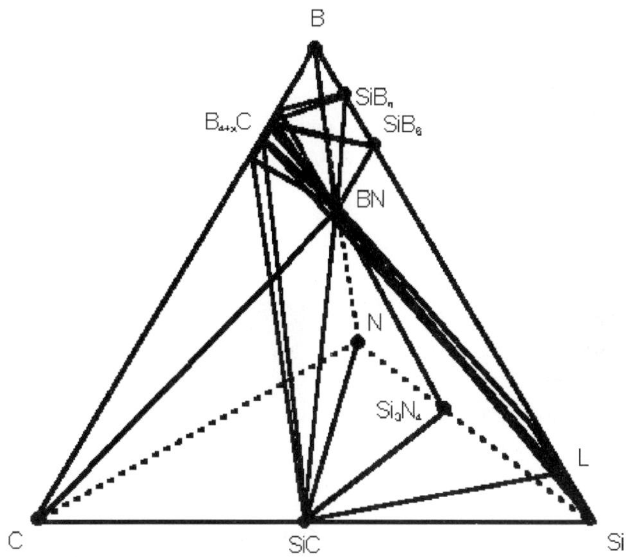

Fig. 1.2 Calculated phase diagrams of the system Si/B/C/N allows for the creation of high-temperature ceramic nanocomposites. The system Si/B/C/N is being investigated with respect to processing new covalent materials. Based on this system, several nanocomposites (SiC/Si_3N_4) have been developed that can, for example, withstand high temperatures (~1500 °C) without degradation or oxidation [10]. (Source [229] used with permission) alternative web site: http://aldix.mpi-stuttgart.mpg.de/E_head.html, used with permission

1.2.1
Nanocomposites by Mechanical Alloying

Mechanical alloying was originally invented to form small-particle (oxide, carbide, etc.) dispersion-strengthened metallic alloys (Figure 1.3) [11]. In this high-energy ball milling process, alloying occurs as a result of repeated breaking up and joining (welding) of the component particles. The process can prepare highly metastable structures such as amorphous alloys and nanocomposite structures with high flexibility. Scaling up of synthesized materials to industrial quantities is easily achieved in this process, but purity and homogeneity of the structures produced remains a challenge. In addition to erosion and agglomeration, high-energy milling can provoke chemical reactions that are induced by the transfer of mechanical energy, which can influence the milling process and the properties of the product. This idea is used to prepare magnetic oxide-metal nanocomposites via mechanically induced displacement reactions between a metal oxide and a more reactive metal [12,13]. High-energy ball milling can also induce chemical changes in nonmetallurgical systems, including silicates, minerals, ferrites, ceramics, and organic compounds. The interest in mechanical alloying as a method to produce nanocrystalline materials is due to the simplicity of the method and the possibility for scaled-up manufacturing.

Displacement reactions between a metal oxide and a more reactive metal can be induced by ball milling [14]. The reaction may progress gradually, producing a nanocomposite powder. In some cases, the reaction progresses gradually, and a metal/metal-oxide nanocomposite is formed. Milling may also initiate a self-propagating com-

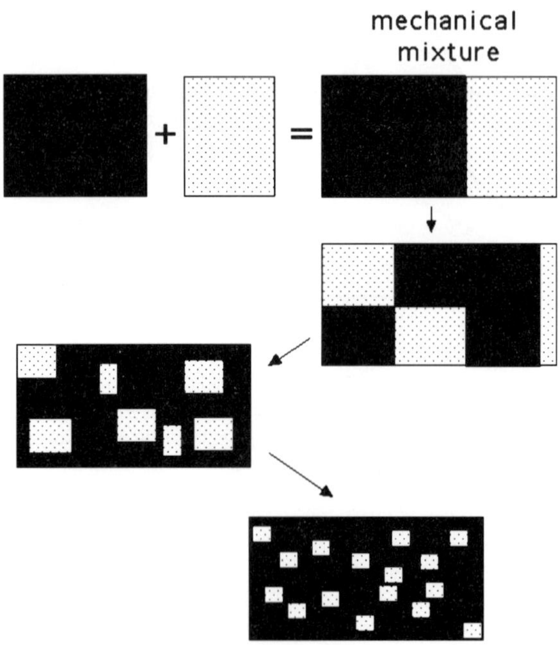

Fig. 1.3 Schematic of the formation process of typical nanocomposite microstructures by the mechanical alloying method. (Source [230, 11] used with permission)

bustive reaction [15]. The nature of such reactions depends on thermodynamic parameters, the microstructure of the reaction mixture, and the way the microstructure develops during the milling process. The mechanical stresses developed during high impact hits can also initiate combustion in highly exothermic systems, melting the reaction mixture and destroying the ultrafine (nanocrystalline) microstructure. Milling mixtures of ceramic and metal powders can induce mechanochemical reactions, and this process is an efficient way of producing nanocermets [16]. Depending on the thermodynamics of the metal/metal-oxide systems and the kinetics of the exchange (displacement) reactions during processing, various nanocomposite systems could evolve. As an example, the reduction of metal oxides with aluminum during reactive ball milling can result in nanocomposites of Al_2O_3 and metallic alloys (Fe, Ni, Cr; particularly binary alloy systems), and such ceramics with ductile metal inclusions produce toughened materials with superior mechanical properties [17]. These nanocomposite materials also have better thermomechanical properties, such as higher thermal shock resistance, due to better metal–ceramic interfacial strength.

Ball milling by direct milling of a mixture of iron and alumina powders has been used to prepare nanocomposites with magnetic phases, such as nanoparticles of iron embedded in an insulating alumina matrix [18]. The average particle size can be reduced to the 10-nm range, as indicated by x-ray diffraction linewidths and electron microscopy. The magnetic properties of this system (e.g., saturation magnetization and coercivity) can be tailored by changing the phase composition, particle size, and the internal stresses accumulated during milling. In this system, the iron nanoparticles were formed with lattice strains of about 0.005; coercivities up to 400 Oe were achieved. The magnetization of the iron particles is 25%–40% less than that expected for bulk iron. Systems of smaller magnetic particles embedded in a nonmagnetic matrix can be prepared by high-energy ball milling [19]. For example, nanocomposites of Fe_3O_4 particles dispersed in Cu have been prepared by ball milling a mixture of Fe_3O_4 and Cu powders directly, as well as by enhanced ball milling-induced reaction between CuO and metallic iron [20]. Both processes result in magnetic semi-hard nanocomposites with a significant superparamagnetic fraction, due to the very small particle sizes of the dispersed magnetic phase. In situ chemical reactions provide a means to control the ball milling process and to influence the microstructure and magnetic properties of the product. Nanocomposite magnets (such as hard magnetic SmCoFe phases in soft magnetic Fe/Co systems), discussed in detail later in this chapter, are routinely prepared by mechanical milling and heat treatment. The metastable nanocrystalline/amorphous structures inherently obtained in mechanically alloyed powders result from repeated deformation and fracture events during collisions of powders with the balls. Plastic deformation in powders initially occurs through the formation of shear bands, and when high dislocation densities are reached, the shear bands degenerate into randomly oriented subgrains. The large surface area of the nanocrystalline grains often helps in the transformation of crystalline into amorphous structures [21]. Deformation-induced defect density and the local changes in temperature due to impacts affect the diffusion coefficients of the several species involved during the milling process. In fact, the final microstructure and stoichiometry of mechanically milled samples often reflects the competing processes of milling-induced disorder and diffusion-limited recovery, rather than being solely dependent on the starting material (e.g., depending on whether the starting mixtures are pre-alloyed or in their elemental forms).

1.2.2
Nanocomposites from Sol–Gel Synthesis

Aerogels, due to their high-porosity structure, are clearly an ideal starting material for use in nanocomposites. Aerogels are made by sol–gel [22,23] polymerization of selected silica, alumina, or resorcinol-formaldehyde monomers in solution and are extremely light (densities $\sim$0.5–0.001 g cc^{-1}) but highly porous, having nanosize pores. In nanocomposites derived from aerogels, the product consists of a 'substrate' (e.g., silica aerogel) and one or more additional phases (of any composition or scale). In the composites, there is always at least one phase whose physical structures have dimensions on the order of nanometers (the particles and pores of the aerogel). The additional phases may also have nanoscale dimensions or may be larger. The systems most commonly made are silica-based nanocomposite systems [24], but this approach can be extended to other aerogel (alumina, etc.) precursors.

Aerogel nanocomposites can be fabricated in various ways, depending on when the second phase is introduced into the aerogel material. The second component can be added during the sol–gel processing of the material (before supercritical drying). It can also be added through the vapor phase (after supercritical drying), or chemical modification of the aerogel backbone may be effected through reactive gas treatment. These general approaches can produce many varieties of composites. A non-silica material is added to the silica sol before gelation. The added material may be a soluble organic or inorganic compound, insoluble powder, polymer, biomaterial, etc. The additional components must withstand the subsequent processing steps used to form the aerogel (alcohol soaking and supercritical drying). The conditions encountered in the CO_2 drying process are milder than in the alcohol drying process and are more amenable to forming composites. If the added components are bulk insoluble materials, steps must be taken to prevent its settling before gelation. The addition of soluble inorganic or organic compounds to the sol provides a virtually unlimited number of possible composites. Two criteria must be met to prepare a composite by this route. First, the added component must not interfere with the gelation chemistry of the aerogel precursor. Possible interference is difficult to predict in advance, but it is rarely a problem if the added component is reasonably inert. The second problem is the leaching out of the added phases during the alcohol soak or supercritical drying steps. This problem can be a significant impediment if a high loading of the second phase is desired in the final composite. When the added component is a metal complex, it is often useful to use a chemical binding agent that can bind to the silica backbone and chelate the metal complex. Many use this method to prepare nanocomposites of silica aerogels or xerogels. After the gel is dried, the resulting nanocomposite consists of an aerogel with metal atoms or ions uniformly (atomically) dispersed throughout the material. Thermal post-processing creates nanosize metal par-

Fig. 1.4 (a) Microstructure of aerogel-encapsulated phase nanocomposite. (b) Left image of three pieces of nanocomposites shows silica aerogel samples that have been coated with silicon nanoparticles by chemical vapor methods. The composites emit red light when excited with ultraviolet light. Right image of six pieces of nanocomposites prepared by adding metal salts or other compounds to a sol before gelation; they show different colors depending on the metal species present. The deep blue aerogel contains nickel; the pale green, copper; the black, carbon and iron; the orange, iron oxide. (Source, the silica aerogel photo gallery [231] used with permission)

ticles within the aerogel matrix. Such composites can have many applications. An example is their use as catalysts for gas-phase reactions or for catalyzed growth of nanostructures.

Vapor phase infiltration through the open pore network of aerogels provides another route [25] to creating various forms of aerogel-based nanocomposites; almost any compound can be deposited uniformly throughout an aerogel. In fact, adsorbed materials in silica aerogels can be modified into solid phases by thermal or chemical decomposition. The same is true for materials that have a porous interior structure, such as zeolites. The nanosize pores within these porous hosts can be utilized for depositing a second phase by chemical or vapor phase infiltration and thermal decomposition. Recently, single-walled carbon nanotubes have been deposited within pores of zeolites to create nanocomposite materials that have unique properties, such as superconductivity [26].

Some examples of nanocomposites (Figure 1.4) that have been created out of silica-based aerogel matrices are the following:

Silica aerogel/carbon composites [27]: These can be made by the decomposition of hydrocarbon gases at high temperatures. The fine structure of aerogels allows the decomposition to take place at a low temperatures (200–450 °C). Carbon loadings of 1%–800% have been observed. The carbon deposition is uniform throughout the substrate at lower loadings, but at higher loadings, the carbon begins to localize at the exterior surface of the composite. These nanocomposites have interesting properties, such as electrical conductivity (above certain loadings) and higher mechanical strength relative to the aerogel.

Silica aerogel/silicon composites [28]: Thermal decomposition of various organosilanes on a silica aerogel forms deposits of elemental silicon. In this case, rapid decomposition of the silane precursor leads to deposits localized near the exterior surface of the aerogel substrate. The nanocomposite, with 20–30-nm diameter silicon particles, exhibits strong visible photoluminescence at 600 nm.

Silica aerogel/transition-metal composites [29]: Organo/transition-metal complexes can be used to deposit metal compounds uniformly through the aerogel volumes. The compounds can be thermally decomposed to their base metals. These intermediate composites, due to the disperse nature of the metallic phase and hence their high reactivity, can be converted to metal oxides, sulfides, or halides. The loading of the metallic phase can be changed by repeated deposition steps. The nanocomposites contain crystals of the desired metal species with sizes in the range of 5–100 nm in diameter.

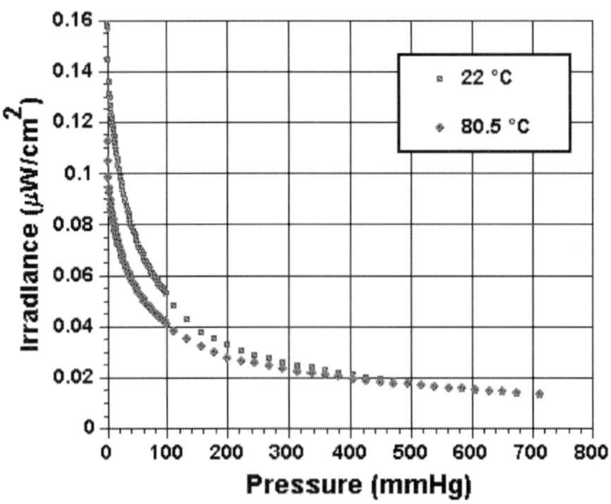

Fig. 1.5 Photoluminescence intensity (irradiance) vs. oxygen pressure (concentration gives a similar plot) at two temperatures measured with a prototype sensor made of silica aerogels. The photoluminescence intensity is indirectly proportional to the amount of gaseous oxygen within the aerogel. The quenching of photoluminescence by oxygen is observed in many luminescent materials. Source [232] used with permission)

The chemical structure of the silica (or other oxide) backbone of an aerogel can also be easily modified. For example, silica aerogel surfaces can be partially reduced by hydrogen. The resulting composite consists of thin interior surface layers of oxygen-deficient silica (SiO_x). This material exhibits strong visible photoluminescence at 490–500 nm when excited by ultraviolet (330 nm) light. The chemical process used to change the surface characteristics of the aerogel does not alter the physical shape or optical transparency of the original structure. This composite is the foundation for the aerogel optical oxygen sensor [30] (Figure 1.5), which is based on the fact that the intensity of photoluminescence is indirectly related to the oxygen concentration in the nanocomposite.

1.2.3 Nanocomposites by Thermal Spray Synthesis

Thermal spray processing is a commercially relevant, proven technique for processing nanostructured coatings [31]. Thermal spray techniques are effective because agglomerated nanocrystalline powders are melted, accelerated against a substrate, and quenched very rapidly in a single step. This rapid melting and solidification promotes the retention of a nanocrystalline phase and even amorphous structure. Retention of the nanocrystalline structure leads to enhanced wear behavior, greater hardness, and sometimes a reduced coefficient of friction compared to conventional coatings.

Figure 1.6 shows a generalized thermal spray process [32]. To form the starting powders, conventional powders can be cryomilled to achieve a nanocrystalline structure [33–35]. Under the right conditions, for example, Fe alloyed with Al, precipitates form, and these precipitates stabilize the nanoscale grain structure to 75 % of the melting temperature of the pure metal. Pure metals (except for aluminum) require some alloying before the nanocrystalline structure is stable at elevated temperatures [36]. For WC/Co and Cr_3Cr_2/NiCr, the hard particles are broken into nanometer-size grains, and they are embedded in the binder [37, 38]. Other systems have also been milled for thermal spraying, such as steel [39] and NiCr/Cr_3C_2. In all cases, there appears to be some nitrogen or oxygen contamination.

The nanoscale powders, prepared by various techniques, must be agglomerated so that grains on the order of 50 nm can be introduced into the thermal spray gun. Unlike sintering of ceramics, this agglomeration does not prevent full densification. A reasonably narrow particle size distribution ensures uniform heating. Nanocrystalline feedstock is generally injected internally (inside the torch), but powders can be injected externally. The type of flame or jet produced depends on the thermal spray technique, and within each technique, gas heating and gas flow parameters can control the velocity and temperature profile. The temperature and velocity profile, combined with the spray distance (the distance from the end of the nozzle to the substrate), control the temperature that the powders reach. Successive impact of particles in a molten or viscous state on the substrate or on previously deposited layers of material forms a coating.

The ability to maintain the nanocrystalline structure during processing and upon consolidation is critical to improving its properties because it is the nanoscale micro-

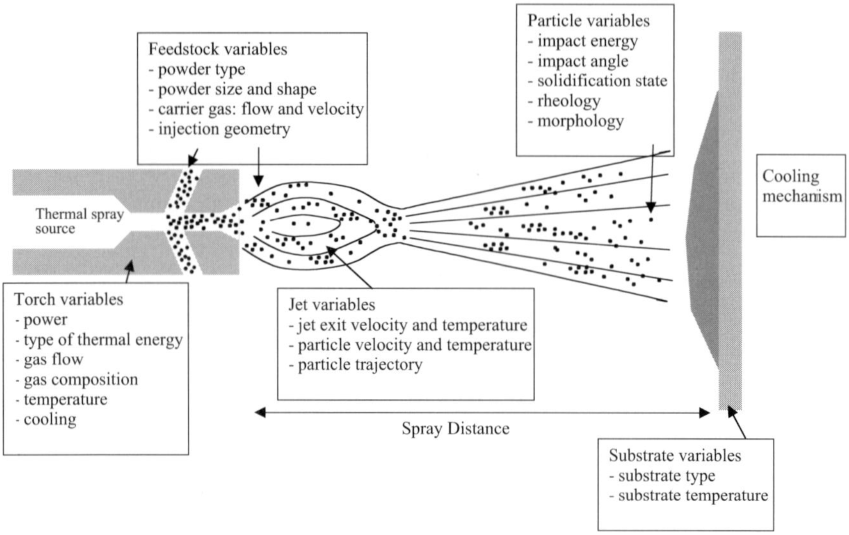

Fig. 1.6 Schematic for a generalized thermal spray process, showing the different variables used. The qualities of the coatings (bonding to the substrate, microstructure of the coating, hardness, wear resistance, etc.) are affected by a multidimensional parameter space

structure that leads to the unique properties. Several parameters are critical: (a) The thermal stability of the agglomerated powders: nanocrystalline materials can experience grain growth at temperatures well below the temperatures observed for conventional materials. The high surface area drives this growth. (b) The degree of melting that occurs in flight: this can be controlled by the spray distance, the temperature of the jet, and the velocity of the jet, and optimal parameters are determined primarily by experiment. (c) The cooling rate: a high cooling rate leads to high nucleation and slow grain growth, which promotes the formation of nanocrystalline grains. The systems that tend to maintain their nanocrystalline structure even at elevated temperature are apt to have impurities or a second phase that stabilize the grain structure [40]. For example, cryomilling often results in nanoscale particles (oxides, nitrides, or oxynitrides) [41] that fix the grain boundaries. In addition, significant impurities or excess solute atoms at the grain boundaries also limit grain growth [42, 43].

Plasma spraying and high velocity oxy fuel (HVOF) processes are the most widely used thermal spray methods for producing nanocrystalline and nanocomposite coatings. In plasma spraying, an electric arc is used to ionize an inert gas to produce a highly energetic thermal plasma jet with gas temperatures and velocities of approximately 11 000 K and 2000 ms^{-1}. Vacuum plasma spraying and low-pressure plasma spraying have been used to effectively process WC/Co nanocomposite coatings. Use of HVOF involves an internal combustion chamber in which fuel (hydrogen, propylene, acetylene, propane) is burned in the presence of oxygen or air (HVAF). This results in a hypersonic gas velocity. The particle velocities are higher than the 800 ms^{-1} achieved with plasma spray, and the thermal energy is lower (it may reach 3000 K), which reduces superheating and particle vaporization. The high speed and low temperatures result in more strongly adhering and more homogeneous coatings with lower oxide content.

WC/Co coatings are of great interest because they already exhibit excellent wear properties. Nanostructuring further increases the wear resistance and decreases the coefficient of friction. Thermally sprayed WC/Co coatings, however, do not always exhibit improved properties. WC/Co coatings sprayed via HVOF exhibited decreased wear resistance due to decomposition of the carbide phase during spraying [44]. Nanostructured powders reach temperatures almost 500 ° higher than their conventional counterparts. Vacuum plasma spraying, however, resulted in coatings with significantly improved wear resistance and lower coefficient of friction, presumably because the Ar atmosphere prevented oxidation of the carbide phase [45].

Cr_3C_2/NiCr composites are also used in applications where wear resistance is required, but they have an added advantage over WC/Co, which has excellent corrosion resistance. Nanostructuring of these coatings has also resulted in improved hardness and scratch resistance, as well as reduced coefficient of friction. The improved homo-

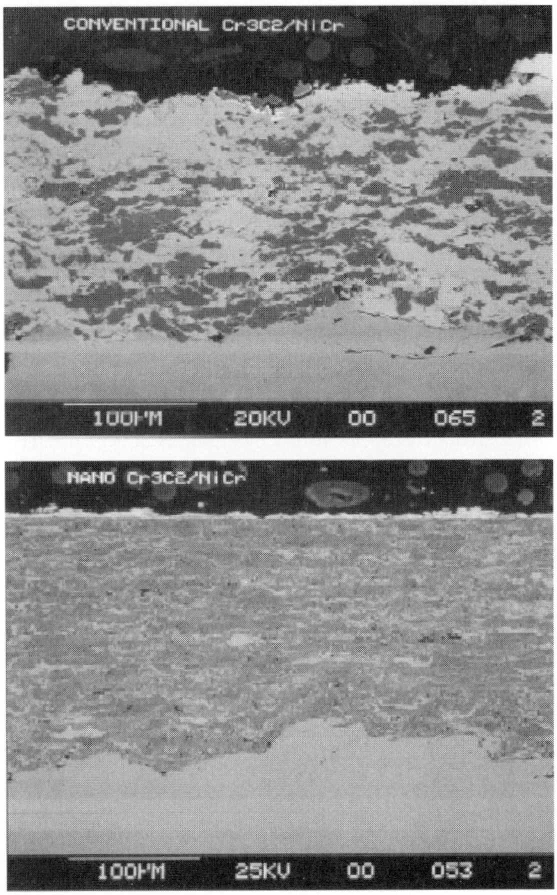

Fig. 1.7 Microstructures of thermal-sprayed Cr_3C_2/NiCr coatings. Top micrograph shows conventional coating and bottom micrograph shows nanocomposite microstructure. A uniform, dense microstructure is observed in the nanostructured coatings, compared to an inhomogeneous microstructure in the conventional coating. (Source [37] used with permission)

geneity of these structures, as well as a high density of Cr_2O_3 nanoparticles (formed by oxidation during the thermal spray process), compared to conventional materials, cause the improved properties. Figure 1.7 shows an example of the improved homogeneity of nanostructured coatings. Ceramics such as alumina/titania and zirconia have also been thermally sprayed, and the nanostructured powders have lead to submicron final grain sizes in the coatings. Key to achieving excellent properties is minimizing the degree of melting [46] so as to maintain the nanostructure in the final coating. On the other hand, significant deformation or splatting of the particles is required upon impact, to assure a large surface contact between the particles [47]. Thus, some melted particles lend well to continuous, good-quality coatings.

1.3
Metal Matrix Nanocomposites

During the past decade, considerable research effort has been directed towards the development of in situ metal-matrix composites (MMCs), in which the reinforcements are formed by exothermal reactions between elements or between elements and compounds [48]. With this approach, MMCs with a wide range of matrix materials (including aluminum, titanium, copper, nickel, and iron), and second-phase particles (including borides, carbides, nitrides, oxides, and their mixtures) have been produced. Because of the formation of stable nanosized ceramic reinforcements, in situ MMCs exhibit excellent mechanical properties.

MMCs are a kind of material in which rigid ceramic reinforcements are embedded in a ductile metal or alloy matrix. MMCs combine metallic properties (ductility and toughness) with ceramic characteristics (high strength and modulus), leading to greater strength to shear and compression and to higher service temperature capabilities. The attractive physical and mechanical properties that can be obtained with MMCs, such as high specific modulus, strength, and thermal stability, have been documented extensively. Interest in MMCs for use in the aerospace and automotive industries and other structural applications has increased over the past 20 years. This increase results from the availability of relatively inexpensive reinforcements and the development of various processing routes that result in reproducible microstructure and properties.

The family of discontinuously reinforced MMCs includes both particulates and whiskers or short fibers. More recently, this class of MMCs has attracted considerable attention as a result of (a) availability of various types of reinforcement at competitive costs, (b) the successful development of manufacturing processes to produce MMCs with reproducible structure and properties, and (c) the availability of standard or near-standard metal working methods, which can be utilized to fabricate these composites. The particulate-reinforced MMCs are of particular interest due to their ease of fabrication, lower costs, and isotropic properties. Traditionally, discontinuously reinforced MMCs have been produced by several processing routes such as powder metallurgy, spray deposition, mechanical alloying (MA), and various casting techniques. All these techniques are based on the addition of ceramic reinforcements to the matrix materials, which may be in molten or powder form. For conventional MMCs, the reinforcing phases are prepared separately prior to the composite fabrication. Thus, conventional MMCs can be viewed as ex situ MMCs. In this case, the scale of the reinforcing phase is limited by the starting powder size, which is typically on the order of micrometers to tens of micrometers and rarely below 1 μm. Other main drawbacks that must be overcome are interfacial reactions between the reinforcements and the matrix and poor wettability between the reinforcements and the matrix due to the surface structure of the reinforcements as well as to contamination.

The properties of MMCs are widely recognized to be controlled by the size and volume fraction of the reinforcements as well as by the nature of the matrix/reinforcement interfaces. An optimum set of mechanical properties can be obtained when fine, thermally stable ceramic particulates are dispersed uniformly in the metal matrix. Efforts have been made to meet such requirements and have led to the development of novel composites – in situ MMCs in which the reinforcements are synthesized in a metallic matrix by chemical reactions between elements or between element and compound during fabrication of the composite. Compared with conventional MMCs produced by ex situ methods, in situ MMCs exhibit the following advantages: (a) formation of reinforcements that are thermodynamically stable in the matrix, leading to less degradation in elevated-temperature services; (b) reinforcement/matrix interfaces that are clean, resulting in strong interfacial bonding; (c) the formation of reinforcing particles of a finer size with a more uniform distribution in the matrix, which yields better mechanical properties.

The great potential that in situ metal matrix nanocomposites offer for widespread applications has resulted in the development of a variety of processing techniques for production during the past decade. Using these routes, in situ composites with a wide range of matrix materials (including aluminum, titanium, copper, nickel, and iron) and second-phase particles (including borides, carbides, nitrides, oxides, and their mixtures) have been produced. Particularly attractive among the several techniques available for synthesizing MMCs are the solidification processes in which the reinforcing particles are formed in situ in the molten metallic phase prior to its solidification. What makes them so attractive is their simplicity, economy, and flexibility. The judicious selection of solidification processing techniques, matrix alloy compositions, and dispersoids can produce new structures and affect a unique set of useful engineering properties that are difficult to reach in conventional monolithic materials. Specifically, the solidification conditions that are present during processing play an important role in dictating the microstructure and the mechanical and physical characteristics of these structures. Microstructure refinement arising from rapid solidification processing (RSP) offers a potential avenue for alleviating solute segregation and enhancing dispersion hardening by substantially reducing the size of the reinforcing phases and modifying their distribution in the matrix. As example, RSP of Ti/B or Ti/Si alloys accompanied by large undercoolings and high cooling rates is very effective in producing in situ Ti-based nanocomposites containing large volume fractions of reinforcing particles [49]. These particles are formed in situ in Ti/B or Ti/Si alloys either upon solidification or, subsequently, by controlled decomposition of the resulting supersaturated solid solutions.

More recently, several workers have used the RSP route [50] to fabricate in situ TiC particulate-reinforced Al-based composites. In their work, master material ingots were prepared by melting a mixture of Al, Ti, and graphite powder in a graphite-lined induction furnace under an argon atmosphere, followed by direct chill cast. Chill block melt spinning was used to prepare rapidly solidified samples in ribbon form. These ribbons were further milled into powders (100 ± 250 μm), which were subsequently canned and degassed and then extruded into rods. In situ formed TiC particles of 40 ± 80 nm were reported to be distributed uniformly in the aluminum matrix with a grain size of 0.3 ± 0.85 μm. The authors reported that RSP can significantly refine the microstructure of the composites. For TiC/Al composites, the microstructure is often characterized by the presence of agglomerated TiC particles. These particles, with a size of 0.2 ± 1.0 μm, accumulate at the Al subgrain or the grain boundaries. The larger particles have polyhedral morphology, and the smaller ones are round or globular. In comparison, typical rapidly solidified microstructures consist of a uniform, fine-scale dispersion of TiC particles with a size of 40 ± 80 nm in an Al supersaturated matrix of 0.30 ± 0.85 μm grain size. One main advantage of RSP is its ability to produce alloy compositions not obtainable by conventional processing methods. Furthermore, RSP materials have excellent compositional homogeneity, small grain sizes, and homogeneously distributed fine precipitates or dispersoids.

The homogeneity of composite materials is crucially important to high-performance engineering applications such as in the automotive and aircraft industries. A uniform reinforcement distribution in MMCs is essential to achieving effective load-bearing capacity of the reinforcement. Nonuniform distribution of reinforcement can lead to lower ductility, strength, and toughness of the composites. Nanoscale ceramic particles synthesized in situ are dispersed more uniformly in the matrices of MMCs, leading to significant improvements in the yield strength, stiffness, and resistance to creep and wear of the materials. For example, in situ fabrication of TiC-reinforced Al, Al/Si, and Al/Fe/V/Si matrix composites by the RSP route is far more effective in

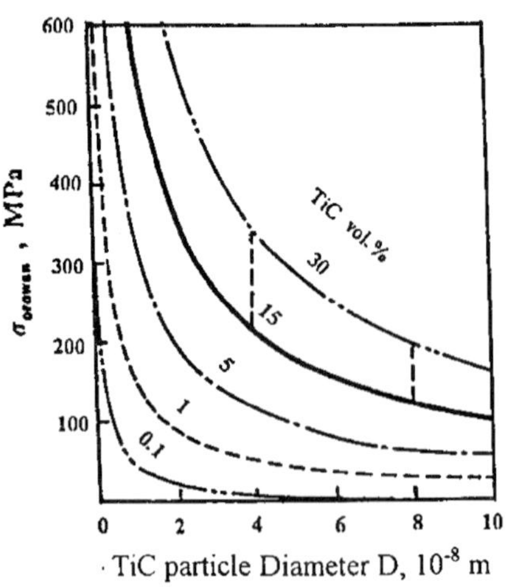

Fig. 1.8 Increase in strength (σ) of in-situ fabricated TiC-reinforced Al nanocomposites with increasing volume fractions or decreasing diameters of dispersed-phase (TiC) particles. (Source [50] used with permission)

Tab. 1.1 Typical magnetic properties of nanocrystal/amorphous composites and amorphous alloys. Alloy 1: $Fe_{73.5}$ Cu_1 Nb_3 $Si_{13.5}$ B_9 (at. %). Alloy 2: $Fe_{73.5}$ Cu_1 Nb_3 $Si_{16.5}$ B_6 (at. %).
Core loss: 100 kHz, 0.2 T [52].

Alloy	t (μm)	B_s (T)	B_r /B_s (%)	H_c (A m^{-1})	μ_c (1 kHz)	Core loss (kW/m^{-3})	λ_s ($\times 10^{-6}$)	Curie temp. (K)
Alloy 1	18	1.24	54	0.53	100000	280	+2.1	843
Alloy 2	18	1.18	58	1.1	75000	280	~ 0	833
Fe/Si/B/M	20	1.41	16	6.9	6000	460	+20	631
Co/Fe/Si/B/M	18	0.53	50	0.32	80000	300	~ 0	453

improving the tensile properties of these composites, due to the formation of a refined microstructure. The in situ composites exhibit excellent strength at room temperature and elevated temperatures. The values of strength (σ) increased with increasing volume fractions or decreasing diameters of dispersed-phase (TiC) particles (Figure 1.8). When the volume fraction of dispersed particles is about 15 – 30 vol % and the particle diameters 40 – 80 nm, the values of σ are 120 – 270 MPa and 200 – 350 MPa, respectively (Figure 1.8).

Nanocrystalline materials in general are single- or multi-phase polycrystals with grain sizes in the nanometer range. Owing to the extremely small dimensions, many properties of nanocrystalline samples are fundamentally different from, and often superior to, those of conventional polycrystals and amorphous solids. For example, nanocrystalline materials exhibit increased strength or hardness, improved ductility or toughness, reduced elastic modulus, enhanced diffusivity, higher specific heat, enhanced thermal expansion coefficient, and superior soft magnetic properties in comparison with conventional polycrystalline materials [51]. Crystallizing completely amorphous solids under proper heat treatment conditions can result in formation of nanocrystalline materials. However, controlled crystallization of amorphous alloys can

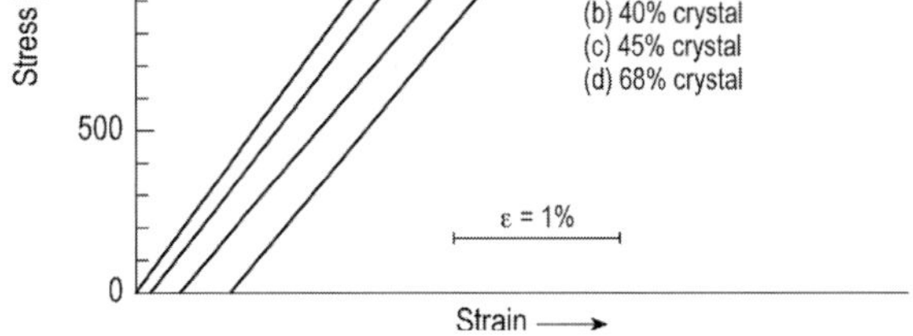

Fig. 1.9 Compressive stress–strain curves of amorphous and partly crystallized $Zr_{57}Al_{10}Cu_{20}Ni_8Ti_5$ alloy nanocomposite (a) as-cast, (b) 40 vol. % nanocrystals, (c) 45 vol. % nanocrystals and (d) 68 vol. % nanocrystals. The sample containing a volume fraction of 40 % nanocrystals (b) seems to provide the best compromise between strength and ductility. (Source [53] used with permission)

be used to obtain partially crystallized materials with nanosized crystallites embedded in the residual amorphous matrix. This special nanocrystal/amorphous nanocomposite structure allows the materials to exhibit excellent mechanical or magnetic properties. The Fe/Cu/Nb/Si/B alloys are a good example of this type of material. The Fe/Si/B/*M* (*M*: additives) alloys properly prepared by annealing amorphous alloys and having bcc Fe solid solution and 10-nm-diameter nanostructures embedded in the residual amorphous matrix, show excellent soft magnetic properties (Table 1.1). The nanocrystal/amorphous composite shows high saturation flux density, low magnetostriction, and excellent soft magnetic properties. This nanocomposite, therefore, is expected to find use in many kinds of magnetic devices such as choke coils and transformers.

Production of bulk nanostructured composites with amorphous matrices has been carried out by die casting and mechanical alloying and subsequent consolidation at elevated temperatures in Zr-based alloys [53]. The distribution of finely dispersed nanocrystals increases the flow stress significantly. For example, a $Zr_{57}Al_{10}Cu_{20}Ni_8Ti_5$ sample containing a volume fraction of 40% nanocrystals (Figure 1.9, curve b) seems to provide the best compromise between strength and ductility.

1.4 Bulk Ceramic Nanocomposites for Desired Mechanical Properties

Over the past half century, ceramics have received significant attention as candidate materials for use as structural materials under conditions of high temperature, wear, and chemical attack that are too severe for metals. However, one characteristic of these ceramics that has prevented them from being widely used is their inherent brittleness. Thus, significant scientific effort has been directed towards making ceramics more flaw-tolerant through design of their microstructure. An important example of high-toughness structural ceramics is *self-reinforced* silicon nitrides, which were first developed during the 1970s [54]. These materials have high toughness and room-temperature strength, along with good resistance to corrosion and oxidation. However, their high-temperature ($>1000\,^{\circ}C$) strength is compromised by low creep resistance and the occurrence of subcritical crack growth. These phenomena are caused by the softening of a glassy phase that is located at the grain boundaries as the temperature is increased. A possible way to overcome this problem is the fabrication of Si_3N_4/SiC nanocomposites [55]. Also, several approaches have been used to improve the processability (the sinterability of these materials is rather poor) and high-temperature properties of monolithic silicon nitrides, with limited success.

Considerable attention has been devoted to 'functionally graded nanocomposite materials', for which gradually varying the dispersion (nanoparticles) to matrix ratio in chosen directions continuously changes the material. An example of such a material is SiC dispersions in a C (pyrolytic graphite) matrix, which has served well as thermal barriers on the space shuttle due to its excellent resistance to oxidation and thermal shock. One route to preparing these composite systems is by chemical vapor deposition using multicomponent gas reactions. For example, 10–100-nm sized SiC dispersions can be prepared with pyrographite using precursors of $SiCl_4/C_3H_8/H_2$ or $SiCl_4/CH_4$ in CVD. The entire range of compositions from carbon to SiC has been prepared by this method [56]. Changing the deposition conditions can control the morphologies of the second phase and hence the microstructure of the composites; this control influences the mechanical properties of the material. Nanocrystalline carbide-embedded composites, particularly those with amorphous or diamond-like carbon matrices, can be considered for tribological applications. The challenges here are to minimize formation of the sp^2 soft graphite-like phase during synthesis and to retain a high fraction of the hard a-C matrix. Pulsed laser deposition (in which laser-ablated plumes from graphite are intercepted by a low-energy metal plasma produced by magnetron sputtering) [57] at near room temperature has been used successfully to create ~10 nm TiC/TiCN nanocrystals embedded uniformly in a-C with diamond-like characteristics. Hardness values as high as 60 GPa, coefficient of friction values as low as 0.1, and high toughness values have been achieved in these films, which have high tribological value and possible applications in surface-protection coating technologies. The two-phase heterogeneous structure in the nanocomposites provides crack deflection mechanisms, reducing the tendency toward easy brittle failure in these hard composites.

Significant interest was generated in 1991 when Niihara [58] reported large improvements in both the fracture toughness and the strength of materials with a unique microstructure: ceramics with nanometer-range particles (20–300 nm) embedded within a matrix of larger grains and at their grain boundaries. These ceramics reportedly showed up to 200% improvement in both strength and fracture toughness, better retention of strength at high temperatures, and better creep properties. Materials like SiC, Al_2O_3, ZrO_2 and Si_3N_4 are excellent candidates for demanding structural applications due to their mechanical and thermomechanical properties (Figure 1.10) [59]. The incorporation of fine SiC and Si_3N_4 particles (smaller than 300 nm) in an alumina

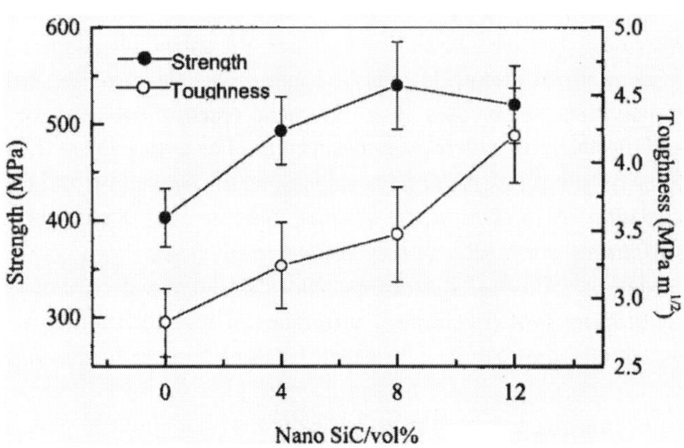

Fig. 1.10 Mechanical properties of a SiC/zirconia-toughened mullite nanocomposite as a function of nanosized SiC content. (Source [59] used with permission)

(Al_2O_3, a structural ceramic material) matrix first demonstrated the concept of structural nanocomposites. The dispersion of these particles improved the fracture toughness from 3 to 4.8 MPa $m^{1/2}$ and the strength from 350 to 1050 MPa at only 5 vol. % additions of SiC [60]. The bending strength of such composite was also measured to be $\sim$ 1 GPa. Further improvements in strength to about 1500 MPa were achieved by annealing the samples at 1300 °C. The high strengths were maintained to about 1000 °C. For Si_3N_4/SiC nanocomposites, several processing techniques have been used, including conventional powder processing, sol–gel processing, and polymer precursor processing; the SiC particles can originate from admixed commercial SiC powders, SiCN powders produced by plasma synthesis, in situ reaction pyrolysis of carbon-coated Si_3N_4 particles, and pyrolysis of a polycarbosilane based on SiCN precursor. An example of conventional processing is co-milling of the solid precursor powder materials followed by hot pressing, which produces nanocomposites possessing both inter- and intra- granular SiC particles [61]. Intragranular particles are most effective in toughening the material, because they are mainly responsible for crack deflection and crack impediment. Intergranular particles are often detrimental (initiating cracks), but they could provide some advantages by grain refinement during processing. The toughness of the composite depends on the volume fractions of these two types of dispersoids, and controlling these fractions precisely is challenging. In polymer-based processing, a mixture of Si_3N_4 powder, sintering additives, and polymethylphenylsilane is pyrolyzed at $\sim$1000 °C in Ar and sintered in nitrogen. Another possibility is preparing an amorphous Si/N/C powder by crosslinking and pyrolysis of polymethylsilazane [62]. Whereas the conventional process leads to a micronanostructure with nano-sized SiC dispersions mainly inside Si_3N_4 grains, the polymer-processing route results in atto (nano-nano) structures. Hybrid polymer/powder processing can also be applied to other composite microstructures such as Al_2O_3/SiC. Coating a silicon-containing polymer (polycarbosilane) onto alumina powder, followed by pyrolysis, can result in a finer SiC nanophase [63]. Novel, superior processing routes can be used to prepare well dispersed ceramic nanocomposites, for example Al_2O_3/SiC. Colloidal consolidation and reaction sintering is one such process [64]: micro-size particles of the two phases are colloidally dispersed and consolidated to form uniform compacts. The SiC phase particles are then oxidized to reduce them to nanometer-size cores. The interfacial reaction between the oxidized SiC particles and the alumina matrix produces mullite. The advantage of this process is that particle sizes need not be brought down to nanoscale by milling and can be controlled well by oxidation. Also, due to the volume increase that occurs during reaction sintering, shrinkages often seen during sintering are small.

Materials fabricated from polymer-derived powders made by hot pressing have yielded the best mechanical properties. These techniques are expensive and limit the shapes that can be fabricated. Gas-pressure sintering or pressure-less sintering are the most attractive processing techniques. However, to date, research on gas-pressure sintering has used mixed powders, which result in poor powder dispersion, agglomeration of SiC, and changes in the glass phase chemistry due to reaction with SiC. Thus, the focus for future processing rests on routes that produce commercially viable powders with uniform dispersion of SiC particles (with controlled size and volume fraction) that can be fabricated into dense components by gas-pressure or pressure-less sintering. The effects of reinforcement on the mechanical properties of silicon nitride nanocomposites is not well understood. As with the Al_2O_3/SiC nanocomposites, the presence of the SiC at the grain boundaries restricts grain growth, resulting in the formation of a matrix of finer Si_3N_4 grains. In fact, at higher volume percent of reinforcements, grain growth is reported to be severely restricted, resulting in fine α-Si_3N_4 grains and a composite showing superplastic (large strain to failure) behavior. The 'nanosize' strengthening and toughening effect due to thermal expansion mismatch proposed for Al_2O_3/SiC composites needs to be critically examined for Si_3N_4/SiC composites: the difference between the thermal expansion coefficients of Si_3N_4 and SiC is small and in the opposite sense as in Al_2O_3/SiC. The presence of a liquid phase in Si_3N_4 and its composites further complicates matters: reactions between the reinforcements and the liquid phase could alter its composition and quantity, thus changing the sintering behavior and creep resistance. Thus, a more fundamental understanding of the effect of nanosized SiC reinforcements on the behavior of Si_3N_4 matrix composites is required. In particular, systematic studies of the effects of reinforcement size and volume fraction on the microstructure, processing, and properties of these nanocomposites are needed. Such studies exist for composites in which the reinforcement size is in the several-micron range, and in these studies, both the reinforcement size and volume fraction significantly affect the fracture toughness and strength [65].

For unreinforced Si_3N_4 ceramics, there is sufficient evidence that creep in the liquid phase-sintered materials is controlled by grain boundary transport and sliding. Thus, the incorporation of fine reinforcements at the grain boundaries may improve creep resistance [66]. The amount and viscosity of the intergranular phase due to chemical reactions and accelerated crystallization, the hindrance of grain-boundary sliding due to fixing by SiC particles, and the obstruction of easy diffusion paths by the SiC particles explain the better creep resistance of the nanocomposites [67]. For Al_2O_3/SiC nanocomposites, the better creep resistance (5 vol. % SiC nanoparticles reduces the tensile creep of Al_2O_3 by 2–3 orders of magnitude) is explained by SiC nanoparticles fixing the grain boundaries, resulting in less grain-boundary sliding, a small viscoelastic contribution to creep, and enhanced grain-boundary strength, allowing plastic deformation of grains through dislocation motion [68]. The presence of non-bridged grain boundaries causes a rearrangement of the microstructure due to initial grain sliding. But this initial sliding process is stopped easily as boundaries fixed by the SiC nanoparticles are encountered; hence, only lattice mechanisms based on dislocation motion or, to a lesser extent, viscoelastic mechanism, contribute to creep. However, in general, the studies on nanocomposites and the exact mechanisms of creep resistance have not been entirely conclusive, and systematically investigating the effects of various microstructural parameters as a functions of particle size, oxidation, and volume of the dispersed phase is desirable. An additional factor that needs careful control is the volume fraction of reinforcements on the grain boundary as opposed to within the grain. The erosion properties of such composites also deserve mention. The residual surface stress induced by grinding and polishing nanocomposites (e.g., Al_2O_3/SiC) and monolithic alumina are quite different, and studies show that the residual surface stress in the nanocomposite is more sensitive to surface treatment than that in the corresponding monolithic structure. Direct observations by transmission electron microscopy suggest that deformation micromechanisms, from twinning in the alumina to dislocation generation in the nanocomposite, dominate the plastic deformation induced by the surface treatments.

Crack-tip bridging by particles is a primary mechanism of strengthening ceramic nanocomposites. Small brittle particles (for example, silicon carbide particles in Al_2O_3/SiC composites) cause crack tip bridging at small distances behind the cracks [69]. Residual stresses around the particles cause the strengthening mechanism to operate effectively even at small volume fraction loading of SiC. The increase in the nanocomposite strength due to the reduction in the critical flaw size is achieved by the fine dispersion of nanoparticles [70]. Intergranular fracture, which is responsible for fracture in monolithic alumina, is suppressed in the nanocomposite, because the crack extension along grain boundaries is suppressed by particles that strongly bond the matrix–matrix interfaces. Hence, transgranular fracture is a common mode of fracture in nanocomposites. Tensile stresses in the intragranular particles induce the cracks to extend to the particles, leading to particle bridge toughening.

Finally, dispersing metallic second-phase particles into ceramics is not only suitable for improving the mechanical properties of ceramics, but also presents a wide variety of advantages for addition of new functions such as magnetic, electric, and optical properties, due to the size effect of nanosized metal dispersion. Granular films can be made with a ceramic phase embedded with nanosize metal granules [71] (e.g., Fe/Al_2O_3, Fe/SiO_2). Such films display unusual and often enhanced transport, optical, and magnetic properties. An important parameter that affects the physical behavior of granular films is the percolation threshold of the embedded metal phase. Percolation effects in nanocomposites are discussed later in this chapter. A dramatic manifestation of this percolation effect is the change, by many orders of magnitude, of the electrical resistivity (insulator-to-metal transition) as the percolation threshold is passed. The percolation threshold is generally found to be in the range of 0.5–0.6 metal volume fractions for nanocomposite films [72]. The mechanical properties can also undergo changes near the percolation threshold. The obvious explanation for this is that a change occurs in deformation behavior as the embedded phase forms connected networks that spread through the matrix phase. Conventional powder metallurgical methods and solution chemical processes like sol–gel, and coprecipitation methods can also be used to prepare composite powders for ceramic/metal nanocomposites such as Al_2O_3/W, Mo, Ni, Cu, Co, Fe; ZrO_2/Ni, Mo; MgO/Fe, Co, Ni and so on. They are sintered in a reductive atmosphere that gives homogeneous dispersion of metallic particles within the ceramic matrices. The microstructural refinement by nanodispersion and the plasticity enhance the fracture strength, toughness, and/or hardness.

In summary, the advantage of ceramic nanocomposites lies not only in the mechanical strength of the composite material but also in other mechanical properties such as fracture toughness, hardness, and creep resistance. The degree of improvement in these properties depends on the composite systems involved. Although there are some generalities in the strengthening and toughening mechanisms in such composites, such as crack deflection and crack-tip bridging by the dispersed particles, the actual size, location, and volume fraction of particles strongly influence the final outcome with respect to mechanical behavior.

1.5
Thin-Film Nanocomposites: Multilayer and Granular Films

Thin-film nanocomposites are films consisting of more than one phase, in which the dimensions of at least one of the phases is in the nanometer range. These nanocomposite films can be categorized as multilayer films, in which the phases are separated along the thickness of the film, or granular films, in which the different phases are distributed within each plane of the film (Figure 1.11). Multilayered thin-film nanocomposites consist of alternating layers of different phases and have a characteristic thickness on the order of nanometers. These films are usually used for their enhanced hardness, elastic moduli, and wear properties. The elastic moduli is higher in multilayered thin films than in homogeneous thin films of either component. The supermodulus effect [73] is observed in some metallic systems, by which, at certain characteristic thicknesses (typically $\sim$2 nm, corresponding to a bilayer consisting of one layer of each phase) of the film, the elastic modulus increases by more than 200 %. The most satisfactory explanation of this effect assumes an incoherent interface between the adjacent layers [74], suggesting that atoms are displaced from their equilibrium positions and that, during loading, all the layers undergo compression, which results in a higher resistance to deformation. The increase in hardness of the multilayer nanocomposites has been explained [75] by considering that, when ultrathin films of materials with different dislocation line lengths are stacked, the strength approaches the theoretical limit. The dislocations cannot move from layer to layer, due to the difference in dislocation line lengths, and the films are thin enough that independent dislocation sources do not become operative. Conventional thin-film deposition techniques (sputtering, physical vapor deposition, CVD, electrochemical deposition, etc.) can produce multilayer nanocomposites, and the excellent flexibility of these techniques creates extremely thin films of uniform compositions.

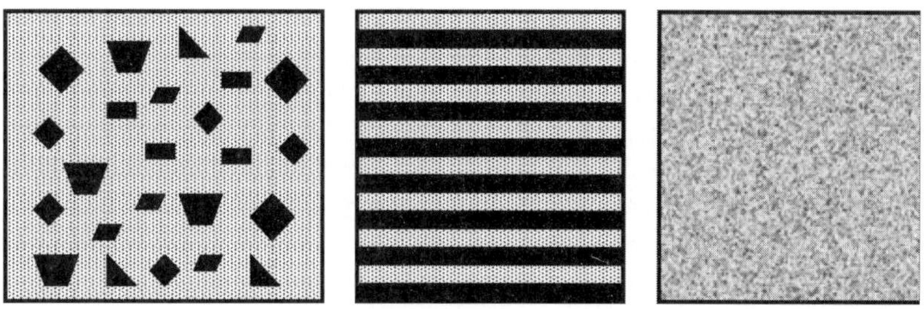

Fig. 1.11 Schematic of possible microstructures for nanocomposite and nanostructured coatings: isotropic dispersed multiphase microstructure (e.g., TiC/amorphous carbon), multilayered microstructure (e.g., TiN/TiC) for nanocomposite coatings, and homogeneous alloyed microstructure as possible homogeneous nanostructured coatings (e.g., NiCoCrAlY alloy)

Granular nanocomposite films are those that contain both phases (metal and ceramic) in the same layer of the film and have no abrupt interfaces across the film thickness as in multilayer films. Here, one phase can be in the nanosize range (similar to dispersions), or both phases can have nanocrystalline grains distributed contiguously and laterally in the film. Granular nanocomposite coatings are also common; in these, the matrix phase is a polymer and the dispersed phase is inorganic. Granular films in which at least one distributed phase has electrical/magnetic properties are mainly used in electrical and magnetic applications [76]. In certain systems consisting of metal and ceramic particles (such as iron oxide/silver and alumina/nickel), changing the fraction of the phases present can alter the magnetic properties. At small volume fractions of the metal component, the material exhibits ferromagnetic behavior. Beyond a certain volume fraction of the metal phase, the ferromagnetic ordering gives way to superparamagnetism, because, once the metal particles attain percolation, the film behaves essentially like the metallic phase. An important parameter that critically affects the properties of granular films is the percolation threshold of the metal, which typically corresponds to a metal volume fraction of 0.5 – 0.6. A dramatic observation at percolation is the large (several orders of magnitude) change in electrical resistivity of such films, going from an insulating ceramic-like behavior to a conducting-metal behavior. Similar techniques (such as CVD or electrochemical methods) as used in multilayer film growth can also be used to prepare homogeneous nanostructured composite films. As an example, nanostructured AlN/TiN composite films were made by CVD using high-speed deposition of gaseous precursors ($AlCl_3$, $TiCl_4$, NH_3) to form insoluble solid mixtures [77]. In the films so generated, the grain sizes corresponded to 8 nm (AlN) and 6 nm (TiN). The grain size range allows the films to have better ductility and greater toughness compared to bulk AlN/TiN composites.

1.6 Nanocomposites for Hard Coatings

Improved wear resistance, good high-temperature stability, and improved friction properties are important characteristics of good coatings for use in applications such as cutting tools. Most widely used coatings are made from TiN, TiC, TiAlN, CrN, diamond-like carbon (DLC), WC/C, MoS_2, Al_2O_3, etc. For improved coatings in which lower friction, increased life time, increased toughness, higher thermal stability, and in some cases, environmental (biomedical, for example) compatibility are needed, new types of materials are being considered, including nanocomposite materials [78]. Nanocomposite structures such as multilayers or even isotropic coatings can be made from nanoscale entities with properties superior to single-phase materials. This approach of using nanocomposites is an alternative to using specific alloying elements in single-phase coating materials (to improve properties such as hardness) and provides far better flexibility in tailoring multifunctional coatings.

Nanocomposite coatings usually consist of two or more phases combined as multiple layers [79] or as homogeneous isotropic multiphase mixtures. Classical multilayer (3 to many layers) coatings have total thicknesses of several micrometers, and the many layers are typically used to provide toughness (by crack deflection at the many interfaces) and properties such as oxidation resistance (due to, e.g., TiAlN layers). Building multilayer structures also provides better tribological properties [80]. Gradient layers are also often introduced to counterbalance the vast differences in thermal expansion coefficients between multiple layers, which would cause internal stresses and delamination at the interfaces between layers. In nanoscale multilayer coatings, when the thickness of each layer is in the nanometer range, superlattice effects can increase the hardness and other properties of the coatings. The increased hardness of such coatings (e.g., TiN/VN), where the superlattice period is a few nanometers, can be orders of magnitude higher than that of the corresponding base materials [81, 82]. The hardness increase in these nanoscale multilayer films results mainly from hindering dislocation movements across the sharp interfaces between two materials having vastly different elastic (particularly shear modulus) properties and lattice mismatch (coherency strain) [83]. One interesting point to note is that, if the individual layers are very thin (less than 3–5 nm typically), the hardness increase can disappear, because the strain field around the dislocations falls mainly outside the particular layer. Also, the sharpness of the interface also affects the hardening mechanisms, because high-temperature interdiffusion between layers can decrease the sharp variation in shear modulus. Another possible reason for increased hardness and improved chemical stability is chemical differences intentionally inserted at the interfaces so as to induce strains and electronic interactions; for example, TiAlN/CrN multilayer structures are more efficient than TiAlN films [84]. The hardness effect resulting from lattice mismatch disappears at multilayer periods greater than 10 nm, due to lattice relaxation (Figure 1.12). Hence, for practical applications, the thickness of the layers in these coatings is designed based on the above considerations so as to obtain the optimum hardness values and wear properties with appropriate temperature stability. In fact, for many coating applications (e.g., coatings for cutting tools), toughness at high operating temperatures and chemical stability are more crucial than hardness alone [85]. Commercial multilayer coatings with multilayer periods in the nanoscale range do exist; for example, WC/C coatings used in the cutting tool industry. When the multilayer nanoscale coatings are deposited by periodic variation of the deposition (e.g., sputtering) conditions, a templating effect is often observed. In building superlattice structures from different materials of different structures, the layer first deposited can force the next layer (of the different material) to adopt the crystallographic structure of the first layer; this occurs, for example, in TiN/Cr_2N coatings in which Cr_2N is forced into the structure of the TiN (fcc) underlayer [86] and in TiN/AlN, in which the AlN (originally wurtzite) is forced into the NaCl structure of the TiN.

In addition to nanoscale multilayer coatings, it is also possible to fabricate isotropic nanocomposite coatings consisting of crystallites embedded in an amorphous matrix, with grain sizes in the nanometer range. These coatings generally have one phase that is hard (load bearing; e.g., transition metal carbides and nitrides) and a second phase that acts as a binder and provides structural flexibility (amorphous silicon nitride, amorphous carbon). The formation of these composite structures involves phase separation between two materials (which show complete immiscibility in solid solution),

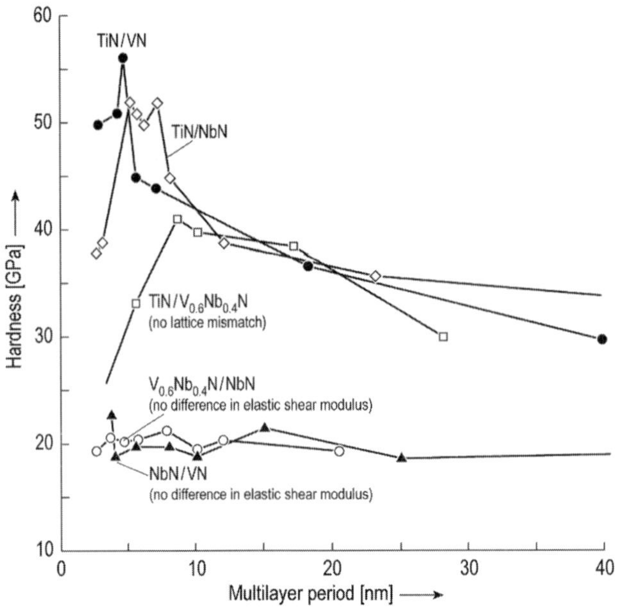

Fig. 1.12 Hardness of several multilayered nanocomposite coatings as a function of multilayer period in the coatings. The individual layers in these coatings can differ in inherent elastic properties (modulus) as well as in structure and lattice spacing. (Source [78] used with permission)

which are often codeposited, for example, by sputtering or plasma deposition. Unlike in multilayer composite systems, the possible material compositions and particle sizes in nanocomposite coatings are restricted by material properties and deposition conditions. Typically, these nanocomposite coatings are deposited by plasma-assisted chemical vapor deposition (PACVD) or physical vapor deposition (PVD). Very hard (~50–60 Gpa) coatings of nanocomposites have been made from a TiN/a-Si_3N_4 system, using PACVD from $TiCl_4$, $SiCl_4/SiH_4$, and H_2 at about 600 °C [87]. The nanocomposite contains nanocrystalline TiN (4–7 nm) in a matrix of amorphous Si_3N_4. The gas phase nucleation (uncontrollable rates), chlorinated precursors (unreacted species remain in the process, contaminating the films), and high processing temperatures are disadvantages in this process. PVD processing can be used to prepare the same coatings by sputtering Ti and Si targets in nitrogen gas at room temperatures. The disadvantage is that the films are of inferior quality, and the hardness is lower than the PACVD-deposited nanocomposite coatings. In addition to improved hardness, nanocomposite coatings also show better oxidation resistance in comparison to TiN coatings.

It is interesting to note and analyze the high hardness of these nanocrystalline particles/amorphous matrix coatings. Typically in single-phase nanocrystalline materials, the major increase in hardness comes from the lack of plastic deformation because the dislocations face far more barriers to mobility. However, hindered dislocation alone cannot explain the superior hardness properties of these nanocomposite coatings.

Since the nanocrystallites within the amorphous matrices are only a few nanometers large, dislocation formation simply does not occur, and hence, plastic deformation is largely quenched. Plastic deformation occurs through a pseudo-plastic deformation in which the nanocrystals move against each other. Since this process requires higher energy, resistance to plastic deformation in these materials is relatively high. Crystalline carbides in amorphous carbon matrices are a good example of this category of nanocomposites and can provide hard, low-friction coatings [88]. Ti/C systems are most promising in this regard. Advances in laser-assisted deposition techniques (e.g., magnetron sputtering-assisted pulsed-laser deposition) have aided the fabrication of hard composites with nanocrystalline and amorphous phases [89]. The volume fraction and particle size of the coatings can be adjusted to obtain optimum properties in terms of toughness and hardness. The design of appropriate particle size allows for the optimum generation of dislocations and micro- and nano-cracks, which result in a self-adjustment in composite deformation from hard elastic to plastic at loads exceeding the elastic limit. Thus, compliance of the coatings is improved and catastrophic brittle failure is avoided. Such load-adaptive nanocomposites based on optimal design of the composite microstructure are extremely useful in applications subject to wear.

Hard coatings such as TiN and Ti-C-N/DLC have significant advantages in aerospace systems. Apart from their hardness, toughness, and low friction properties, these coatings also resist attack by corrosive fluids that are used (as engine oils or lubricants) in aircraft engines and parts. Magnetron sputter-assisted pulsed-laser deposition is a good candidate technique for depositing such coatings in different configurations, such as functional-gradient, multilayer, and granular nanocomposites [90]. The tailoring of the different architectures in creating appropriate coatings that blend the various tribological properties is key to creating high-performance hard coatings. Nanostructured coatings provide several pathways for reducing stress in hard coatings and terminating cross-sectional dislocations, with possibilities for controlling dislocation movement by dispersion strengthening or lattice mismatching between alternate layers, as discussed above.

Other systems that create excellent nanocomposite-based, low friction, hard coatings are based on carbide particles (Ti, Ta, Nb, etc.) in DLC matrices [91–93] (Figure 1.13). Several techniques, such as PVD, pulsed-laser deposition, and reactive magnetron sputtering, are used to deposit these coatings. These typically contain larger crystalline particles (10–50 nm) surrounded by thick amorphous carbon coatings (5 nm). The particle sizes are large enough to allow dislocations but are too small for crack propagation. The larger grain separation allows incoherency strains to develop and, under loading, cracks to originate between crystallites, which allow pseudoplastic deformation. Thus, their hardness is also much higher ($\sim$30 Gpa) than that of single-crystalline carbide materials, and these coatings have, in addition, much higher toughness. These types of coatings can also be modified by introducing other elements; for example, W or Cr for creating optically absorptive coatings for solar energy converters [94] and materials such as MoS_2 (TiN/MoS_2, TiB_2/MoS_2) for lubricating coatings [95].

Metal carbide/ductile metal systems are considered for cutting tools, because the carbide phase provides hardness and the metal provides toughness. For example, composites such as WC/Co, WC/TiC/Co are commonly used in cutting and forming tool

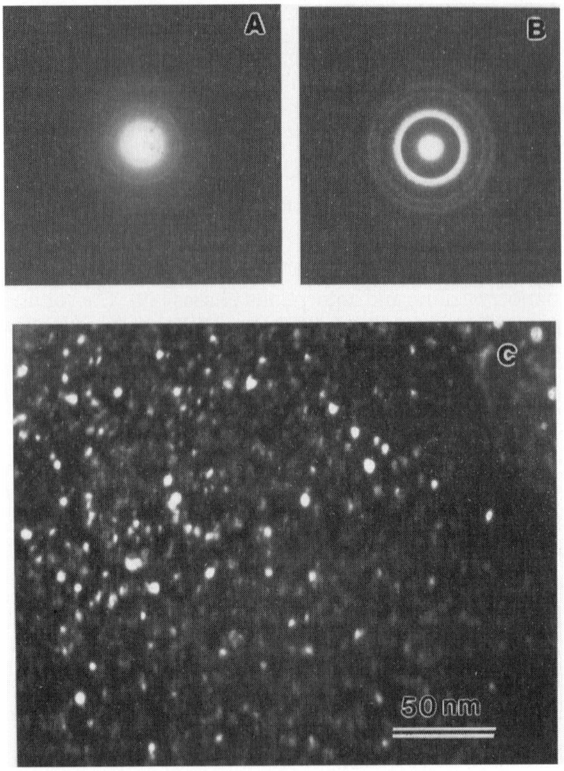

Fig. 1.13 Plan-view transmission electron microscope examinations of W-DLC layers: (a) electron diffraction pattern of a W-DLC with 9.7 at. % W, (b) electron diffraction pattern of a W-DLC with 32.0 at. % W, (c) dark-field micrograph of the W-DLC layer with 9.7 at. % W, taken with WC rings. (Source [93] used with permission)

applications, and these can be synthesized by various routes [96]. These nanocomposites, in which the particle or grain sizes of the component phases are in the nanometer range, have much better mechanical properties (strength, hardness, toughness). Typically, reductive decomposition of W- and Co-containing salts, followed by gas-phase carburization (with CO/CO_2) are used to prepare nanocomposites such as WC/Co. These processes produce carbon-deficient metastable carbide phases with inferior mechanical properties. Alternative approaches, in which a polymer precursor such as polyacrylonitrile is used as the carbon source during the chemical synthesis and subsequent heat treatment to obtain high-quality nanocomposites (particle sizes ~50–80 nm), have been developed [97]. TaC/Ni nanocomposites are interesting from two aspects: excellent thermal stability and outstanding mechanical properties [98]. They are used as surface coatings for protection against wear and corrosion. These nanocomposites can be prepared by devitrification of sputtered amorphous films of Na/Ta/C of nonstoichiometric composition. The grain size of the matrix Ni is about 10–30 nm, and the TaC particles (~10–15 nm) are uniformly distributed in the matrix. Even at 700 °C, no grain growth is observed, suggesting excellent thermal stability for these nanocrystalline duplex-phase composites. The measured hardness of the composite (12 GPa) matches that of conventional WC/Co nanocomposites at a much reduced volume fraction (35 %) of the composite.

Chemical and physical vapor deposition (CVD and PVD, respectively), and laser ablation have been used to prepare a variety of superhard nanocomposites made of nitrides, borides, and carbides. In these systems, the hardness of the nanocomposite significantly exceeds that given by the rule of mixtures in bulk. For example, the hardness of nanocrystalline $M_nN/a\text{-}Si_3N_4$ (M = Ti, W, V, etc.) nanocomposites with the optimum content of Si_3N_4 (close to the percolation threshold) reaches 50 GPa, although that of the individual nitrides does not exceed 21 GPa. For a binary solid solution, such as $TiN_{1-x}C_x$, the hardness increases monotonically with increasing x from that of TiN to the that of TiC, following the rule of mixtures. Recently, superhardness has been achieved in coatings consisting of a hard transition-metal nitride and a soft metal that does not form thermodynamically stable nitrides, such as nanocrystalline M_nN-M' (M = Ti, Cr, Zr, M' = Cu, Ni) [99]. However, these systems have low thermal stability and their hardness decreases upon annealing to $\geq 400\,°C$.

The generic concept for the design of novel superhard nanocomposites that are stable up to high temperatures ($\sim 1000\,°C$, which is very important in industrial applications) is based on thermodynamically driven segregation in binary (and ternary) systems. These systems display immiscibility and undergo spinodal decomposition even at such temperatures. In these systems, any small local fluctuation in the composition of the mixed phase decreases the free energy of the system, thus leading to spontaneous segregation and, as a result, a nanocomposite that remains stable against grain or particle coarsening develops. For many hard coatings (e.g., PVD deposited films, in which the compressive stress is typically 4 to 6 GPa or even greater than 10 GPa), the measured enhancement of the apparent hardness and elastic modulus is due to high biaxial compressive stress. It is important to perform annealing experiments in such systems to verify what the real, 'intrinsic' hardness of the film is. The observed high resistance of superhard nanocomposites to crack formation has been studied in terms of conventional fracture mechanics scaled down to dimensions of 1–2 nm. An analysis of indentation curves measured on superhard nanocomposites in terms of Hertzian elastic response shows that they are indeed strong materials. The stress concentration factor based on nanoscale flaws is low, and therefore, the stress needed to propagate such a small nanocrack is very high. The propagation of such nanocracks in 3D nanocomposites involves much deflection and branching of the plane of the cracks, which hinder growth of the nanocracks. However, the self-organization of these systems, due to thermodynamically driven spinodal segregation, results in a very low concentration of intrinsic flaws. Thus, the remarkably high resistance of these nanocomposites to crack formation can be understood in terms of a high threshold for the initiation of larger microcracks, which may lead to their catastrophic growth.

Carbon–carbon nanocomposites: Carbon–carbon composites have received much attention due to their tremendous potential for applications in the aerospace, automobile, and energy industries. This material's attractive properties include low specific weight, low thermal expansion, high thermal shock resistance, and high strength over large temperature ranges. They are usually made with carbon fiber reinforcement, with the matrix being disordered or partially graphitized carbon infiltrated from various resins, pitches, and gaseous (hydrocarbon) precursors. The main limitation of

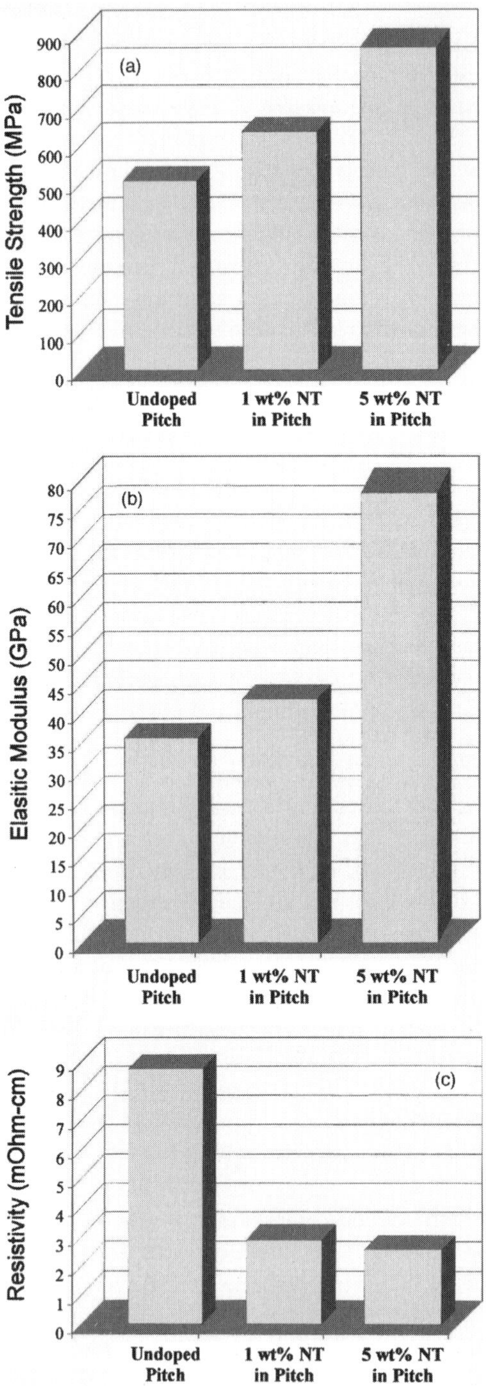

Fig. 1.14 Comparison of the tensile strength (a), modulus (b), and electrical resistivity (c) of composite carbon fibers with 1 and 5 wt. % nanotube loadings with the corresponding values in unmodified isotropic pitch fibers. Each data point represents an average value for strength, modulus, or resistivity obtained from 12 composite fibers. Standard deviations of experimental values are strength, 15%; modulus, 15%; and resistivity, 20%. (Source [100] used with permission)

these composites is their susceptibility to oxidation, but surface modification or coating of the fibers or addition of appropriate ceramic refractory fillers that shield against oxygen attack at high temperatures alleviates this problem. The advent of several novel forms of nanocarbons has led to attempts to create nanocomposites with nanocarbon (e.g., nanotubes) components. In a recent experiment, single-walled carbon nanotubes (5% by weight) were uniformly dispersed in an isotropic petroleum pitch matrix to make nanotube-based carbon nanocomposite fibers (Figure 1.14), with better electrical and mechanical properties than isotropic pitch-based carbon fibers [100]. The tensile strength increased nearly 100%, and the electrical conductivity increased fourfold. The advent of the new carbon nanostructures suggests that novel carbon–carbon composites can be prepared with properties far superior to those of conventional carbon–carbon-fiber-based composites. The challenges lie in achieving proper dispersion of these nanostructures and tailoring the interfacial strength between the nanocarbons and the matrix.

1.7
Carbon Nanotube-Based Nanocomposites

The mechanical behavior of carbon nanotubes is exciting, since nanotubes are seen as the ultimate carbon fiber ever made [101]. The most important application of nanotubes, based on their mechanical properties, will be as reinforcements in composite materials. Chapter 2 includes a detailed discussion of nanotube-based polymer composites; here, we briefly discuss some developments regarding ceramic/metal-matrix-based nanotube composites. The nanotube reinforcements promise to increase the fracture toughness of the composites by absorbing energy through their highly flexible elastic behavior during deformation, which will be especially important for nanotube-based ceramic matrix composites. Possible applications are in lightweight armor or conductive durable ceramic coatings. An increase in fracture toughness on the order of 10% has been seen in nanotube/nanocrystalline SiC ceramic composite fabricated by a hot-pressing method at 2273 K (25 MPa in Ar for 1 h) [102].

Nanoscale ceramic powders with carbon nanotubes provide another opportunity for creating dense ceramic-matrix composites with enhanced mechanical properties. The strength and fracture toughness of hot-pressed α-alumina is typically much greater than that of conventional grain-size polycrystalline alumina. Addition of carbon nanotubes to the alumina results in lightweight composites with even greater strength and fracture toughness (Figure 1.15). The mechanical properties of such composites depend strongly on the processing method and surface treatment of the carbon nanotubes. Sintered alumina has high strength, hardness, and fracture toughness. An exciting possibility, as well as a processing challenge, is incorporating carbon nanotubes into an alumina-matrix composite to improve these properties. Alumina (γ-phase) matrix composites with 5–20 vol. % MWNT (multiwalled nanotubes) have been fabricated [103]. The γ-phase alumina powder was transformed to α-alumina (with mean particles sizes close to 60 nm) during sintering at 1300 °C. The MWNT were lightly oxidized at 640 °C in air, which removed the disordered carbonaceous material and

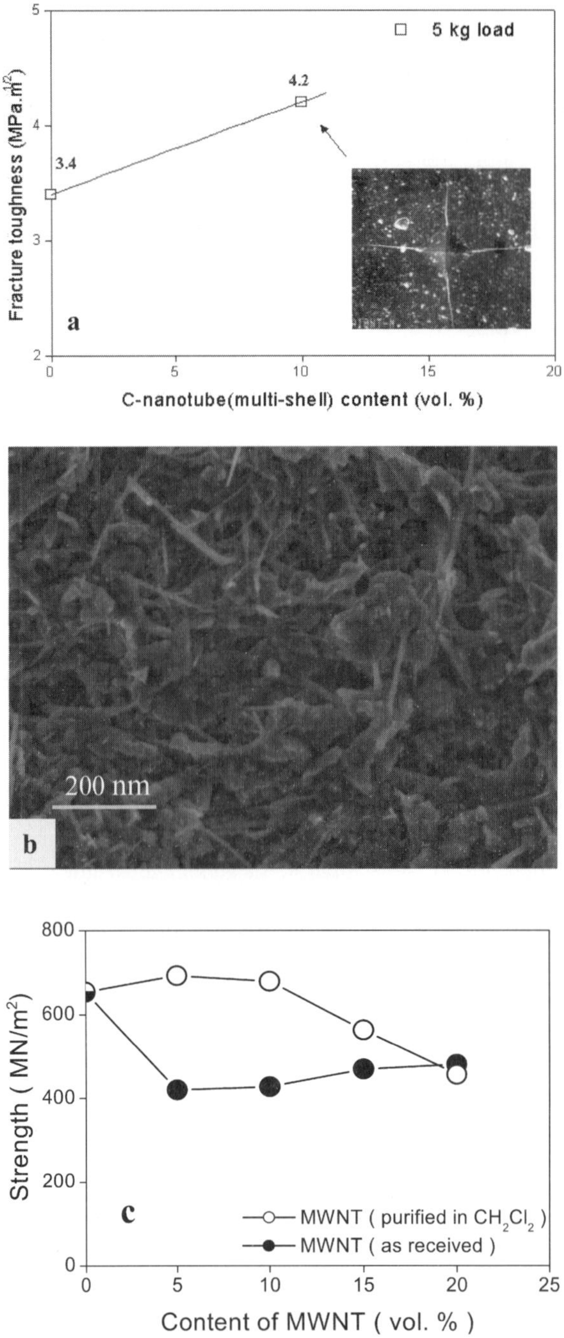

Fig. 1.15 (a) Fracture toughness of multiwalled nanotube (MWNT)/alumina nanocomposite compared with that of sintered nanophase alumina. The inset in (a) shows an indent and resulting cracks. The microstructure of the nanocomposite is shown in (b). (c) Diametral strengths of alumina-matrix nanotube composites hot-pressed in Ar at 1300 °C and 60 MPa for 1 h and containing various amounts of MWNT. (Source [103])

made it easier to disperse the nanotubes. The alumina- MWNT mixtures were dispersed in organic solvents via ultrasound, dried, and then sintered by hot pressing in a graphite die at 1300 °C and a pressure of 60 Mpa. The structures of the MWNT remained the same before and after processing, without any visible degradation. The density of the sintered composites was > 97% of the theoretical density.

The diametral strengths of alumina-MWNT composites with different MWNT contents indicate great improvements in fracture toughness with addition of nanotubes. The bulk alumina made from nanoparticles had a strength of 654 MPa, compared to typical strengths of 200 – 350 MPa for sintered bulk alumina, and was even comparable to strengths reported for single-crystal alumina (sapphire) of 350 – 1000 MPa. With the addition of purified, well-dispersed nanotubes, the strength first increased at 5 and 10 vol. % MWNT and then decreased for higher vol. percents of MWNT. The fracture toughness of bulk alumina and alumina/MWNT composites shows that, for 5 vol. % MWNT composites, the toughness increased by about 25% to about 5.1 MPa $\sqrt{m}$. The toughness values were higher than those reported for single-crystal alumina (sapphire) and polycrystalline alumina. The hardness of the composite depends on how well the nanotubes are dispersed in the matrix and how the nanotube surface is modified (with functional groups via oxidation) to produce anchors to the matrix; for the best value, the Vickers hardness is 20.4 GPa compared to 18.4 GPa for bulk alumina. From the general trends in the results of a few scattered reports in the literature, the strength and fracture toughness of bulk structures hot-pressed from nanophase ceramic powders is much higher than the typical strength and toughness of conventional polycrystalline ceramics. The addition of small amounts of nanotubes to nanophase ceramics to form dense nanocomposite improves the diametral strength and fracture toughness. However, the real improvements occur when the nanotubes are purified and well dispersed in the matrix, so that large aggregates do not create macrodefects in the structures, leading to easy fracture and lower strength. Vacuum hot pressing also removes entrapped gases in the composite powder mixture, preventing formation of residual stresses and reduction of strength. These processing improvements show the great promise of nanophase alumina/ MWNT composites for light-weight, high-strength, high-toughness materials and in applications such as impact-resistant body armor.

It is widely perceived that carbon nanotubes will allow the construction of composites with extraordinary properties. Metal-matrix composites, as discussed before, are a new range of advanced materials and normally, the liquid-phase fabrication method is an efficient process because of its simple processing. If carbon nanotubes can remain stable within some suitable high-strength metal matrix, outstanding nanocomposites may be obtained. The rapid solidification technique allows the extension of alloying levels and more refined microstructures, which can lead to enhanced mechanical and physical properties. In addition, rapid solidification processing may help carbon nanotubes to survive the process because of the controllable melting step, which is the most important issue to be solved in fabricating nanotube-reinforced composites. Rapid solidification processing has achieved the successful synthesis of nanotube/ $Fe_{80}P_{20}$ metallic glass nanocomposites [104]. Their properties show that the addition of nanotubes greatly enhances the thermal stability of the glass matrix and increases

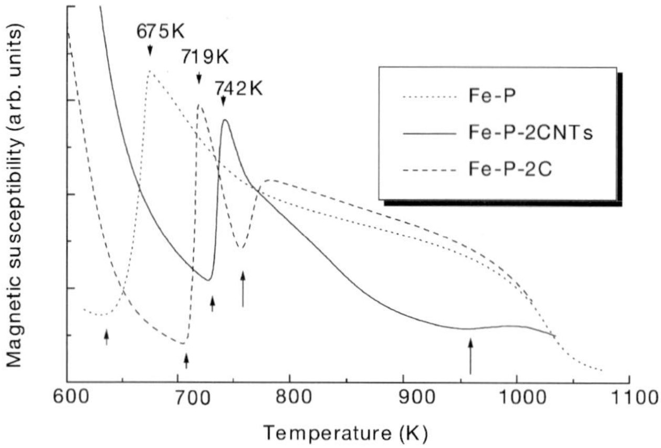

Fig. 1.16 Magnetic susceptibility vs. temperature for various amorphous alloys in nonisothermal heat treatments. The temperatures at the end of the first transition are indicated. The first peak in the plots corresponds to the first crystallization process. The onset temperature of crystallization for amorphous Fe/P/2% CNTs is about 100 K higher than that for amorphous Fe/P alloy at the same heating rate. (Source [105] used with permission)

the low-temperature electrical resistivity by 70% (Figure 1.16; Table 1.2). The activation energy of crystallization for the nanotube composite is higher than those of both the $Fe_{80}P_{20}$ metallic glass and the metallic glass containing 2% carbon. The onset temperature of crystallization for the nanotube composite is almost 100 K higher than for the original glass, according to magnetothermal analysis (MTA).

Tab. 1.2 The onset temperatures at a heating rate of 5 K min^{-1} and activation energy of crystallization for various amorphous metallic glass nanocomposite materials containing carbon (C) and Carbon Nanotubes (CNTs) [105].

Materials	Activation energy (kJ mol^{-1})	Onset T (MTA, K)
Fe/P	354.0 ± 0	630
Fe/P/2C	460.9 ± 29.8	709
Fe/P/2CNTs	574.1 ± 35.4	729

1.8 Functional Low-Dimensional Nanocomposites

Recent years have seen a drive towards creating heterogeneous nanostructures, which could serve as multifunctional materials and essentially satisfy the definition of functional nanocomposites. These stand out from the bulk nanocomposites that are the main subject of this book, but will have important applications as nanoscale sensors, electronic and optical nanodevices, nanoprobes, nanoelectromechanical systems, unique reinforcements, and drug-delivery media. These are particularly interesting for layered material systems in which atomic layers of different material compositions (e.g., graphite, dichalcogenides, clays, etc.) can be assembled into different geometries on the nanoscale. Some excellent examples, which are described below, are endohedral fullerenes (metal/carbon systems), filled carbon nanotubes, boron/carbon/nitrogen nanostructures, and nanosize coaxial cables from a wide range of layered materials.

What are the various synthetic strategies for creating such composite nanoscale structures? One simple, elegant idea is to use infiltration of nanoporous structures, commonly used in bulk composites for creating heterogeneous materials. At the nanoscale, this problem is not trivial, especially if the process occurs through the infiltration of a liquid or molten phase in which the viscosity of the infiltrating phase is high. A very useful example of such a nanocomposite structure is nanotubes and nanowires built into porous alumina templates. The alumina template can be prepared by anodization of Al metal in acidic solvents, and it is possible to create templates with straight, random pores that have diameters of 20–50 nm and are several micrometers long, running through the thickness of the alumina template [106]. Once the templates with pores are prepared, they can be filled (via electrochemical deposition) with nanowires (such as Au, Co, etc.; Figure 1.17) by changing the electrochemical conditions and introducing metal-containing solvents into the electrolyte [107]. The structure of the metal wires can be tailored (amorphous or crystalline) by changing the electrochemical deposition conditions. The distinct difference between nanowires created by other techniques and by template-based synthesis is that the latter technique offers the possibility of creating well-separated individual nanowires with adjustable diameters. Structures such as carbon nanotubes can also be deposited in the pores defined by such templates, by infiltration of gaseous hydrocarbon precursors (e.g., acetylene, benzene) and subsequent annealing [108]. Electrochemical techniques for materials processing (from highly anisotropic wires inside porous templates to particle-dispersed thin-film nanocomposites) provide a powerful alternative technique to vapor phase deposition and solid state reaction for the synthesis of nanomaterials and nanocomposites [109]. Their main advantage is control of the process so as to modulate dimensions and compositions. Their disadvantages are mainly poor crystallinity (being a low-temperature deposition process) and contamination with impurities from the electrochemical baths.

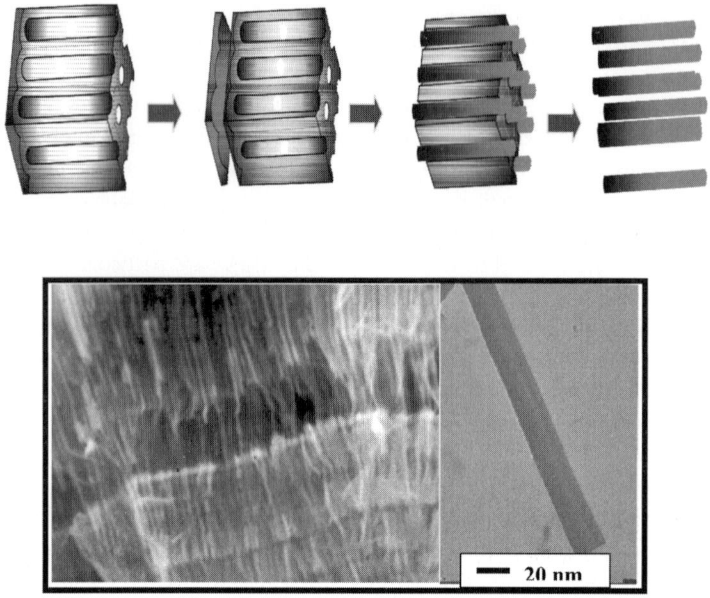

Fig. 1.17 Nanowires grown in alumina templates. Top schematic shows how pores in a straight-pore anodic alumina template are filled with metal by electrodeposition. Scanning electron micrograph of Co nanowire-filled template is shown in lower image; an individual nanowire removed from the template is shown (right). (Courtesy Prof. Mamoun Muhammed, The Royal Institute of Technology Stockholm, Sweden)

1.8.1
Encapsulated Composite Nanosystems

Here, we describe the creation and characterization of composite carbon/metal-based nanostructures (systems consisting of combinations of a layered material and a solid phase, which have been studied by several groups in recent years). These are unique nanocomposite systems with promise for many interesting applications.

Graphite-encapsulated metal/ceramic structures are interesting systems, and the generation of metal/carbon nanosystems requires knowledge of the interactions of metal atoms and the graphite lattice. The experimental study of the behavior of individual metal atoms in a graphitic environment is difficult and has been successful in only a few cases. In 1993, Ruoff et al. [110] performed carbon-arc discharge experiments using C anodes filled with metals and metal carbides and obtained carbon nanotubes and polyhedral carbon nanoparticles that contained both encapsulated metals and metal carbides. Polyhedral graphitic particles generally contain a hollow core, which is given shape by the outer faceting of the particle, and metal nanoparticles can easily fill these cores during formation (Figure 1.18). Metal surfaces are expected to act as nucleation sites for graphitic carbon to surround. The expected tendency of preexisting graphitic filaments is to wrap around metallic nanocrystallites and close by

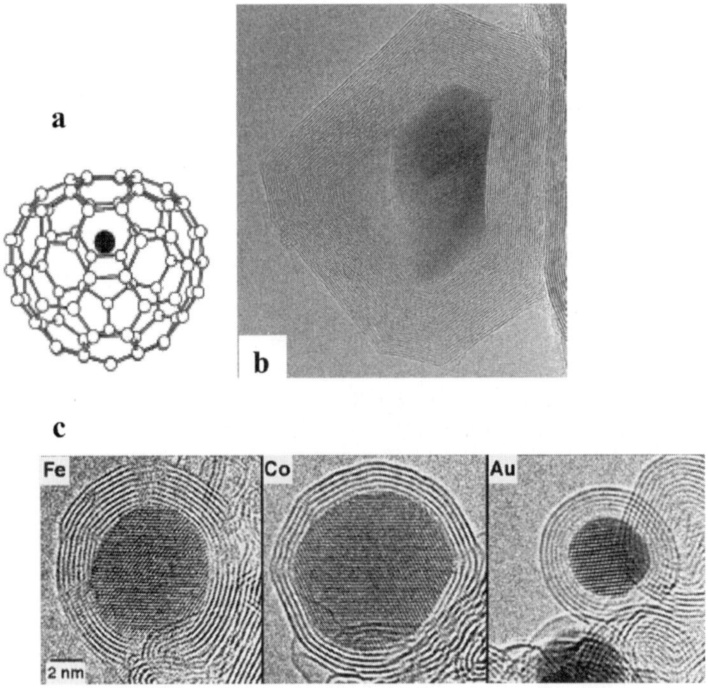

Fig. 1.18 (a) Schematic of an endohedral fullerene. Structural model of a La atom in C_{82} fullerene (b) Encapsulated nanoparticle of lanthanum carbide inside a carbon polyhedral particle (approximately 50 nm in diameter) produced in an arc discharge with a composite carbon/La electrode. (c) Encapsulated nanoparticles of metals (Fe, Co, Au) in spherical graphitic layers (carbon onions). These structures were produced by electron irradiation of carbon soot containing the metal species

saturating all carbon dangling bonds. The properties of encapsulated metal crystals have been the subject of many experimental studies [111]. Graphite-coated ferromagnetic metal crystals should show a reduced magnetic coupling that depends on the thickness of the graphite shell [112], with potential applications in magnetic recording media. Guerret-Piecourt et al. [113] sought general principles relating to the nature and structure of the filling materials, for efficient filling of nanotubes made via the electric-arc method (Figure 1.19). Several metals and/or their compounds have been encapsulated in carbon cages and studied; e.g., Ti, Cr, Fe, Co, Ni, Cu, Zn, Mo, Pd, Sn, Ta, W, Gd, Dy, and Yb. In later experiments, the authors also managed to fill hollow carbon nanostructures with semiconducting structures such as Se, S, Sb, and Ge by similar processes [114]. S, present as impurity in the carbon, was later reported to play a major role in the effective formation of the filled carbon nanotube/nanoparticle structures. The above technique has made it possible to encapsulate a large number of carbides nanocystallites [115, 116] (LaC_2, YC_2, CeC_2, Gd_2C_3, TiC, V_4C_3, ZrC, TaC, MoC, NbC, HfC). With pure metals such as Fe, Ni, or Co, which strongly interact with carbon, encapsulating nanowires inside nanotubes is challenging. However, experiments have shown that onion-like nanoparticles of metal-containing carbon clusters can be produced relatively easily with the arc method. Here, the metal or metal carbide encapsulate is protected by the C layer coating, which effectively insulates the magnetic metal particles from oxidization upon exposure to the environment and retains its magnetic properties. It is noteworthy that high melting point refractory metal carbides such as TaC, NbC, and MoC, which exhibit superconducting transitions at low temperatures, have been inserted into nanotubes and graphitic nanoparticles [117]. Bulk magnetic susceptibility measurements of encapsulated TaC, MoC, and NbC revealed diamagnetic responses and superconducting transitions between 10 K and 14 K, respectively. The structure and state of these encapsulated nanostructures affects their physical properties; for example, a shrinkage in the lattice parameters for the (fcc) MoC encapsulated structure appears to decrease the Tc by 3.8 K.

Magnetic alloy (e.g., FeCo) nanoparticles encapsulated in carbon layers can also be made in radio frequency (RF) torch reactors, and acetylene was used to generate a large number of carbon-coated particles [118]. The carbon lamination provides several functions for these nanoparticles that have low magnetocrystalline anisotropy and large saturation magnetization (making them valuable as soft magnetic materials). These functions include oxidation protection, prevention of coarsening and particle coalescence during growth, shielding interparticle magnetic interactions, and even reducing eddy current losses in high frequency environments. This technique can readily produce particle sizes anywhere from 5 to 50 nm. Other techniques, such as the tungsten arc, blown arc, or electric arc generation, can be used to create encapsulated nanostructures of similar dimensions.

The first experiments on filling of nanotubes date to 1993, where the filling was done via capillary forces, which pushed molten metals and compounds into the cavities of nanotubes during oxidative opening of their tips (Figure 1.20) [119]. Since then, the most efficient filling of hollow nanostructures to create heterogeneous nanomaterial systems has been achieved by chemical treatment. Tsang et al. [120] filled larger

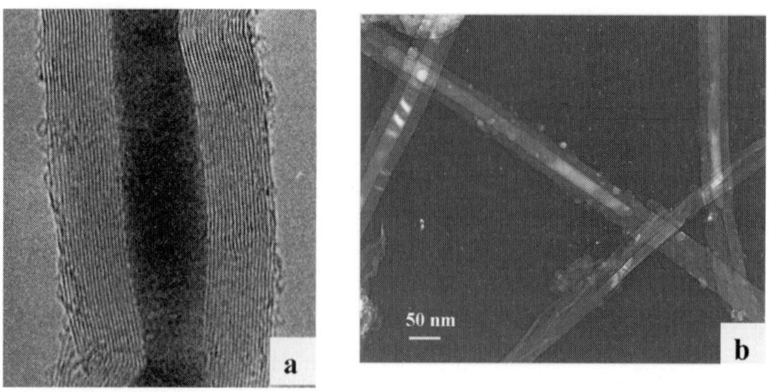

Fig. 1.19 (a) High-resolution transmission electron micrograph of a Mn metal-filled carbon nanotube made by electric arc discharge with electrodes of carbon filled with Mn (b) Encapsulated nanotubes with Fe carbides, formed during a chemical vapor deposition process involving hydrocarbon and metalorganic precursors

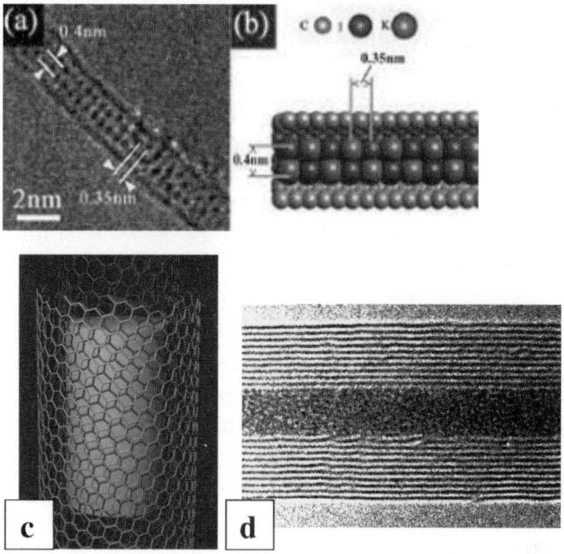

Fig. 1.20 Top: (a) High-resolution transmission electron micrograph of a KI crystal in a 1.4-nm diameter SWNT, (b) its structural representation. Encapsulation of halides and other materials within SWNTs enables the study of 1D crystal structures formed on the smallest scale possible. Bottom: (c, d) Schematic of a filled MWNT and a corresponding transmission electron micrograph of Pb oxide nanowire-filled nanotube, formed by capillary filling of the molten material inside nanotube cavities. In (d) the horizontal lines indicate graphite planes which have a separation of 0.34 nm. (Source, top image and schematic [233] used with permission)

multi-walled carbon nanotubes oxidatively by opening the ends in nitric acid and then filling with solutions containing metal oxides (e.g., oxides of Ni, Co, Fe, U, Mo, Sn, Nd, Sm, Eu, La, Ce, Pr, Y, Zr, Cd) using capillary force. They were then dried to form encapsulated solid materials inside the nanotube cavities. Several other material systems such as Pd, Ag, Au, AuCl, proteins, and enzymes have been introduced into nanotubes with this method. Recently it was reported that single-walled nanotubes (SWNTs) can be opened selectively and filled by wet-chemistry techniques [121]. Treatment with concentrated HCl leads to the selective opening of SWNTs at their tips, and foreign materials may be drawn in, in a similar fashion as MWNTs; the much smaller ($\sim$1 nm) cavities in SWNT make the filling more challenging. Usually, the nanowires formed inside the nanotubes are single crystals. Alternative strategies to fill nanotube cavities with materials have been explored. For example, exposing SWNT nanotubes filled with AgCl/AgBr to light or an electron beam can reduce the encapsulated compounds, resulting in extremely small Ag nanowires [122]. Similarly, constrained chemical reactions within narrow channels can be used to create and characterize nanowire structures of many materials. These promise to be ideal structures for studying 1D properties (electrical, thermal, optical) in materials for applications such as quantum wires and magnetic storage arrays. Characterization of these structures is often challenging, but recent progress in highly spatially resolved microscopy and spectroscopy techniques have enabled researchers to carefully study the structure–property relationships in confined nanocomposite wires. SWNTs containing fullerenes in their inner hollow are another unique form of nanocomposite structure with interesting electronic properties. C_{60} insertion into SWNT could occur during their formation in the arc or laser process. Very recently, researchers have managed to introduce C_{60} inside SWNTs in high yield (Figure 1.21) by thermally annealing C_{60} powders over SWNTs above 600 °C under vacuum [123]. This class of hybrid materials may form the basis of functional devices for applications in electronics, biomedicine, and microelectromechanical systems. Similarly, endohedral fullerenes [124] (or fullerenes containing single or multiple metal atoms, which in their own right can be considered novel nanocomposite systems; e.g., $La_2@C_{80}$, $Gd@C_{82}$, etc.) have been introduced into SWNTs through the vapor phase. These structures are breaking new ground because creation of nanocomposite structures with discrete atomic species in their cores are resulting in the discovery of novel properties of 1D nanoscale materials.

Recent theoretical studies suggest that the presence of a metal nanowire inside a nanotube greatly alters the mechanical properties of the nanotube, for example, suppressing tube-buckling instability. Increasing tube diameter increases the bending strength; however, in contrast to hollow tubes, there is no decrease in the maximum deflection before buckling. Analysis of the principal bending vibrational mode shows a lowering of the frequency, associated with increased tube inertia. Remarkably, metal-filled tubes exhibit strong damping, whereas unfilled single-walled and multi-walled tubes show no damping of oscillations (Figure 1.22). Simulations of multi-walled nanotubes revealed the lack of damping behavior; the dissipation of the mechanical energy is intrinsically associated with tube filling by a dissimilar material [125]. These studies demonstrated the benefits of filling tubes with solids for modifying the bending strength and flexibility, suggesting implications for nanotube-based elements in micromechanical devices or nanoprobes. Experimental studies on these unique 1D nanocomposite systems are eagerly awaited, since a variety of them have already been fabricated.

Condensed-phase electrolytic processing provides an alternative way to fabricate nanotubes with or without encapsulated metals. The technique involves passage of an electric current through graphite electrodes immersed in molten mixtures of ionic salts (e.g., LiCl, $SnCl_2$) under inert gas atmospheres [126]. The graphite crucible anode

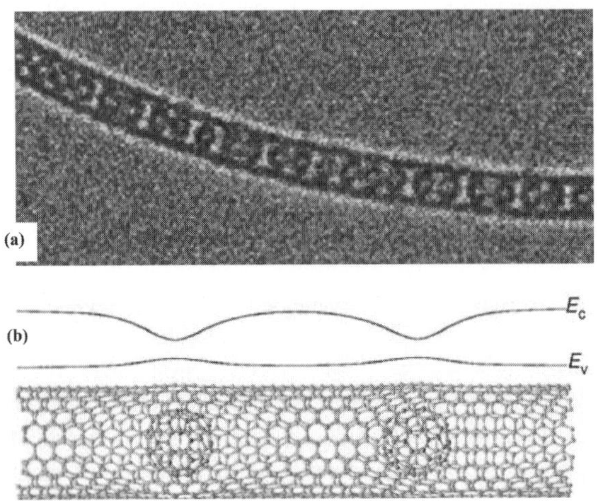

Fig. 1.21 (a) Transmission electron micrograph of a $La_2@C_{80}@SWNT$ assembly. Each C_{80} molecule appears as a dark circle, and the SWNT (1.4 nm diameter) itself appears as two parallel lines. (b) Effect of inserted Gd metallofullerenes (GdMFs) on the topography and band structure of a single-walled nanotube (SWNT). The illustration of a (11,9) GdMF-SWNT. Schematic representation of estimated modulation of conduction (Ec) and valence (Ev) band edges is also shown. (Source: top image [234], bottom part [235] used with permission)

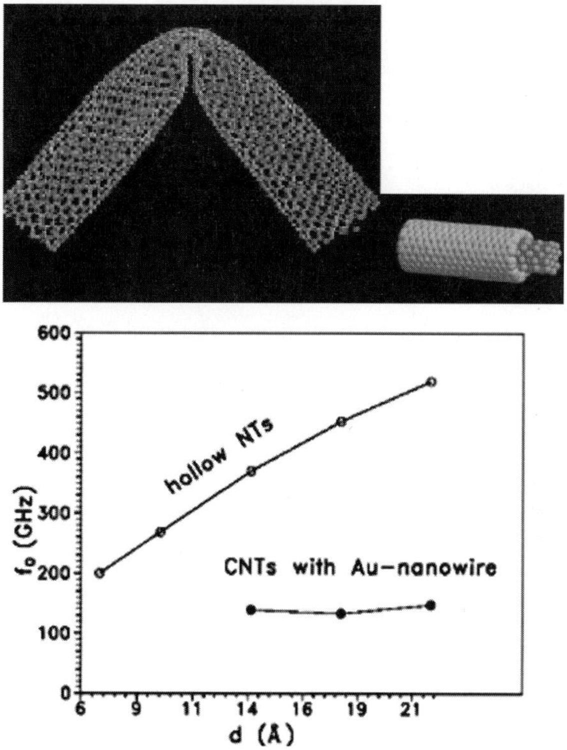

Fig. 1.22 Damping in pure vs. metal-filled nanotubes, simulation results. Top: bending of filled nanotubes; filling makes the nanotubes stiffer to bending and buckling. Oscillatory frequency (f_o) of vibration, as a function of tube diameter for filled and hollow tubes as function of tube diameter. The filled nanotube composite structures show high damping during vibrations. (Source [125])

contains the electrolyte salt, which is heated by an external furnace. When the salt mixture melts, the cathode (graphite rod) is immersed in the electrolyte. During electrolysis, the graphite electrode interacts with the molten electrolyte, forming nanostructures. If pure LiCl is used as the electrolyte, the carbonaceous material formed consists of tangled carbon nanotubes, some of which contain the carbide of the salt (i.e., LiC). Adding small amounts of low-melting metals such as Sn, Pb, and Bi to the electrolyte forms nanowires of the respective metals. The formation of carbide nanowires is based on the reduction of metal ions on the graphite cathode surface and intercalation of the Li between the graphite layers, followed by the formation of Li_2C_2. Carbon then precipitates in the form of nanotubes. When Sn, Pb, and Bi are added to the electrolyte, the cations derived from these metals are discharged at the electrode surfaces by the electrons provided by oxidation of the carbide. This ultimately forms metallic nanowires surrounded by graphitic sheets in the nanotube form. This technique has also generated carbon-encapsulated Sn/Pb nanowires [127].

Pyrolysis of hydrocarbons over metal catalysts at elevated temperatures also produces filamentous structures containing metal and carbon [128]. If the catalyst concentrations are high (>10–20%), encapsulation of the metal catalysts can produce nanowire composites. Pyrolysis of organic compounds or other carbon sources such as C_{60} in the presence of transition metal species (e.g., metallorganics such

Fig. 1.23 Scanning electron micrograph showing the fibrous structure of carbon nanotube/V_2O_5 nanocomposite. The individual nanotubes (seen as tiny filaments) are finely coated with subnanometer layers of the oxide, uniformly on the surface, modifying the nanotube surface properties. The structure was made by mixing nanotubes and oxide powder and annealing.

as ferrocene) leads to nanotubes containing these metals or their alloys [129]. Metal nanowires over carbon structures can also be prepared via substitution. Recent work suggests that suspending SWNTs and exposing them to metal vapor can be used to fabricate nanowires of Pt, Au, Ti, etc. Hollow nanotubes of the metal are formed around nanotube templates [130]. Composite nanotube-based nanocomposites consisting of nanotubes and other materials can also be created by coating nanotube templates by capillary effects. Nanoscale skins of layered oxides such as V_2O_5 can be uniformly coated onto nanotube surfaces [131] (Figure 1.23). It is also possible to create composite nanowire systems by CVD techniques, in which individual layers of laminated materials such as MoS_2 and WS_2 are coated onto multi-walled carbon nanotubes [132]. The electronic properties of these coated tubes are dominated by their outermost shells (oxides or sulfides, which are generally semiconducting or insulating). However, the inner tube structure (carbon nanotubes) may dominate the mechanical properties. In this way, it will be possible to tailor both the mechanical and electronic properties of these 1D systems. Physical property characterization of these nanocomposite systems is challenging, but much excitement exists about uncovering a rich range of properties inherent to these nanoscale multiphase systems.

Multiple-phase nanowires, constituting different structures along the wire length, may also become important in future nanotechnology. The advantage here will be the ability to form controlled junctions, which could have semiconducting properties (metal–semiconductor junctions), novel deformation behavior (whereby toughness may be increased when they are used as reinforcements), photoluminescence (if

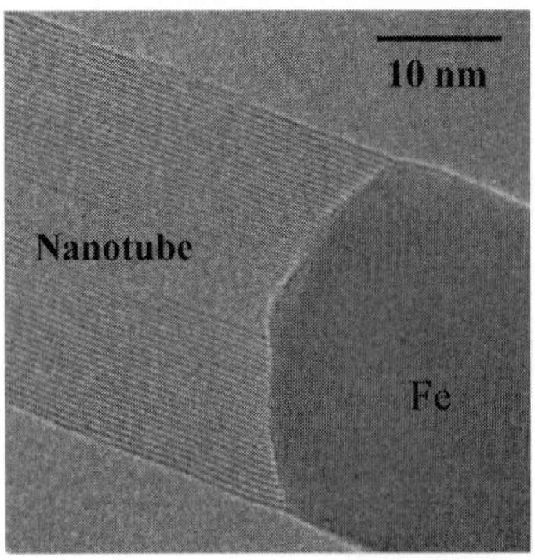

Fig. 1.24 High resolution transmission electron micrograph of a heterojunction formed between an Fe nanowire and a multiwalled carbon nanotube. The structure was formed in the vapor phase during pyrolysis of hydrocarbon and ferrocene. Similar structures can also be made by stepwise electrodeposition inside alumina nanoporous membranes.

phases are semiconducting with distinct n- or p- character), or thermoelectric-junction properties. Carbon nanotube–metal junctions have been made by using excess catalyst (Fe) in the chemical vapor deposition process or by electrochemical deposition in porous alumina templates (Figure 1.24) [133, 134], but the properties of these multiple-phase nanowires have not been evaluated. Controlled solid-state reactions have been used to form heteronanowires of single-walled carbon nanotubes and nanorods of silicon carbide and transition-metal carbides (TiC) [135]. The reactions are carried out via contact annealing with a metal (or Si) solid surface and nanostructures such as nanotubes. The structures created are some of the smallest heterojunctions that have ever been made and will play a large role in the fabrication of future hybrid nanodevices. Similarly, techniques such as laser ablation have been used to form coaxial cables containing heterocompositional layers. For example, high resolution transmission electron microscopy and spatially resolved electron energy loss spectroscopy have characterized wires that consist of β-silicon carbide sheathed inside amorphous silicon oxide surrounded by graphitic shells made of C or h-BN [136] (Figure 1.25). These coaxial nanocables of semiconductor-insulator-metal/semiconductor structures could find applications in nanoscale electronics. The formation of such multiple-layer nanocable structures is challenging and has to be controlled through the self-assembly of various elemental and molecular species in the vapor phase during relatively short growth times, and the directed assembly of such complex structures is one of the most crucial steps in the creation of useful low-dimensional functional nanocomposite structures.

Recent work has also demonstrated that semiconductor nanowires can be controllably grown in cross-bar array architectures, and, if the crossing nanowires can be doped into n- or p- type, then light emission can result at the crossing junctions when a voltage is applied across the wires. High-temperature laser-assisted deposition

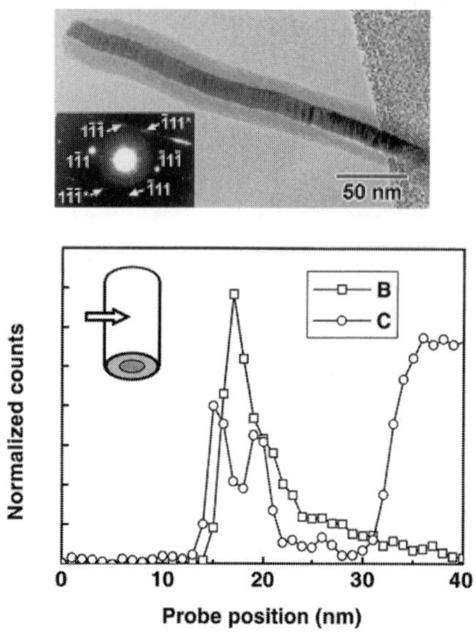

Fig. 1.25 Transmission electron micrograph of nanocable structure ($SiC/SiO_2/BN/C$) showing a crystalline core and an amorphous layer. Inset: the selected area diffraction pattern indicates that the crystalline core is β-phase SiC with the <110> axis parallel to the electron beam. Bottom: profiles of B and C across half a nanocable from EELS measurements, showing a phase separation of graphite and BN layers. (Source [136] used with permission)

has been used to grow single-crystal indium phosphide (InP) nanowires using catalytic gold particles [137], either in p-doped or n-doped form. Superlattice nanowires can also be made with alternating blocks of different semiconducting materials (e.g., Si and SiGe) on a single wire [138]. A hybrid pulsed-laser ablation/chemical vapor deposition method can be used to custom-create various heterostructures on individual nanowires. These structures systematically contain junctions and interfaces between the multiple blocks present in the same wire, enabling them to be useful in miniaturized transistors, LEDs, and lasers.

1.8.2
Applications of Nanocomposite Wires

Of several applications proposed for composite nanowire structures, only one is discussed here. Magnetic recording exploits the hysteresis behavior of ferromagnetic materials: the residual magnetization of a tape or magnetic disc is a record of the maximum field intensity the material experienced with respect to the magnitude and direction of the applied field. For long-term storage, usable magnetic materials should have a reasonably high coercivity, because external magnetic fields should not alter the magnetization. However, the coercive force should not be too large if the medium is to be reusable, since the magnetic fields used during recording should be capable of remagnetizing the material for new data recording. Recently, feature sizes on magnetic disks made using complicated e-beam lithography protocols have shrunk to the sub-100 nm range [139]. Ferromagnetic Fe nanowires could provide an ideal system for recording, but the tendency to oxidize is strong in nanoscale metallic systems. The encapsulation of Fe nanowires within carbon nanotubes will assure that the metal is maintained in the reduced state, due to the protective C coating. These encapsulated nanowires exhibit magnetic coercivities larger than those associated with pure Ni and Co nanowires [140]. The coercive field decreases linearly with increasing temperature, revealing the presence of anisotropy. The small size, anisotropy, and the single (isolated) domain nature of the encapsulated Fe crystals appear to be responsible for this enhanced coercivity. The magnetic properties of encapsulated Fe nanowires do not degrade with time, because they are sealed within the graphitic tubes. These nanowire arrays may have significant potential in magnetic data storage device applications (e.g., quantized magnetic disks) due to their size and anisotropic behavior, with the possibility of drastically increasing the attainable recording density. Film structures of this kind, based on single-domain elements, exhibit the best available storage densities (e.g., 65 Gb in^{-2}) [141]. The type of magnetic nanocomposite particles should also find applications in the fabrication of fine-particle magnets for use in magnetic inks and as toners in xerography.

1.8.3
Applications of Nanocomposite Particles

Several strategies are being pursued for creating nanoparticles containing multiple shell structures (for multifunctionality), with each shell composed of different materials. Template-based approaches are effective in this situation, for which an inner removable template particle (silica, polymer beads) can be used to coat shells of other materials (e.g., a metal) via multi-step colloidal or vapor-phase assembly and can later be easily removed to create empty shells. These could even be filled with different materials to produce multiple-shell composites [142]. Creating uniform coatings on particle templates by colloidal self-assembly is based on the concept of self-assembled organic molecular species. The two ends of the molecules to be joined have specific functional groups (e.g.. thiols, amines, carboxylic groups) that can be targeted for specific interactions with the template and the clusters that are used to make the coatings. Uniform, dense packing of the molecules around the templates leads to close packing

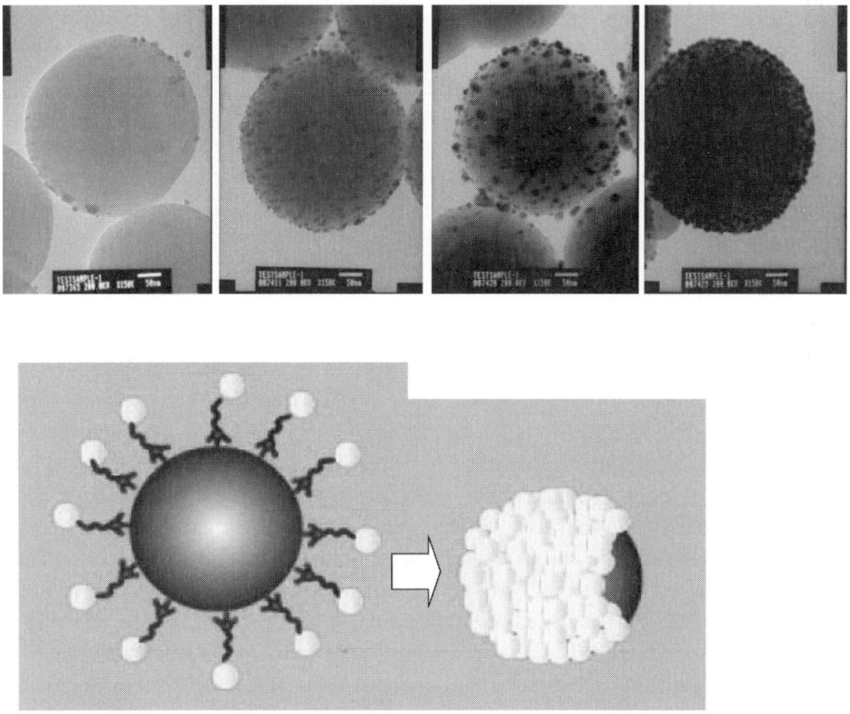

Fig. 1.26 Fabrication of multilayer nanoparticle structures by colloidal synthesis. Top: micrograph of silica spheres being continually coated with gold nanoparticles until a uniform coating is produced. The process of attachment of gold particles (shown in the schematic below) occurs via connectors of self-assembled monolayers of bifunctional organic molecules (e.g., mercaptopropyltrimethoxysilane). Multiple-layer structures can be fabricated by colloidal assembly techniques. (Courtesy Prof. Mamoun Muhammed)

of the clusters that form a porous but space-filled shell around the template (Figure 1.26). Noninteracting metal-coated magnetic particles (SiO_2/Au, Fe_3O_4/Au, NiO/Co, etc.) or coated semiconducting particles (PbS/CdS) are examples of such composite particle structures. These structures can have applications in magnetic recording, as multilayered catalyst materials, or in drug delivery. In such applications of drug delivery systems [143], for which a biocompatible outer layer and a drug-containing inner core are necessary, the multilayered particle approach will be crucial. As an alternative to colloidal templating, structures such as PbS-coated CdS nanocomposite particles (a few nanometers in diameter) may be synthesized by ion displacement in inverse microemulsions and could be useful in nonlinear optical applications. The observed large refractive nonlinearity in these nanocomposite particles may be attributed to the optical Stark effect and to strong interfacial and internanoparticle interactions.

1.9
Inorganic Nanocomposites for Optical Applications

Nonlinear optical effects, such as nonlinear optical absorption and second- and third-order optical nonlinearities, can be used to make optical limiters, optical modulators, etc. Although many organic materials have high optical absorption and nonlinearity, their thermal and optical stabilities are poor. Often, it helps to create hybrids or composites (organic/inorganic, inorganic/inorganic) that have acceptable optical properties and stability. Quantum-confined nanoparticles have been extensively used in the fabrication of such composites because of their novel optical and electronic properties. The organic-matrix nanocomposites for optical applications are described in chapter 2, which briefly discusses some of the organic materials that contain dispersed nanoparticles in various hosts.

Recent advances in controlling the fabrication and dispersion of semiconductor nanoparticles in polymer and ceramic matrices have suggested possible uses for such nanocomposites in optical applications. A good example of an optically functional ceramic nanocomposite is GaAs nanocrystals embedded in SiO_2 matrix. The interest in the novel optical properties of semiconductor nanocrystals [144] has resulted in strategies to package them as nanocomposites. A variety of techniques, such as colloidal synthesis, self-assembly, and electrochemistry, can be used to produce the semiconductor nanoparticles; however, ion coimplantation (Ga+, As+) is an efficient way of creating well-dispersed nanocomposite materials (e.g., GaAs/SiO_2) [145]. Typically, the ions are implanted at fluxes of $\sim 10^{16}$ cm^{-2} into 100 nm silica films on Si substrates and annealed at appropriate temperatures to create nanocrystals of GaAs (several nanometers in size) in the matrix (Figure 1.27). Photoluminescence studies show an efficient, broad luminescence band in the visible and near-infrared spectral regions due to quantum confinement in GaAs nanocrystals and defect states in SiO_2.

Nanoparticles in matrices have interesting photoluminescence properties, due to the effects of quantum confinement on their optical properties. Indirect band semiconductors such as Si and Ge have very poor luminescence efficiency (hence, efficient

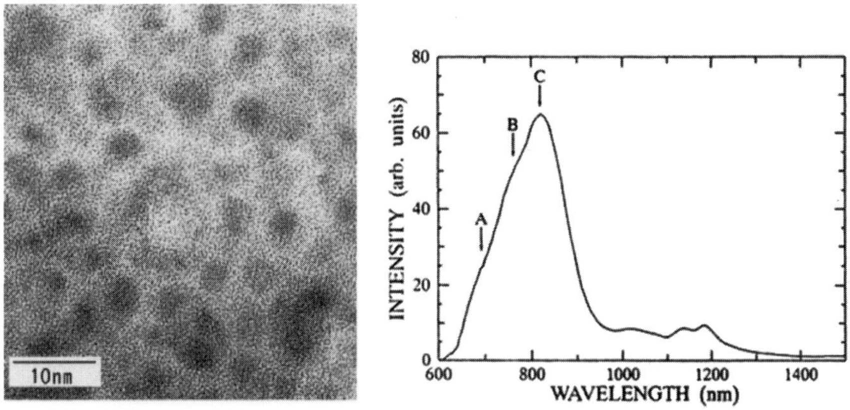

Fig. 1.27 Transmission electron micrograph, left, GaAs nanocrystals embedded inside SiO_2 glass matrix. Sequential ion implantation followed by thermal annealing was used to form GaAs nanocrystals in SiO_2 films. Right, efficient, broad photoluminescence band observed in the red and near infrared spectral regions. The efficient luminescence is attributed to both quantum confinement states in GaAs nanocrystals and defects in SiO_2. (Source [145] used with permission)

light emitters cannot be fabricated directly from them), because the band center transition is optically forbidden. But, by making the particle sizes smaller than the exciton Bohr radius (4.9 nm for Si, 24.3 nm for Ge), the resulting confinement produces an increase in oscillator strength, shifting the luminescence to higher energies; this allows tailoring novel optical materials, which are otherwise impossible. To take full advantage of the particle size effect, the particles must be separated, and the best possible way of accomplishing this is to disperse the particles in a matrix. Understanding and minimizing the interaction of the host matrix and the particles that create interface states are crucial to many of the optical applications. Embedding nanoparticles in different matrices and studying the optical properties provides a way to decipher interface effects. Ge quantum dots have been studied in various hosts, and one way to minimize host interaction is to embed them in oxygen-free environments, such as AlN; AlN/Ge multilayer structures deposited by pulsed-laser deposition show a blue-shifted photoluminescence peak [146]. Similarly, indium oxide (InO) nanoparticles dispersed (by solvent-phase impregnation) within the pores of mesoporous silica (prepared by sol–gel technique) show multiple photoluminescence peaks related to the size and structure of the particles [147]. Amorphous InO particles ($<$6 nm) show a photoluminescence peak that is down-shifted compared to slightly larger crystalline particles. For larger aggregates of InO particles, no photoluminescence is observed. Si nanoparticles themselves are interesting from the point of view of photoluminescence, and porous Si, which contains small particle domains, has been studied in great detail. The problem with porous Si is the structural nonuniformity, lack of reproducibility in emission and aging, and environmental degradation problems. Incorporating light-emitting Si structures in a matrix that is chemically inert and has a wide band gap suitable for quantum confinement can overcome this problem. Nanocomposites of Si nanoparticles ($\sim$5 nm) embedded in polycrystalline diamond matrix have been prepared and studied. The room temperature photoluminescence behavior of such materials in the 1.6–2.5 eV range shows a strong increase in emission efficiency from the Si nanostructures. The studies conducted on this nanocomposite reveal that selecting the size of the embedded Si particles makes it possible to tune the luminescence frequency in the yellow–green spectral region.

Nanocomposites of nanosized metal particles in transparent dielectrics can also be applied as nonlinear optical materials in photonic devices. These materials are characterized by large third-order optical nonlinearity (χ^3) and fast response times, which are important for device applications such as optical computing, real time holography, and phase conjugators. To incorporate metal nanoparticles into dielectrics, several techniques such as ion implantation, sol–gel techniques, and sputtering can be employed. It is even possible to prepare graded layers of metal nanoparticle distributions (with different particle sizes and interparticle separations) by implantation, and these structures can be carefully tailored to produce interacting and noninteracting nanoparticle layers that produce different optical response in the plasma resonance frequencies. Several of these composite systems have been studied, for example, Au/SiO_2, Ag/SiO_2, Au/Al_2O_3, Au/TiO_2 with metal concentrations varying from 15%–60%, for nonlinearity and plasma resonance frequency shifts. High values of susceptibilities ($\chi^3 \sim 6 \times 10^{-7}$ esu, compared to low values of $\sim 10^{-12}$–10^{-11} esu for glass) have been reported for composites containing an optimized fraction of the metal component [148, 149]. Groups II–VI semiconducting nanocrystals (e.g., CdS) prepared in glass hosts have been studied in great detail, because of their large optical nonlinear response and small carrier lifetimes. Enhancement of carrier recombination rates is observed in nanocrystals with large surface-to-volume ratios, and the enhancement results from increases in the density of surface states and fast surface recombination and capture due to multiphoton emission. Surface recombination in nanocrystals embedded in glass matrices exhibits thermally activated nonradiative recombination, which is enhanced at reduced particle size. The semiconductor/glass interfaces in the composites may give rise to deep traps responsible for photoactive phenomena.

In recent years, nanoparticles have been used to make transparent nanocomposite structures having high refractive indices. Polymers containing inorganic particles in a range of 1–100 nm (nanocomposites) are interesting in this regard. In contrast to composites having particles in the micron size range, nanocomposites do not scatter light and are interesting for optical applications. Preparation of nanocomposites with refractive indices over the entire range of <1 to >3, which is by far the lowest and highest ever achieved for a polymer composite, has been possible. Transparent polymeric materials can be coated with surface layers (~100 nm thick) of UV-absorbing nanocomposites to inhibit degradation of the polymer by UV light [150]. High-refractive-index transparent materials are mainly used for improving the optical coupling efficiencies in photonic devices. The refractive indices of polymers vary between 1.3 and 1.7, those for inorganic semiconductors vary between 2 and 5, and for high bandgap semiconductors this value is <3. The technological challenge in creating near-transparent high-refractive-index organic/inorganic nanocomposites lies in the creation of nanoparticles in the size range 20–40 nm, to produce high-refractive-index nanoparticles with low absorption coefficients in the visible range. Examples of such composites are PbS, InP, GaP, Ge, and Si particles in a gelatin matrix. The particle sizes, as well as the loading fraction of the particles, affect the refractive index and absorption coefficient. Very small particles show reduced refractive index. But typically there is a critical size above which the value of refractive index reaches the bulk value (for PbS this is $\sim$25 nm). Hence, to obtain predictable values of refractive index, it is important to have good dispersion of particles. The particles useful for creating these nanocomposites can be produced through a variety of techniques, ranging from mechanical attrition [151] (followed by centrifugation to obtain near-uniform Si particles) to colloidal processing. Nanocomposites can then be prepared by mixing and spin coating (or casting) aqueous solutions of gelatin with suspension of nanoparticles in appropriate organic solvents.

1.10
Inorganic Nanocomposites for Electrical Applications

The incorporation of particles of a nonpiezoelectric phase into a piezo material is unlikely to improve properties such as the strain per unit field or the charge per unit force, but the dielectric loss of the nanocomposite material under high alternating electric fields is an interesting issue. High-drive electrical applications are often restricted by reduced efficiency, due to heating caused by the large dielectric loss at high fields. A clear distinction has been shown between *weak* and *strong* field behavior in materials such as $BaTiO_3$. Above a critical field level, the dielectric constant and dielectric loss increase dramatically, due to hysteretic domain wall motion and the resultant reorientation by spontaneous polarization. It is possible that incorporating nanosized particles (SiC of small volume percentages, $\sim$2.5 %) into an electroceramic matrix ($BaTiO_3$) may hinder domain wall motion sufficiently to reduce dielectric losses at high fields [152]. Accordingly, the microstructure requires the particles to be situated intragranularly. This concept is similar to that used in the development of hard ferromagnetic materials, for which heat treatment is used to produce a fine dispersion of particles within the microstructure, which increases the resistance to domain wall motion. In addition to the possible effect of the particles on piezoelectric properties, additional issues that require examination include the influence of any secondary phases produced by solid-state reactions between the matrix and nanoparticles; in a $BaTiO_3$/SiC system, heat treatment and hot pressing can form new phases such as $Ba_2TiSi_2O_8$. When such reaction phases are present in the composite, the permittivity of the system is lowered. Addition of silicon carbide leads to a change in the microstructure of the material and a decrease in grain size. This outcome has been reported in structural systems, which show corresponding improvements in mechanical properties. In addition, the fracture mode was transformed from transgranular for monolithic $BaTiO_3$ to intergranular for the nanocomposite. Many studies have been undertaken to fabricate piezoelectric particle-dispersed ceramic nanocomposites from other materials: $Pb(Zr, Ti)O_3$ nanocomposites have been prepared from high-purity PZT powder and small amounts of oxides (Al_2O_3, MgO, etc.) [153]. Small additions of a second phase improve the mechanical properties, such as hardness and fracture toughness, of the PZT. The nanocomposite's reduced grain size during processing with the second phase is responsible for the improved mechanical properties. The piezoelectric properties of these composites remain essentially unchanged, although for the MgO composites, the electromechanical coupling factor (which is important for actuator applications) becomes larger.

Nanocomposites can also be useful for such applications as electrical contact materials, particularly to replace the toxic Ag/CdO contacts being used now. The reactive milling process has been used for manufacturing Ag/SnO_2 nanocomposites. A high-energy ball mill controls the reaction to obtain nanosized SnO_2 particles in an Ag matrix [154]. Then this powder is hot pressed into electrical contacts with superior erosion resistance and good thermal and electrical conductivity. Such cermet-like nanocomposites (in different configurations) have applications in other areas, like sensors. Many reports have appeared on gas-sensing technologies based on the measurement of changes in the electrical resistances of metal oxides (such as SnO_2, TiO_2) as the environment is changed. The sensitivity and selectivity of these metal oxide sensors can be affected by deposition of discontinuous metal films on their surface. Metals such as Pd, Pt, etc. have been used as surface activators. Similarly, incorporating metal clusters into bulk oxides to form nanocomposites also makes it possible to tailor the effective bulk electrical response of sensor materials. This incorporation allows the range of resistivity changes caused by changes in the carrier concentration at the surface or bulk of the metal oxide sensor material to be altered.

Although ceramic/metal composites are exciting mostly due to improved mechanical properties, the temperature-dependent electrical behavior of metal in ceramic nanocomposites is quite interesting [155]. As examples, Ag and Pb particles have been prepared in silica gels by electrochemical methods. DC electrical characterization of these materials shows surprising changes from semiconducting to metallic behavior as a function of temperature and depending on particle size. The electron conduction mechanism in these nanocomposites is complicated but, in general, occurs via two conduction mechanisms: electron tunneling between metal particles and metallic conduction due to the presence of metal percolated paths. The particle size and microstructure of the composite can be changed to adjust these properties [156]. The variable conductivity of these composites can be used in applications, such as electromagnetic induction shielding.

Percolation Effects and Transport Phenomena in Composite Systems

To understand nanocomposites in terms of conduction processes, we need a basic understanding of the bulk materials as well as of the composite filler. This is where certain issues arise. When dealing with organic or inorganic insulators and semiconductors, we may rely on a form of charge transport through the medium in terms of variable range-hopping conductivity when we add fillers to the system [157]. However, the conductivity is determined by the dominant charge carrier, because the carrier's mobility in hopping from one site to the next is the dominating feature of the conduction mechanism. A detailed study of hopping and its effect on conductivity has been summarized earlier [158, 159].

For metal filler systems in insulating matrices, the dominating feature is the type of filler, with particular regard for its size and shape [160]. Changing the electrical transport properties of these insulators with nanosize fillers is necessary, because we want to retain the innate properties that make the insulators useful. We also want to induce electronic properties that allow us to make quantitative determinations without altering the chemical and/or physical nature of the dominant medium (insulator). For the simple percolative effect, we must look at the insulating/metallic mix. In general, for bulk transport properties, the conductivity follows a power law behavior:

$$\sigma = \sigma_0 (p - p_c)^t \quad (\text{for } t = 2) \tag{1}$$

where σ denotes the conductivity and p represents the mass fraction [161]. In the past, this behavior has been shown to agree with 3D behavior, by Monte-Carlo simulations for thick films [162]. We do not deal with this behavior in detail here, except to note that the study of superconductivity in percolating films is relevant for the interpretation of critical temperature and heat capacity data [163]. A more detailed application of percolation theory deals with the effect of varying the number of interconnections present in a random system [164]. Fourier et al. [165] proposed a model in bulk materials, based on the Fermi–Dirac distribution, that describes the critical insulator-to-conductor transition:

$$\log(\sigma_c) = \log(\sigma_{nc}) + \frac{\log(\sigma_f) - \log(\sigma_{nc})}{1 + \exp[b(p - p_c)]} \tag{2}$$

where σ_c is the composite conductivity, σ_f is the filler conductivity, σ_{nc} is the nanocrystalline conductivity, p is the mass fraction, and b is an empirical parameter that leads to the change in conductivity at the percolation threshold p_c.

On a macro scale, we look at transport properties in terms of 3D processes. The dimensionality of the charge responds to its environment in the x, y, and z directions, as well as to the field that affects it. The simple phrase 'the path of least resistance' is very true for 3D transport. For years, many organic and inorganic materials were examined for their conduction dimensionality on the nanoscale.

In examining composites for their electrical (and electronic) properties, consideration of the volume or mass fraction addition (where we add a metallic component to our insulator) is crucial. However, where this volume addition occurs, the structural phenomena have a large bearing on the composite's transport properties. Alternatively, the metal particles can be distributed randomly, but their proximity (path length) and size can be varied, depending on the technique for adding them. Some clusters will aggregate and therefore nucleate to form larger grain sizes, but other deposition techniques result in islands of metallic clusters. This heterogenity has a dramatic effect on the conduction process as well as on the dimensionality of the conduction pathway. Variable ranges for the critical mass fraction of added filler (5%–50%) required to reach the percolative threshold have been reported. Why should we care about these values and how are they important? The reason is both complex and simple. To influence transport phenomena, both the shape and size of the filler are crucial. In addition, the manner of deposition and the resulting pathway and mean distance between islands of particles or colloids also directly influence the dimensionality of the charge carriers. We should be aware of certain effects of various fillers. The electrical conductivity of very thin metal films can be much lower than in a bulk material. Closer examination of the thin films reveals a discontinuous structure, where the metal appears as clusters or islands. Detailed examinations have been made of the transition of materials from metallic ballistic transport to insulating, indicating that it is percolative in effect.

As mentioned above, in recent years, the size-dependent properties of nanocrystalline solids have attracted much attention. By gaining an in-depth knowledge of these materials, we can start to use nanocrystalline fillers as components in emerging complex composites. Small changes in the physical size of a simple nanocrystalline material can lead to large changes in its physical properties [166]. In general, to understand such fillers, we must examine many complexities before expanding the investigations into macro- or even nanoscale composites. Work in this field is still inadequate for a complete understanding of percolation effects and dimensionality when nanofillers are used in composites (Figure 1.28) [167]. However, recent studies of carbon nanotubes as a filler in a conjugated polymer have led to some interesting results. Carbon nanotubes were used as fillers in thin-film nanocomposites, and altered conductivity was measured. What makes this effect interesting is that a reasonably well characterized material (the carbon nanotube) was used as a filler in a semiconducting polymer – one whose low conductivity despite conjugation falls in the insulating range [168]. The effect at percolation was to change the conductivity of the composite by 10 orders of magnitude.

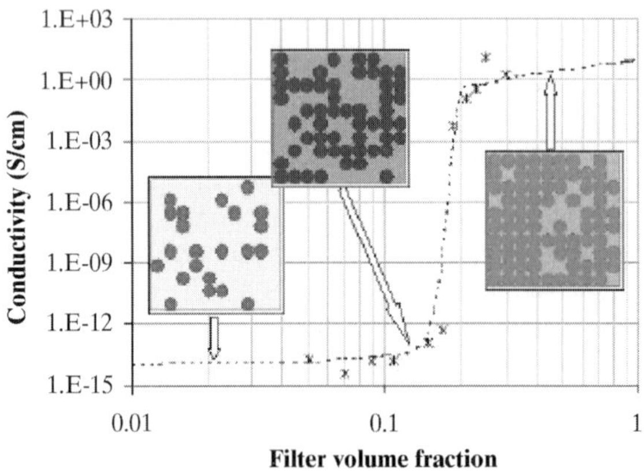

Fig. 1.28 Filler fraction dependence of the conductivity of particle/polymer matrix composites, demonstrating the percolation transition. Insets are schematic microstructures reflecting the extent of particle clustering, at and above the transition. (Source [167] used with permission)

1.11 Nanoporous Structures and Membranes: Other Nanocomposites

Nanoporous materials are materials with pores that have dimensions in the nanoscale. In many ways, these materials can be considered nanocomposites, because they have one phase (empty space) that is either randomly dispersed or ordered within a solid matrix. We have already discussed examples of some of these structures, such as hollow carbon nanotubes with straight hollow channels or porous alumina templates made by electrolytic methods. Examples of random pore distribution are zeolites, mesoporous silica, and porous silicon, all of which have important applications. In this section, we discuss in a little more detail the importance of a few select structures in the realm of nanocomposites. An exhaustive treatise of porous materials and membranes is beyond the scope of this book but has been dealt with elsewhere [169].

The transformation of a chemically homogeneous material into a porous structure by some form of chemical reaction controlling pore size is challenging, because chemical homogeneity and selectivity during such processes do not appear to go hand in hand. However, several strategies can be used (Figure 1.29). Examples include zeolites (pore diameter ~0.3–0.4 nm), Al foils converted to alumina having straight pores (diameter ~7–250 nm) via an electrochemical self-organization, compact polymeric membranes and also mica plates and glassy metals treated by irradiation or particle (e.g., ion) beams that also develop straight pores ('chemical drilling'), porous carbon structures generated by carbonization of appropriate polymeric precursors and subsequently subjected to an activation reaction (diameter 0.3–4 nm), and cyclodextrin polymers [170] (diameter ~0.8 nm).

The pore sizes in porous structures and membranes range from 0.3 nm to 10 μm. Pore dimensions below 0.3–0.4 nm may also be considered, but these are nearly the dimensions of ordered or disordered interstices in solids. However, the upper limit of 10 μm covers the pores of most membranes. Pore size uniformity (and distribution) is obviously important for applications such as separations on any size scale [171]. In addition, the physical and chemical character of the pore surfaces can play a significant role at smaller pore sizes, especially when gases or vapors are to be separated. In general, the fabrication of a defect-free porous membrane addresses two problems: one is the creation of pores of the desired local geometry (size, void fraction, and pore coordination number); the other is the achievement of mechanical robustness while repeating the local arrangement over macroscopic distances.

Polymeric membranes useful at lower temperature (20–100 °C) and for large-scale routine applications are an important class of porous materials [172]. Ceramic membranes, compared with their polymeric counterparts, are usually more fragile, more expensive, and harder to fabricate. On the other hand, ceramic membranes tend to withstand harsher separation conditions (e.g., high temperatures, solvents, and corrosive or fouling-favoring environments) and can often be cleaned easily. Carbon membranes are based on appropriate polymeric or pitch precursors, and they are intermediate (in character and properties) between polymeric and ceramic membranes [173]. The properties of carbon membranes are closer to the properties of their ceramic counterparts: high temperature and chemical stability (except in oxidative environ-

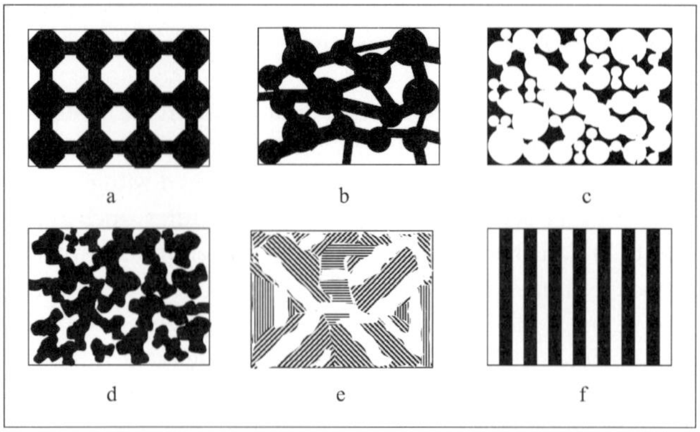

Fig. 1.29 2D schematics of a selection of 3D bicontinuous structures commonly encountered in nanoporous materials. Except for structure (e), the void part in indicated by black. (a) Fully ordered pore network, as in zeolites (pore size 0.4–1 nm). (b) Generic nanoporous material with an imprecisely tailored porous network. (c) Structure generated by compression/mild sintering of solid, nearly round nanoparticles. Pores down to a few nanometers are achievable. (d) Structure typically generated by spinodal decomposition (a Corning silica-based product with 4 nm pores) (e) Nanoporous carbon with a reference nanodomained graphitic structure. Black lines represent graphitic planes and slit-shaped pores correspond to missing graphitic material. (f) Nanoporous material with tubular pores in a parallel semiordered arrangement, anodic alumina (with a minimum pore size in the 10-nm range), and various mesoporous zeolites (pore size in the range of several nanometers). (Courtesy Dr. Konstantinos Beltsios, Institute of Physical Chemistry, NCSR, Demokritos, Greece)

ments at temperatures >350–400 °C). Current interest in carbon membranes focuses mostly on those appropriate for gas separations; at least for air separation, these membranes exhibit a combination of permeability and selectivity superior to those known so far for polymeric materials. Examples of applications include separation of CO_2 from CH_4 in landfill gas, production of oxygen-enriched air, and separation of propylene from propane. As a consequence of some of the routes for their preparation and/or their final structure, BN and SiC membranes constitute a bridge between carbon and generic ceramic membranes. Because BN is superior to C in terms of oxidative resistance, developing BN versions of C membranes is potentially interesting. In analogy to the fabrication of porous carbon structures from pitch and other precursors, appropriate organic precursors can be found for BN fibers. Deposition of a precursor organosilicon polymer or oligomer [174] (e.g., polycarbosilane) on porous ceramic substrates, followed by pyrolysis, can be used to fabricate silicon carbide or Si-O-C containing ceramic top layers of composite membranes. Pores from <1 nm to mesopores can be generated, depending on the precursor and the processing. In very recent years, self-assembled single-walled carbon nanotube-based arrays have been proposed for membranes (Figure 1.30). The membrane action is based on quantized transport of water through nanotubes [175] (<1 nm pore size), mimicking water transport in physiological structures through tubules of similar dimensions.

Several forms of well-defined nanostructured carbons (Figure 1.31) exist, and these forms have important practical applications, such as development of electronic, catalytic, and hydrogen storage systems. These materials can be synthesized by thermal chlorination of various carbides. It has so far been possible to synthesize nanoporous carbon, nanoparticles with multi-walled, band-like lamellar graphitic carbon, and turbostratic carbon. The nanoporous amorphous carbon consists of very high-surface-area carbon with a tunable, very narrow, pore size distribution and is suitable in applications such as supercapacitors and specific adsorbents. Template-based (e.g., quartz, silicates) synthesis techniques can also be used to prepare such materials. Structures such as mesoporous aluminum silicate or nanoporous silica have well-ordered arrays of open structures randomly connected by pores (pore size <3.5 nm). The pores can be filled with organic solvents, heated, then pyrolyzed at high temperature; finally, the template is etched off to produce nanoporous carbons [176].

Semiconductor and metal porous membranes have specialized applications [177]. The most important example of metal membranes is hydrogen-separating palladium-based membranes [178], which work on the basis of selective diffusion of gaseous species through the metal lattice. Composite membranes also exist, in which one or more top or buried layers are added to a bulk membrane material. At least one of the layers plays an active role in separation or provides defect healing. Composite materials, consisting of a matrix material (most often a polymer) and a dispersion of a

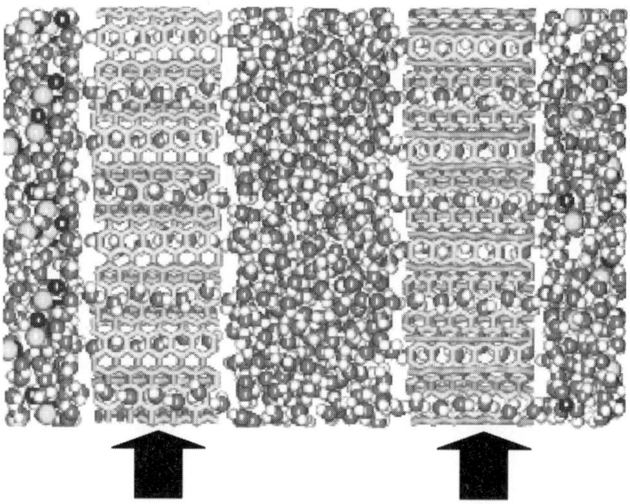

Fig. 1.30 Above: schematic of a possible multiple layer membrane consisting of self-assembled nanotubes (along the arrows), packed together densely. Between the membrane layers is a water-salt mixture. Simulation suggests that osmotic pressure can drive selective transport of species across the membranes, as shown by molecules moving through the nanotubes. Such porous architectures can find use in many applications ranging from bioseparations to drug delivery. (Courtesy Prof. Shekhar Garde, Department of Chemical Engineering, Rensseloer Polytechnic Institute, Troy, New York, USA)

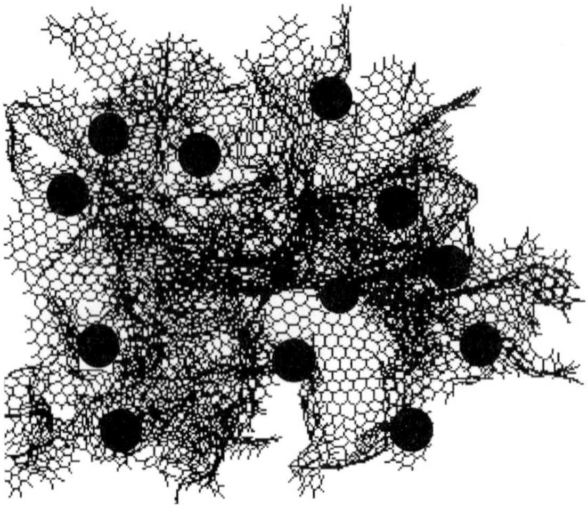

Fig. 1.31 Atomic-level schematic of nanoporous carbon structure with loaded metal particles. High surface area of porous carbon materials are ideal for holding dispersed metal catalyst particles for use in heterogeneous catalysis. (Source: [236] used with permission)

standard nanoporous material (such as zeolites or microporous carbon), can also be useful in catalysis and separation applications.

Straight-pore alumina membranes (which were briefly described above, in the section on preparation of templates for the growth of nanowires and nanotubes) are prepared by anodic oxidation of thin aluminum foils, in the electrolysis of a solution of oxalic, phosphoric, chromic, or sulfuric acid [179]. The electrolyte is chosen among those having some capacity for dissolving alumina. As electrolysis progresses, self-organization occurs; the aluminum oxide formed in the anode undergoes spatial redistribution, resulting in a honeycomb-like structure, with straight pores distributed in a nearly hexagonal arrangement. Pore sizes in the 10 – 250 nm range, porosities on the order of 0.2 – 0.3, and film thickness in the 10 – 100 μm range are usually obtained. For separation applications requiring small pores, it is preferable to develop asymmetric versions of anodic alumina membranes. In addition to asymmetry achievable by depositions, changing the conditions during growth, especially through voltage changes, makes asymmetry possible. In addition to the anodic porous alumina structures created, novel approaches being tested include, for example, the use of a master pattern (made of, e.g., SiC) with nanosize features to imprint well-defined surface textures that guide the formation of nearly defect-free pore channels with high aspect ratio over a wide area [180]. Long-range ordered channel arrays with dimensions on the order of millimeters and a channel density of 10^{10} cm^{-2} can be obtained, with aspect ratios of over 150 for individual channels.

Although the limited thickness achievable by the aluminum anodization approach makes the membranes produced unlikely candidates for extensive use as gas-separation membranes, a substantial amount of research has been dedicated to generating model asymmetric ceramic structures with additional functional layers deposited for specific functions; oriented microcrystals of porous phosphate structures and zeolite-like solids (which have tailorable pore sizes of 2 – 20 nm) can be grown on the surface of straight-pore alumina membranes to provide added functionality in separation applications [181, 182]. Straight-pore alumina membranes have been used as templates for generating a porous metal membrane by electroless plating [183]. Dissolution can remove the original and intermediate templates.

Polymeric nanoporous materials and nanocomposite materials have improved the technology of polymeric membranes drastically. In a complete reversal of roles for organic and ceramic matter in the building of organized gas- or vapor-selective materials, it has been possible to infiltrate the channels of zeolite-like materials with dimensionally constrained active organic (and metal-containing organic) phases. The architecture of the inorganic zeolite can be used to build other porous structures, including porous carbons. For example, a zeolite structure exposed to propylene gas at a high temperature, after carbonization and acid washing, yields a microporous/nanoporous carbon material [184].

1.12 Nanocomposites for Magnetic Applications

1.12.1 Particle-Dispersed Magnetic Nanocomposites

Magnetic nanocomposites in which magnetic species are dispersed within nonmagnetic or magnetic matrices are practically useful. Magnetic recording, giant-magneto resistance, and magnetic refrigeration are some important areas in which magnetic nanocomposites have relevance. Nanocomposite magnets with both particle/matrix systems being magnetic have received much attention, because they may have high remanence (associated with exchange coupling at interfaces separating hard and soft magnetic phases) and large energy product, $(BH)_{max}$, relative to conventional magnetic materials [185]. The remanence enhancement can be as high as 0.8 M_s, the saturation magnetization, compared to the isotropic value of 0.5 M_s. Many reports have appeared on the enhanced magnetic properties of nanocomposite magnets made of hard magnetic phases such as $Nd_2Fe_{14}B$ and $Sm_2Fe_{17}N_3$ within soft magnetic matrices (e.g., $\alpha-Fe$) [186]. The effect of the strength of intergrain exchange interaction is important in these materials and influences properties of nanocomposite magnets such as coercivity and maximum energy product [187]. Decreasing the grain size of the material and reducing the intergrain exchange interaction can increase the maximum energy product to as high as 200 kJ m^{-3} [188]. In general, the coercivity also increases (and peaks at some value) with a decrease of the exchange constant compared to the intragrain exchange interaction. Nanosize magnetic particles typically have grain sizes smaller than the critical single domain particle size. Also typically in magnetic nanocomposite films, the coercivity increases with grain size if the grains are isolated (no interaction), but when the grains start contacting and exchange interaction kicks in, the coercivity falls rapidly with grain size. The coercivity is highest at percolation, where the grains just start touching each other. In a very recent study, exchange-coupled nanocomposites were fabricated by nanoparticle self-assembly. Here, both

FePt and Fe_3O_4 particles were incorporated as nanometer-scale building blocks into binary assemblies, which were subsequently annealed to obtain FePt/Fe_3Pt nanocomposites, in which FePt is a magnetically hard phase and Fe_3Pt a soft phase. An optimum exchange coupling, and therefore an optimum energy product, is obtained by independently tuning the size and composition of the individual building blocks [189].

The improvements in magnetic properties in nanocomposites can be explained as follows. A nanocomposite magnet can be considered as hard and soft magnetic grains connected by magnetically interacting springs. The intergrain interaction has a tendency to align the magnetization in each grain in the same direction. When a magnetic field is applied in the opposite direction, reversal of magnetization occurs only in the soft grain (until a critical field value is reached). The magnetization in these grains is reversed when the field is removed, resulting in high values of reversible susceptibility. However, for high applied fields, which reverse the magnetization even in the hard grains, the magnetization in the soft grains does not reverse upon removal of the field. This effect is most notable in magnets (Nd/Fe/B) having grain sizes smaller than 50 nm. The strength of interaction between the grains is crucial in determining the properties of the nanocomposite magnets; in addition, the magnetic properties also depend on the grain size, the fraction of soft and hard phases present, and the shape of the grains.

Different synthesis routes can be used to produce magnetic composites. Melt spinning/splat cooling of alloys accompanied by heat treatments such as fast annealing have been used to prepare alloy ribbons of nanocrystalline and nanocomposite magnets (e.g., containing $Nd_2Fe_{14}B$, Fe_3B, and α-Fe). The most successful technique has been mechanical alloying of two-phase mixtures (e.g., α-Fe/$Sm_2Fe_{17}N_3$), which typically results in two-phase mixtures consisting of a nanocrystalline soft magnetic matrix and an amorphous hard magnetic phase. Annealing and crystallization at higher temperatures lead to nanocomposite magnets that show superior magnetic behavior. The evolution of the microstructure and grain morphology during mechanical milling and subsequent heat treatments affects the magnetic properties and has to be controlled precisely. Cosputtering from multiple targets is another method capable of fabricating uniform nanocomposite structures. Uniform thin-film nanocrystalline cobalt grains (10–20 nm) surrounded by graphitic layers have been prepared by this technique followed by post-annealing treatments [190–192]. The Co/C system is immiscible, with a negative interdiffusion coefficient, and because the carbides in this system are metastable, they decompose into the elemental forms. Hence, uniform films of segregated magnetic and encapsulant layers can be fabricated for controlling the magnetic interaction between the different Co grains present in the films. Such films may prove useful for high-density magnetic recording, because carbon provides good isolation, so as to reduce the intergrain exchange interaction. The thickness of the film deposited also needs to be controlled for preparing optimal recording media, and this thickness should be comparable to the grain size (~10–15 nm) within the films for best performance.

1.12.2
Magnetic Multilayer Nanocomposites

The increase in storage density of hard disk drives and the parallel decline in cost per megabyte (MB) of storage is happening rapidly [193]. Along with disk performance and price, sputter equipment is evolving in both design and efficiency. Current standard processing modules involve heating, dc or RF sputter deposition, and cooling. Further improvements and required modifications of these modules, as well as new processing steps such as substrate cleaning, multilayer deposition, and plasma chemical vapor deposition of carbon are necessary in the future. Magnetic nanocomposites are the end result of many of those improvements.

1.12.2.1 Microstructure and Thermal Stability of Layered Magnetic Nanocomposites

Granular solids consist of small metal granules embedded in an immiscible medium, which may be insulating or, less frequently, metallic [194]. Like other artificially structured solids, granular solids have intricate structure on the nanometer scale and extra degrees of freedom within which the physical properties can be manipulated to achieve tailored materials for applications and for explorations of physical phenomena. The relevant extra degree of freedom for granular solids is the granule size. In magnetic granular solids, all such granules are single domains exhibiting 'hard' magnetic properties. On the other hand, for samples with large metal volume fractions, the metal granules form an infinite network, exhibiting metallic conductivity and 'soft' magnetic properties, because the conducting paths and magnetic closure structure are readily facilitated. For most common metals (e.g., Fe, Au, Co, Cu), the ultrafine solid granules can be made as small as 10 to 100 nanometers. This is the size range in which finite-size effects, single-domain magnetic properties, and other phenomena are observed. Work on this type of nanostructured film began with the pioneering studies of Abeles et al. [195]. These films have attracted considerable attention in recent years because of theoretical suggestions dealing with their physical properties and also because of practical applications, among which the use of ferromagnetic nanocomposites as data recording media stands out. For this application, superior magnetic performance can be tailored by manipulating the size and volume fraction of the particles. Other advantages also derive from the structure of the film, in that the insulating portion of the film provides chemical and mechanical protection to the metal particles.

A variety of deposition methods can be used to produce granular, layered magnetic solids. Of these, sputtering is the most versatile. Sputtering is most often applied via a single homogenous target, or by cosputtering. Sequential deposition is also applicable if layer formation can be avoided. Generally, because the nanostructures of granular metal solids are strongly influenced by the processing conditions, deposition parameters such as deposition rate, sputtering pressure, and substrate temperature must be tightly controlled. Multilayer thin films with an under layer and a cover layer (total film thicknesses are usually in the range of several micrometers) have been prepared with multiple-gun dc and RF sputtering systems. The multilayers were deposited onto glass or silicon substrates. Flowing high-purity gas (Ar, N_2, etc.) was used in sputtering. The sputtering parameters, including gas pressure, the power rates of the sputtering guns, and the target-to-substrate distance, were chosen to produce films with maximum coercivity. Heat treating the as-deposited films in a furnace or rapid thermal annealing yielded nanocomposite films with random nanostructure. Generally, high coercivity is obtained only after heat treatment.

As an example, cobalt nitride films, CoN, in pure form and also as a nanocomposite in a boron nitride or silicon nitride matrix were generated by reactive sputtering of cobalt metal, cobalt boride, or cobalt silicide as targets, respectively, in nitrogen plasma [196]. Cobalt nitride decomposes into the elements when heated under vacuum at 500 °C. The nanostructure of the composites was preserved in the heat treatment, thus creating a fine dispersion (<10 nm) of cobalt particles in a ceramic matrix. The magnetic properties of the nanocomposites were studied in detail; the precursor cobalt nitride is paramagnetic but the cobalt dispersions, having dimensions smaller than single magnetic domain size, show characteristics typical of systems such as superparamagnetism and, at temperatures lower than the blocking temperature, marked hysteresis. The coercive fields at 5 K for the BN and Si_3N_4 nanocomposites are 3250 and 850 Oe, respectively.

The problem of long-term stability of written bits in magnetic data storage devices places a fundamental limit on the possible increase in area density for conventional magnetic recording. Because decreased magnetic grain size is required for the improved media signal-to-noise ratio that is needed to support higher area densities, thermal energy becomes significant relative to the switching energy, and the playback signal amplitude decays over time. Thermal stability will ultimately limit the maximum area density achievable with conventional longitudinal recording. The key aspects of the media microstructure contributing to thermal stability are the grain size and grain size distribution, alloy composition, alloy segregation, lattice defects, and strain. Random nucleation processes occurring during media deposition create grain size distributions. Continued increases in area density will require tighter grain size distributions and improved microstructural control of very thin magnetic layers [197].

Controlling the media grain size and tightening the grain size distribution are probably the most difficult, yet most important, aspects of media processing. Due to the exponential dependence of the decay on $K_\mu V/kT$, ($K_\mu V$ is the product of the uniaxial anisotropy energy density and the switching volume), small changes in V determine the difference between stable and unstable bits, with the smallest grains in the distribution decaying first. Thermal decay experiments have shown that the data cannot be well explained without taking the contribution of the grain size distribution into account. Slow lateral grain growth rates or high nucleation rates favor smaller average grain sizes. Materials or growth conditions that limit surface mobility reduce the grain size. Because a minimum number of atoms are needed to form a critical nucleus, a narrower grain size distribution is expected as the grain size approaches the size of the critical nucleus. To significantly improve the grain size distribution, one needs a regular array of nucleation sites or a new type of media in which the magnetic grains do not grow together to form a fully dense film. Although research on so-called self-assembled quantum dot arrays for media is ongoing, these approaches are still far from being implemented for production media.

1.12.2.2 Media Materials

Area densities in longitudinal magnetic recording have reached 36 Gbits in^{-2} in laboratory demonstrations [198] and 10 Gbits in^{-2} in media designs already in use (12 $\pm$ 4 nm grains) [199]. Roadmaps toward 40–100 Gbits in^{-2} have been proposed [200]. At such densities, tight control of the media microstructure, especially grain size, grain size dispersion, and chemical isolation to break exchange are necessary, in order to keep the media noise within acceptable bounds [201]. Continued grain size scaling to diameters considerably below 11–12 nm should be possible, allowing for densities well beyond the currently perceived 40–100 Gbits in^{-2} limit. This prospect has been the driving force behind industrial and academic research in the area of thin-film hard magnet materials, which promise minimal thermally stable grain sizes down to 2–3 nm and a more than 10-fold potential density gain. However, for extremely high density recording media, typically defined as >100 Gbits in^{-2}, the critical values of relevant magnetic properties will be stringent for achieving low media noise: coercivities of 4 kOe and weak exchange coupling in grains <10 nm are estimates.

High-K_{μ} (uniaxial magnetocrystalline anisotropy) materials are attractive for ultrahigh-density magnetic recording applications (e.g., audio tapes, video cassettes, magnetic memory devices) because they allow smaller, thermally stable media grains (Figure 1.32). Particle-dispersed recording systems are also possible targets for future magneto-optical data storage. For the particulates, high magnetization is crucial, because it determines the strength of the field to be detected during read/write. Metals such as Fe and Co have high magnetization, but one drawback is that they oxidize easily when particle sizes are small. Alloys have advantages in this regard and encapsulation in a medium (matrix) can solve several problems associated with chemical stability of the particulates. The search for higher magnetic recording densities pur-

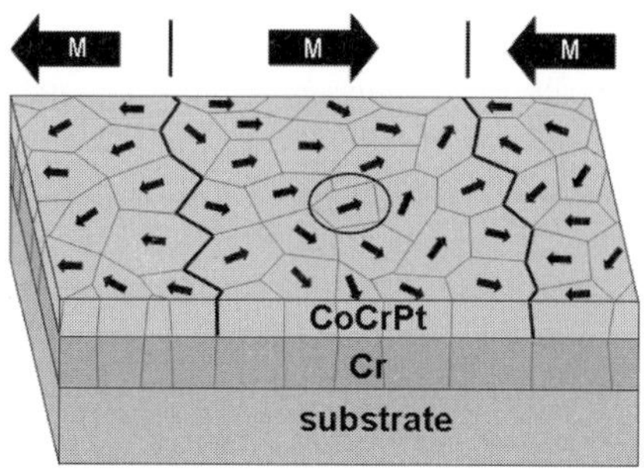

Fig. 1.32 Schematic of multilayer microstructures of typical nanocomposites used in magnetic recording media. The area bound by the thick jagged lines corresponds to the transition between bits (dimension of each bit ~80 × 400 nm). Within each bit, the grains of the magnetic layer, CoCrPt, have uniaxial anisotropy due to the random orientation of magnetic moments. The total magnetic moment is the vector sum of all magnetic orientations (shown by arrows M). The orientation of the Cr layers can induce magnetization in the plane of the film by their epitaxial relationship. The thickness of the films correspond to 20 nm for the CoCrPt layer and ~40 nm for the Cr layer. (Courtesy Prof. M. Shima, Department of Materials Science & Engineering, Rensselaer Polytechnik Institute, Troy, New York 12180, USA.)

sues particle sizes that are <10 nm. With such small particle sizes, high magnetocrystalline anisotropy is needed to avoid thermal and field fluctuations that can destroy the magnetization in recorded locations. When particle sizes become that small, in many material systems the magnetization can become inherently unstable, due to effects such as superparamagnetism. Intrinsic magnetic properties of several potential alternative media alloys can be found in reference [202]. Recent studies have emphasized rare earth intermetallic compounds with high magnetic anisotropy and small grain sizes in the search for the best magnetic properties for high-density recording media.

Nanocomposites, e.g., metal/carbon, metal/oxide, or immiscible metal/metal mixtures, are prospective routes toward generating <10-nm granular structures, needed for ultrahigh density recording. Much work has been reported on nanocrystalline rare-earth transition-metal films, most prominently Co_5Sm- and $Co_{17}Sm_2$-based films. High-density recording studies have been reported on CoSm/Cr structures with coercivities in the 2000–4000 Oe range. Another example is nanocrystalline high coercivity CoPr/Cr films, for which coercivities of about 8000 Oe and grain sizes of 10 nm have been obtained.

Face-centered tetragonal (fct) structures of near-equiatomic composition, CoPt and FePt, have been studied extensively. Both compositions result in large coercivities, close to the theoretical maximum obtained for 12–15-nm-thick epitaxial Fe/Pt (001) films with perpendicular orientation grown on MgO substrates. 'Nanometer-size' CoPt particles (100–300 nm) obtained by dc-magnetron sputtered, and post-annealed CoPt films exhibited coercivities of about 30–37 kOe. Annealed nanocomposite CoPt/C films may be tailored for magnetic recording media applications. Some of the common problems with hard magnet nanomaterials are unfavorably high processing temperatures during annealing or growth, which may adversely affect grain growth. As a consequence, the noise performance is compromised, as was recently found in perpendicular recording studies of polycrystalline ordered Fe/Pt (001) thin films. Codeposited nanophase materials, e.g., CoPt/Ag and CoPt/C, may alleviate some of these problems. CoPt alloys undergo phase transition from fcc to the more anisotropic face-centered tetragonal (fct) phase at higher temperature (835 °C), which makes this material magnetically hard. RF magnetron sputtering from targets containing CoPt alloys and Ag can be used to prepare nanocomposite (CoPt/Ag), and post-annealing treatment embeds the hard magnetic phase. Large coercivities, which can be tailored over the whole range by changing the processing route and hence the microstructure, were achieved for these multilayer films of CoPt fct phases in a fcc Ag matrix, with average grain sizes in the range of 7–100 nm [203]. Different matrix materials (other than Ag) can also be used, serving as the isolation material between the hard magnetic grains, for example, silica and carbon. Carbon should be an ideal isolation material, because it does not react with Co or Pt to form carbides. CoPt/C nanocomposite films have been prepared that show the high coercivity and chemical stability required for high density recording media. Exciting new developments in high density recording media may further derive from chemical synthesis routes, self-assembly of magnetic nanoparticles, and template-based growth of magnetic nanowire arrays; these options are currently being explored for generating monodisperse, ordered nanoparticle arrays.

In addition to the common applications outlined above, magnetic nanocomposite materials also have applications in ferrofluids, imaging, and separation technologies and as magnetic carriers for drug delivery. The challenge here is to make the magnetic nanocomposite particles (e.g., silica-coated iron oxide) resistant to agglomeration and chemical attack. The magnetic particles are often coated with protective and functional materials such as silica, which can be done effectively and easily by sol–gel synthesis or liquid phase coating processes (Figure 1.33). The silica coating provides protection against chemical attack and also provides a shell that can be modified chemically to make functional nanocomposites [204]. The sol–gel process (performed in organic solvent via hydrolysis of tetraethoxysilane) is good for making thin inorganic films with tailored porosity; however, because the coating is porous, chemical protection is not necessarily achieved. Supersaturated monosilicic acid can be used to achieve homogeneous thick coatings on particles. Often, a combination of the two processes is used to obtain uniform coatings of silica on magnetic particles. The key in this technology is to provide a protective coating without any reduction in magnetic properties such as saturation magnetization (important for magnetic carrier technology) and to provide an outer surface for additional easy modification. Thin silica films also ensure the stability of the magnetic nanocomposites against aggregation and metal (Fe) release to the environment, which may contaminate the systems during use. Isotropic ferrofluid dispersion can be obtained for these silica-coated magnetic particles (due to electrostatic repulsion and strong hydration forces between the outer silica surfaces) in the absence of a magnetic field, and the dispersion can be easily switched

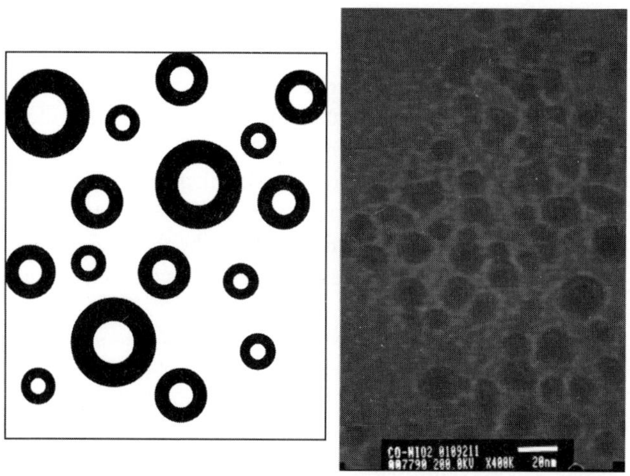

Fig. 1.33 Schematic and micrograph showing the fabrication of core shell structures. Several possible routes can be used to make such structures, for example, vapor phase coating or multi-step precipitation via nanoemulsion processing. The micrograph shows cobalt (core)-nickel oxide shell magnetic nanocomposite particles in the 20 nm size range. Such structures could find applications in magnetic xerography and as ferrofluids. (Micrograph courtesy Prof. Mamoun Muhammed)

to an anisotropic structure in external magnetic fields. These nanocomposites have applications as stable ferrofluids and in magnetic carrier applications. Magnetic particles embedded in polymer matrices (e.g., iron oxides inside a conductive polyaniline matrix) have also been synthesized for the fabrication of conductive, superparamagnetic plastic films [205]. Deposition of conductive polymer chains and magnetic nanoparticles in a layer-by-layer fashion may lead to new properties such as giant magnetoresistance (GMR) and novel organic-based nanostructured GMR materials.

1.13
Nanocomposite Structures having Miscellaneous Properties

Photoelectrochemical properties are inherent to semiconducting nanostructures. Electrochemical deposition from baths loaded with colloidal nanoparticles of the semiconductor (e.g., Ni films incorporating TiO_2, CdS, etc.) can produce metal films containing photoactive semiconducting particles. Much larger photocurrents are produced from these nanocomposite electrodes compared to conventional semiconductor electrodes and contiguous nanocrystalline semiconductor electrodes [206]. Separating the semiconductor nanostructures within a matrix in a nanocomposite system and creating intimate contacts between the particles and the molecular redox media can improve quantum yields in the photoexcitation processes. Controlling the local environments of the semiconductor species in the nanocomposite can modulate electron transfer and carrier mobility at the electrodes. Such photoelectrodes provide an interesting context for studying charge transfer processes at semiconductor–molecular contacts. As noted here, electrochemical deposition to fabricate nanocomposites is a powerful synthetic approach that has been used in several other instances to create nanocomposites with enhanced properties. One example is the formation of Ni/Al_2O_3 films [207], where ~50 nm Al_2O_3 particles are uniformly distributed in a Ni matrix. The only requirement for this synthesis process is a stable inclusion phase (ceramic or semiconductor) and its well-dispersed suspension in the electrochemical solution. The Ni/Al_2O_3 nanocomposite shows enhanced mechanical properties due to reinforcement of the ductile metal matrix by the hard ceramic nanoparticles (dispersion hardening). These can be used as coatings that retain most of the properties of the metal.

Applications of TiO_2 colloids and thin films extend from photovoltaics to catalytic devices and sensors. Nanocrystalline TiO_2 and ordered structures have been synthesized by the sol–gel method. Processing conditions can be controlled to produce various interesting (isotropic to anisotropic) morphologies and structures in the microstructure of the films that can be deposited. Rod-like or truncated pyramidal morphologies are formed at the anatase phase as the temperature is varied during processing [208]. Controlling the base used in the colloidal synthesis can make aligned rod-like structures self-organize into regular arrays. The ordering occurs due to surface charges and forces that can be altered by controlling the basicity and the dielectric constant of the colloidal medium. Thin-film optoelectronic devices with semiconductor and metal nanoparticles are hot candidates for research these days [209]. Alternating layers of

TiO_2 thin films and modified nanoparticles (semiconductor–metal layers) have been prepared for nonlinear optical applications [210]. Layer-by-layer assembly and surface sol–gel processing are used to fabricate these nanocomposites. Surface-stabilized (with organic monolayers such as thiols) gold nanoparticles (4–5 nm) can, in addition, be assembled into layers forming close-packed monoparticulate layers; multilayer films of TiO_2/Au can be prepared this way. The surface hydroxyl groups present on the Au particles act as an active component in the surface sol–gel process [211], which can be used to make metal oxide thin films with nanometer-scale control of film thickness.

The optical properties of small semiconductor particles are interesting, showing quantum confinement effects and shifts in their absorption spectra. Also, dielectric/nanocrystalline semiconductor interfaces enhance carrier recombination to picoseconds, with the possibility of ultrafast optical applications. Applications of semiconductor polymer nanocomposites in optics are discussed briefly in the next chapter. Similarly, metal/semiconductor hybrid nanoparticles produce novel effects such as enhanced optical nonlinearity [212] and shifted plasmon resonances for the metal particles [213]. Such hybrid nanocomposite particles (CdS/Ag) have been prepared in glass matrices by a modified melt–quench process [214]. The hybrid CdS-coated Ag nanoparticles exhibit red-shifted plasmon resonance absorption spectra, which can be tuned by the heat treatment that controls the particle size and other interfacial effects (Figure 1.34). Arrays of nanometer-sized CdSe clusters have been fabricated inside the pores of porous silicon matrices. The purpose here is to increase photoluminescence by enhancing the light-emitting centers by chemical infiltration [215] (using first an aqueous $Cd(COO)_2$ solution, resulting in the formation of CdO, and subsequently selenidation by passage of H_2Se). The nanosize semiconducting clusters (3–5 nm) provide the high-density light-emitting centers in this nanocomposite on top of the intense photoluminescence spectra in the visible range from the porous silicon. Electrochemically depositing metal (Au) within pores of optically transparent nanoporous alumina templates can be used to prepare interesting optical nanocomposites. The plasmon resonance absorption band of the metal nanoparticles as a function of particle size and aspect ratio (nanoparticle vs. nanowires in the channels of the alumina) was studied in such composites [216]. As the particle size decreased, the λ_{max} (wavelength of maximum absorption intensity) approached theoretical values and reached a quasi-static lower limit for particles of $\sim$16 nm, irrespective of the aspect ratio.

Functional nanocomposites in which nanoparticles are deposited inside appropriate substrates can be used as sensors that react by changing optical properties in response to gas exposure. CdO-doped SiO_2 nanocomposites have been prepared by RF cosputtering, and show gas-sensing capabilities [217]. This nanocomposite showed a change in optical transmittance during exposure to nitrogen oxide gas. For such nanocomposites to function, the embedded nanoparticles should change optical transmittance upon exposure to gases, the matrix should have porosity for the gas molecules to diffuse and interact with the particles, and the matrix should be transparent to allow detection of the changing optical signals.

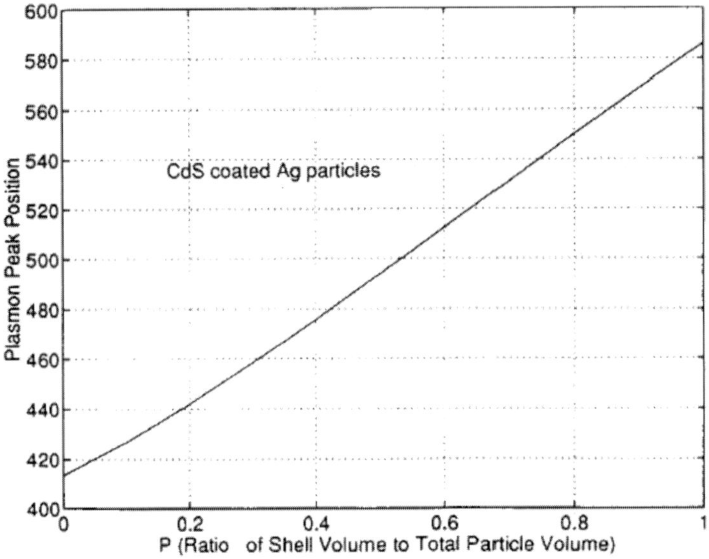

Fig. 1.34 Optical (luminescence) property of CdS-coated Ag particles embedded in a glass matrix nanocomposite. Plot shows the plasmon peak position as a function of the ratio of the CdS shell volume to the total particle volume in the nanocomposite particle. Semiconductor/metal hybrid nanoparticle composites were prepared in a glass host by a modified melt/quench process and examined with respect to their optical properties. The coated hybrid nanoparticles exhibit surface plasmon resonance absorption spectra that are red-shifted relative to that of a dispersion of homogeneous Ag nanoparticles in the same host. (Source [214] used with permission)

Many nanocomposite systems derive their usefulness from the fact that value-added functions can be advantageous, in addition to the primary function of the nanostructured phase. For example, we already discussed the dispersion-induced toughening and strengthening of ceramic nanocomposites (Al_2O_3/SiC, Si_3N_4/SiC, etc.), and, if the dispersoid phase acquire add new functions, such as piezoelectricity and ferroelectricity, then new applications can result. Conventional powder metallurgy techniques (milling, mixing, and hot pressing) can be used to prepare barium titanate-dispersed MgO nanocomposites [218]. The criteria for the formation of well-dispersed nanocomposites are that the dispersoids do not react with the matrix phase during sintering. In these perovskite-dispersed composites, the mechanical properties were not degraded, but new piezoelectric properties were added, enabling possible applications as sensors. Such materials have sensing ability for crack propagation by using electromotive forces or by introducing internal strains via electrical impulses, resulting in increased fracture toughness. The introduction of piezo- and ferro-electricity into structural ceramic nanocomposites will create new materials that exhibit 'intelligent' functions [219]. For transition-metal dispersed-oxide ceramic composites, ferromagnetism is also compatible with its excellent mechanical properties. In addition, good magnetic response to applied stress was found in these ceramic/ferromagnetic-metal nanocomposites, showing the possibility of remote sensing of fracture or deformation of ceramic materials.

Nanocomposites can also be effective as catalytically active membranes (e.g., catalytic converters in automobiles), in which a catalytically active phase is uniformly distributed in a matrix of high porosity. The operating conditions for such materials are quite demanding, because they are used over a wide range of temperatures and chemical environments. Nanostructured γ-Al_2O_3 catalyst supports with a dispersed cerium oxide phase and supporting noble-metal nanoparticle (Pt, Pd, etc.) heterogeneous catalysts are commonly used in automotive exhaust streams. The ceria phase serves as an oxygen-storage component, to buffer sudden changes in the fuel/air ratio; it also stabilizes the alumina structure and the noble metal particle dispersion. Calcination of sol–gels (prepared from Al and Ce chlorides and nitrides), followed by microwave or autoclave heating, have been used to prepare such nanocomposites [220]. The microwave-generated powder contains nanocrystalline ceria and alumina phases, suitable for catalytic applications. Note that, in nanocomposite structures, microstructure evolution and the final dispersion structure and its relation to the matrix decide the functionality of the composite. For example, if the final product in the above alumina/ceria composite is autoclaved instead of microwaved, a coarse microstructure that has no catalytic ability results. Silica/copper oxide nanocomposites have been prepared by a sol–gel process (controlled hydrolysis/polymerization of sodium metasilicate and copper nitrate via a microemulsion technique with anionic surfactants) and show promise in catalytic applications [221]. These nanocomposites have specific surface areas ranging from 200 to 400 $m^2 g^{-1}$ and are mesoporous, with pore sizes in the range 3–6 nm. Nanocomposites are also being considered for use as photocatalysts to decompose environmental pollutants and harmful microorganisms. As an example of such materials, transparent nanocomposite anatase films have been deposited at low temperatures on various substrates by a sol–gel method [222]. For the substrate, materials such as porous SiO_2 work well because, for such applications, a high surface area, transparency to light (including UV), and durability for the photocatalytic activity are required.

Low-dimensional catalytic nanocomposites can also be envisioned. For example, molybdenum carbide has been considered as a substitute catalyst for metals like Ru in several heterogeneous catalysis systems. Sonochemically prepared nanocomposites of molybdenum oxide and carbide dispersed as discontinuous surface coatings on microspheres of silica are being considered as nanocomposite catalysts and fit the profile of the optimal catalyst material, considering that the composites retain the nanoscale dimensions of the active phases, provide effective interaction between the catalyst and the solid support, and have a uniform distribution of the active phase on the support surface [223]. The formation of strong interfacial Si-O-Mo bonds is observed in this material. The use of sonochemical activation is yet another approach to breaking down and dispersing precursor materials into nanocrystalline forms and simultaneously activating surfaces onto which the dispersed phase can be uniformly coated.

Oxidation-resistant coating systems have been developed, based on nanocomposite technology, for applications such as resistive heating elements for industrial melting.

Materials such as Mo, which has several excellent electromechanothermal properties, are used for such applications, but suffer from severe oxidation problems. $MoSi_2$ is a substitute material for such applications, due to its excellent resistance to oxidation at high temperatures. The high temperature strength of this material is a problem, and incorporation of SiC into the $MoSi_2$ matrix provides a compromise. The best solution is to put $MoSi_2$/SiC coatings onto Mo substrates, by techniques such as RF magnetron sputtering of composite targets [224]. As deposited films, they are usually amorphous but can be converted into nanocrystalline form by annealing under vacuum at $>1000\,^{\circ}C$. The high interdiffusion of Si and C in these films can be minimized if the sputtered films are made in a nitrogen atmosphere. Pyrolysis of $MoSi_2$ particles coated with polycarbosilane and subsequent hot-pressing can also be used to prepare bulk $MoSi_2$/SiC nanocomposite materials for high-temperature applications [225]. Uniform dispersions of SiC particles are obtained in the $MoSi_2$ matrix after densification. The room-temperature and high-temperature flexural strengths of this nanocomposite are several times those of unreinforced matrix.

DLC-based nanocomposites possess very interesting properties and hence can be targeted for several applications (a high-profile application being the DLC coating on the Gillette Mach 3 razor blade). Amorphous diamond-like/quartz-like (C:H/Si:O) composites and metals containing such materials constitute a class of DLC-based nanocomposites with unique bulk and surface properties. These nanocomposite materials have excellent diffusion barrier properties, high wear resistance, and low coefficient of friction, in addition to good adhesion to most substrates, variable electrical resistivity, thermal shock resistance, and controllable optical properties. Films of diamond-like nanocomposites can be deposited by plasma-assisted vapor deposition of organosilane precursors and involves C- and Si-containing free radicals and, optionally, transition-metal atoms [226]. These nanocomposites can find applications in tribology, especially in magnetic recording read/write heads. DLC/SiO_2 nanocomposites have been prepared with better optical clarity as thin films [227]. The films of the nanocomposite (Figure 1.35) were deposited at room temperature by RF plasma, from tetraethoxysilane (TEOS) as the precursor. The advantage of this precursor is that it serves as the source for both the DLC and the SiO_2 during a simultaneous deposition process. By introducing O_2 into the plasma, the surface of the deposited film can be made more hydrophilic, changing its wetting characteristics.

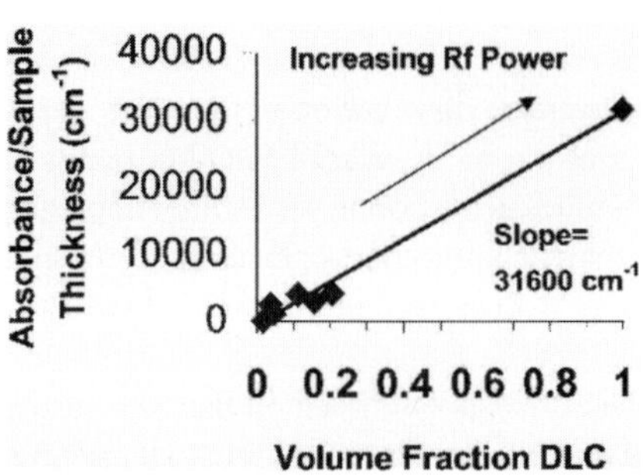

Fig. 1.35 Beer–Lambert plot of normalized absorbance of DLS/SiO_2 nanocomposites as a function of volume fraction of DLC. With increasing RF power the normalized absorbance increases linearly with the volume fraction of DLC in the nanocomposite thin films. The slope is the optical absorption, and the value obtained for the nanocomposite is similar to typical polar absorptions found for organic compounds. (Source [227] used with permission)

1.14
Concluding Remarks on Metal/Ceramic Nanocomposites

This chapter includes examples of how ceramic and metal-based nanocomposite materials can be useful in several applications and how they can be superior to conventional composites and to the separate, independent constituent phases. Thischapter in no way provides an exhaustive list, and the field is still evolving, providing new directions for applications of nanocomposites. Some of the synthesis techniques were discussed, as were challenges involved in fabricating nanocomposites with optimized properties. Synthesis parameters, which determine the microstructure, greatly affect the final properties of the nanocomposite, particularly with respect to tailoring the interface between the constituent phases. In the next chapter we discuss another important class of nanocomposites based on polymer matrices.

Acknowledgements
I greatly appreciate the help of Dr. Bing Qing Wei in putting together parts of this chapter and several of the figures. I also thank Dr. Seamus Curran and Dr. Robert Vajtai for their help in preparing this chapter. I thank the National Science Foundation (in particular the Nanoscale Science and Engineering Center at RPI) for funding support.

References

1 M. Jose-Yacaman, L. Rendon, J. Arenas, M. C. Serra Puche, Science, 273, 223 (1996).
2 K. Niihara , Metal Powder Report, 54(3), 37 (1999).
3 Y. H. Choa, A. Nakahira, K. Niihara, J. Mater. Sci., 35, 3143 (2000).
4 K. Niihara, Y. K. Jeong, T. Sekino, Y.-H. Choa, Ceram. Trans., Adv. Ceram. Matrix Composites IV, Am. Ceram. Soc., Westerville, OH, USA, 96, 185 (1999),
5 J.-F. Yang, T. Sekino, Y.-H. Choa, K. Niihara, T. Ohji, J. Am. Ceram. Soc., 84, 406 (2001).
6 M. Yoshimura, T. Ohji, M. Sando, K. Niihara , Mater. Res. Innovations, 1, 16 (1997).
7 T. Nakayama, T.A. Yamamoto, Y.H. Choa, K. Niihara, Ceram. Trans., 108, 257 (2000).
8 T. Nakayama, Y. Choa, T. Sekino, K. Niihara , J. Ceram. Soc. Jpn, 108, 781 (2000).
9 H. J. Hwang, S. Ueda, K. Niihara, Ceram. Trans., 44, 399 (1994).
10 R. Riedel, H. J. Kleebe, H. Schonfelder, F. Aldinger, Nature, 374, 526 (1995).
11 H. K. D. H. Bhadeshia, Proc. R. Microsc. Soc., London, 35(2), 95 (2000).
12 L. Takacs, Mater. Lett., 13, 119 (1992).
13 M. Pardavi-Horvath, L. Takacs, Scripta Met. Mater., 33, 1731 (1995).
14 L. Takacs, Nanostruct. Mater., 2, 241 (1993).
15 L. Takacs, in *Nanophase and Nanocomposite Materials*, eds. S. Komarnemi, J. C. Parker, G. J. Thomas, MRS Symp. Proc., 286, 413 (1993).
16 K. Niihara, A. Nakahira, T. Sekino, MRS Symp. Proc., 286, 405 (1993).
17 D. Osso, O. Tillement, G. Le Caer, A. Mocellin, J. Mater. Sci., 33, 3109 (1998).
18 M. Pardavi-Horvath, L. Takacs, IEEE Trans. Magn., 28, 3186 (1992).
19 L. Takacs, M. Pardavi-Horvath, in *Nanophases and Nanocrystalline Structures*, eds. R.

D. Shull, J. M. Sanchez, The Minerals, Metals & Materials Society, Warrendale, PA, USA, 1994, p. 135.
20 M. Pardavi-Horvath, L. Takacs, J. Appl. Phys., 73, 6958 (1993).
21 J. S. C. Jang, C. C. Koch, J. Mater. Res., 5, 498 (1990).
22 G. Picaluga (ed), Sol–Gel Preparation and Characterization of Metal–Silica and Metal Oxide–Silica Nanocomposites, Trans Tech Publications, Zurich, Switzerland (2001).
23 J. D. Wright, N.A.J.M. Sommerdijk, *Sol–Gel Materials: Chemistry and Applications*, Taylor & Francis, New York, NY, USA (2001).
24 L. W. Hrubesh, J. F. Poco, J. Non-Cryst. Solids, 188, 46 (1995).
25 A. J. Hunt, M. R. Ayers, W. Q. Cao, J. Non-Cryst. Solids, 185 227 (1995).
26 N. Wang, Z. K. Tang, G. D. Li, J. S. Chen, Nature, 408, 51 (2000).
27 X. Y. Song, W. Q. Cao, M. R. Ayers, A. J. Hunt, J. Mater. Res., 10, 251 (1995).
28 W. Q. Cao, A. J. Hunt, Appl. Phys. Lett., 64, 2376 (1994).
29 M. R. Ayers, X. Y. Song, A. J. Hunt, J. Mater. Sci., 31, 6251 (1996).
30 M. R. Ayers, A. J. Hunt, J. Non-Cryst. Solids, 225, 343 (1998).
31 H. Jiang, M. Lau, V.L. Tellkamp, E.J. Lavernia, Synthesis of nanostructured coatings by high velocity oxygen-fuel thermal spraying, Chapter 3 in *Handbook of Nanostructured Materials and Nanotechnology*, ed. H.S. Nalwa, Academic Press, San Diego, CA, USA, 2000.
32 E. Petrovicova, L. S. Schadler Int. J. Mater. Rev., In press.
33 M.L. Luton, C.S. Jayanth, M.M. Disko, S. Matras, J. Vallone, in *Materials Research Society Symposium Proceedings*, L.D. McCandlish et al., eds, Pittsburgh PA, 1989, p. 79.
34 B. Huang, J. Vallone, M.J. Luton, Nanostruct. Mater., 5, 631 (1995).
35 J. He, M. Ice, S. Dallek, E. J. Lavernia, Metall. Mater. Trans. A, 31, 541 (2000).
36 B. Huang, R.J. Perez, E.J. Lavernia, Mater. Sci. Eng., A 255, 124 (1998).
37 J. He, M. Ice, E. J. Lavernia, Metall. Mater. Trans. A, 31, 555 (2000).
38 J. He, M. Ice, J. M. Schoenung, D. H. Shin, E. J. Lavernia, J. Therm. Spray Technol., 10, 293 (2001).
39 M.L. Lau, E.J. Lavernia, Mater. Sci. Eng., A, 272, 222 (1999).
40 T.R. Malow, C.C. Koch, in Synthesis and processing of Nanocrystalline powder, D.L Bourrell, ed., The Minerals, Metals & Materials Society, Warrendale, PA, USA, 1996, p. 33.
41 M.L. Luton, C.S. Jayanth, M.M. Disko, S. Matras, J. Vallone, in *Multicomponent Ultrafine Microstructures* (L.D. McCandlish et al., eds) MRS Symp. Proc., Pittsburgh PA, USA, 1989, p. 79.
42 B. Fultz, L.B. Hong, Z.Q. Gao, C. *Bansal, in Synthesis and Processing of Nanocrystalline Powder*, D.L Bourrell, ed., The Minerals, Metals & Materials Society, Warrendale, PA, USA, 1996, p. 233.
43 G.J. Fan, W.N. Gao, M.X. Quan, Z.Q. Hu, Mater. Lett., 23, 33 (1995).
44 D.A. Stewart, P.H. Shipway, D.G. McCartney, Wear, 225–229, 789 (1999).
45 Y. Zhu, K. Yukimura, C. Ding, P. Zhang, Thin Solid Films, 388, 277 (2001).
46 E.H. Jordan, M. Gell, Y.H. Sohn, D. Goberman, L. Shaw, S. Jiang, M. Wang, T.D. Xiao, Y. Wang, P. Strutt., Mater. Sci. Eng., A, 301, 80 (2001).
47 R.S. Lima, A. Kucuk, C.C. Berndt, Surf. Coat. Technol., 135, 166 (2001).
48 S.C. Tjong, Z.Y. Ma, Mater. Sci. Engin., 29, 49 (2000).
49 W.O. Soboyejo, R.J. Lederich, S.M.L. Sastry, in *High Performance Composites for the 1990s*, S.K. Das, C.P. Ballard, F. Marikar, eds.), The Minerals, Metals & Materials Society, Warrendale, PA, USA, 1991, p. 127.
50 X.C. Tong, H.S. Fang, Metall. Mater. Trans. A, 29, 893 (1998).
51 K. Lu, Mater. Sci. Eng., 16, 161 (1996).
52 Y. Yoshizawa, S. Oguma, K. Yamauchi, J. Appl. Phys., 64, 6044 (1988).
53 J. Eckert, A. Reger-Leonhard, B. Weiss, M. Heimaier, Mat. Sci. Eng., A, 301, 1 (2001).
54 A.W. Weimer, R. K. Bordia, Composites B, 30, 647 (1999).
55 M. Herrmann, C. Schuber, A. Rendtel, H. Hubner, J. Am. Ceram. Soc., 81, 1095 (1998).

56 Y. Wang, M. Sasaki, T. Goto, T. Hirai, J. Mate. Sci., 25, 4607 (1990).
57 A. A. Voevodin, S. V. Prasad, J. S. Zabinski, J. Appl. Phys, 82, 855 (1997).
58 K. Niihara, J. Ceram. Soc. Jpn., 99, 945 (1991).
59 X. H. Jin, L. Gao L. H. Gui, J. K. Guo J. Mater. Res., 17, 1024 (2002).
60 A. Sawaguchi, K. Toda, K. Niihara, J. Ceram. Soc. Jpn., 99, 510 (1991).
61 Y. Xu, A. Zangvil, A. Kerber, J. Eur. Ceram. Soc., 17, 921 (1997).
62 J. Dusza, M. Steen, Int. Mater. Rev., 44, 165 (1999).
63 M. Sternitzke, B. Derby, R. J. Brook, J. Am. Ceram. Soc., 81, 41 (1998).
64 Y. Sakka, D. D. Bidinger, I. A. Aksay, J. Am. Ceram. Soc., 78, 479 (1995).
65 H.J. Choi, K.S. Cho, J.G. Lee, J. Am. Ceram. Soc., 80, 2681 (1997).
66 K. Niihara, K. Izaki, A. Nakahira, J. Mater. Sci. Lett., 9, 598 (1990).
67 J. Dusza, P. Sajgalik, M. Steen, Key Eng. Mate., 175, 311 (1999).
68 P. Descamps, D. O'Sullivan, M. Poorteman, J. C. Descamps, A. Leriche, F. Cambier, J. Eur. Ceram. Soc., 19, 2475 (1999).
69 T. Ohji, Y.-K. Jeong, Y.-H. Choa, K. Niihara, J. Am. Ceram. Soc., 81, 1453 (1998).
70 J. Perez-Riguero, J. Y. Pastor, J. Llorca, M. Elices, P. Miranzo, J. S. Moya, Acta Mater., 46, 5399 (1998).
71 M. R. Scanlon, R. C. Cammarata, J. Appl. Phys., 76, 3387 (1994).
72 B. Abeles, P. Sheng, M. D. Coutts, Y. Arie, Adv. Phys., 24, 407 (1975).
73 W. M. C. Yang, T. Tsakalakos, J. E. Hilliard, J. Appl. Phys., 48, 876 (1977).
74 R. C. Cammarata, K. Sieradzki, Phys. Rev. Lett., 62, 2005 (1989).
75 J. S. Koehler, Phys. Rev. B, 2, 547 (1970).
76 M. Yu et al., Appl. Phys. Lett., 75, 3992 (1999).
77 Y. J. Liu, H. J. Kim, Y. Egashira, H. Kimura, H. Komiyama, J. Am. Ceram. Soc., 79, 1335 (1996).
78 R. Hauert, J. Patscheider, Adv. Eng. Mater., 2, 247 (2000).
79 W. D. Sproul, Science, 273, 889 (1996).
80 C. Subramanian, K. N. Strafford, Wear, 165, 85 (1993).
81 U. Helmersson, S. Todorova, S. A. Barnett, J. Sundgren, L. C. Markert, J. E. Greene, J. Appl. Phys., 62, 481 (1987).
82 M. Shinn, L. Hultman, S. A. Barnett, J. Mater. Res., 7, 901 (1992).
83 X. Chu, S. A. Barnett, J. Appl. Phys., 77, 4403 (1995).
84 I. Wadsworth, I. J. Smith, L. A. Donahue, W. D. Munz, Surf. Coat. Technol., 94/95, 315 (1997).
85 H. Holleck, V. Schier, Surf. Coat. Technol., 76/77, 328 (1995).
86 M. Shinn, S. A. Barnett, Appl. Phys. Lett., 64, 61 (1994).
87 S. Veprek, S. Reiprich, L. Shizhi, Appl. Phys. Lett., 66, 2640 (1995).
88 A. A. Voevodin, J. S. Zabinski, J. Mater. Sci., 33, 319 (1998).
89 A. A. Voevodin, M. A. Capano, A. J. Safriet, M. S. Donely, J. S. Zabinski, Appl. Phys. Lett., 69, 188 (1996).
90 S. Zabinski, A. A. Voevodin, J. Vac. Sci. Technol., A, 16, 1890 (1998).
91 J. E. Sundgren, B., O. Johansson, S. E. Karlsson, Thin Solid Films, 105, 353 (1983).
92 H. Dimigen, C. P. Klages, Surf. Coat. Technol., 49, 543 (1991).
93 W. J. Meng, B. A. Gillispie, J. Appl. Phys., 84, 4314 (1998).
94 R. Gampp, P. Gantenbein, Y. Kuster, P. Reimann, R. Steiner, P. Oelhafen, S. Brunold, U. Frei, A. Gombert, R. Joerger, W. Graf, M. Kohl, *Optical Materials Technology for Energy Efficiency and Solar Cell Conversion*, SPIE Proc. Series, vol. 2255, 92 (1994).
95 R. Gilmore, M. A. Baker, P. N. Gibson, W. Giassler, Surf. Coat. Technol., 105, 45 (1998).
96 L. E. McCandlish, B. H. Kear, B. K. Kim, Mater. Sci. Tech., 6, 953 (1990).
97 Y. T. Zhu, A. Manthiram, J. Am. Ceram. Soc., 77, 2777 (1994).
98 J. R. Wilde, A. L. Greer, Mater. Sci. Eng., A, 304, 932 (2001).
99 H.J. Choi, K.S. Cho, J.G. Lee, J. Am. Ceram. Soc., 80, 2681 (1997).
100 R. Andrews, D. Jacques, A. M. Rao, T. Rantell, F. Derbyshire, Y. Chen, J. Chen, R. C. Haddon, Appl. Phys. Lett., 75, 1329 (1999).
101 P. M. Ajayan, Chem. Rev., 99, 1787 (1999).

102 R. Z. Ma, J. Wu, B. Q. Wei, J. Liang, D. H. Wu, J. Mater. Sci., 33, 5243 (1998).
103 S. Chang, R. H. Doremus, P. M. Ajayan, R. W. Siegel, Ceram. Eng. Sci. Proc., 21, 653 (2000).
104 Y.B. Li, B.Q. Wei, C.L. Xu, J. Liang, D.H. Wu, J. Mater. Sci., 34, 5281 (1999).
105 B.Q. Wei, Y. Li, Ph. Kohler-Redlich, R. Lück, S. Xie, J. Appl. Phys., In press.
106 H. Masuda, M. Satoh, Jpn. J. Appl. Phys., 35, 126 (1996).
107 X. Y. Zhang, L. D. Zhang, G. W. Meng, F. Philipp, Chem. Mater., 13, 2511 (2001).
108 G. L. Che, B. B. Lakshmi, E. R. Fischer, C. R. Martin, Nature, 393, 346 (1998).
109 P. C. Searson, R. C. Cammarata, C. L. Chien, J. Electron. Mater., 24, 955 (1995).
110 R. S. Ruo, D. C. Lorents, B. C. Chan, R. Malhotra, S. Subramoney, Science, 259, 346 (1993).
111 S. A. Majetich, J. O. Artman, M. E. McHenry, N. T. Nuhfer, S. W. Staley, Phys. Rev. B, 48, 16845 (1993).
112 T. Hayashi, S. Hirono, M. Tomita, S. Umemura, Nature, 381, 772 (1996).
113 C. Guerret-Piecourt, Y. Le Bouar, A. Loiseau, H. Pascard, Nature, 372, 761 (1994).
114 A. Loiseau, H. Pascard, Chem. Phys. Lett., 256, 246 (1996).
115 S. Subramoney, Adv. Mater., 10, 1157 (1998).
116 Y. Saito and T. Yoshikawa, J. Cryst. Growth, 134, 154 (1993).
117 M. Tomita, Y. Saito, T. Hayashi, Jpn. J. Appl. Phys., 32, L280 (1993).
118 J. H. J. Scott, A. A. Majetich, Z. Turgut, M. E. McHenry, M. Boulos, MRS Symp. Proc., 457, 219 (1997).
119 P. M. Ajayan, S. Iijima, Nature, 361, 333 (1993).
120 S. C. Tsang, Y. K. Chen, P. J. F. Harris, M. L. H. Green, Nature, 372, 159 (1994).
121 R. R. Meyer, J. Sloan, R. E. Dunin-Borkowski, A. I. Kirkland, M. C. Novotny, S. R. Bailey, J. L. Hutchison, M. L. H. Green, Science, 289, 1324 (2000).
122 J. Sloan, D. M. Wright, H. G. Woo, S. Bailey, G. Brown, A. P. E. York, K. S. Coleman, J. L. Hutchison, M. L. H. Green, Chem. Commun., 8, 699 (1999).
123 B. W. Smith, M. Monthioux, D. E. Luzzi, Nature, 396, 323 (1998).
124 J. Lee, H. Kim, S.-J. Kahng, G. Kim, Y.-W. Son, J. Ihm, H. Kato, Z. W. Wang, T. Okazaki, H. Shinohara, Y. Kuk, Nature, 415, 1005 (2002).
125 D. Danailov, P. Keblinski, P. M. Ajayan, J. Nanosci. Nanotechn., In press.
126 W. K. Hsu, H. Terrones, M. Terrones, N. Grobert, A. I. Kirkland, J. P. Hare, K. Prassides, P. D. Townsend, H. W. Kroto, D. R. M. Walton, Chem. Phys. Lett., 284, 177 (1998).
127 W. K. Hsu, S. Trasobares, H. Terrones, M. Terrones, N. Grobert, Y. Q. Zhu, W. Z. Li, J. P. Hare, R. Escudero, H. W. Kroto, D. R. M. Walton, Chem. Mater., 11, 1747 (1999).
128 M. S. Dresselhaus, G. Dresselhaus, K. Sugihara, I. L. Spain, H. A. Goldberg, *Graphite Fibers and Filaments*, vol. 3, Springer, Berlin, Germany (1988).
129 C. N. R. Rao, R. Sen, B. C. Satishkumar, A. Govindaraj, J. Chem. Soc. Chem. Commun., 15, 1525 (1998).
130 Y. Zhang, H. J. Dai, Appl. Phys. Lett., 77, 3015 (2000).
131 P. M. Ajayan, O. Stephan, P. Redlich, C. Colliex, Nature, 375, 564 (1995).
132 W. K. Hsu, Y. Q. Zhu, H. W. Kroto, D. R. M. Walton, R. Kamalakaran, M. Terrones, Appl. Phys. Lett., 77, 4130 (2000).
133 M. Terrones, P. M. Ajayan, Unpublished.
134 J. Luo, L. Zhang, Y. Zhang, J. Zhu, Adv. Mater., 14, 1414 (2002).
135 Y. Zhang, T. Ichihashi, E. Landree, F. Nihey, S. Iijima, Science, 285, 1719 (1999).
136 Y. Zhang, K. Suenaga, C. Colliex, S. Iijima, Science, 281, 973 (1998).
137 Lieber Group, Nature, 409, 66, (2001).
138 Y. Wu, R. Fan, P. Yang, Nanoletters, 2, 83 (2002).
139 S. Y. Chou, M. S. Wei, P. R.Krauss, P. B. Fischer, J. Appl Phys., 76, 6673 (1994).
140 T. M. Whitney, J. S. Jiang, P. C. Searson, C. L. Chien, Science, 261, 1316 (1993).
141 S. Y. Chou, Proc. IEEE, 85, 652 (1997).
142 D. K. Kim, M. Mikhaylova, M. Toprak, Y. Zhang, M. Muhammed, *Nanoparticulate Materials*; MRS Proc., 704 (2001).
143 D. K. Kim, Yu Zhang, M. Muhammed, W. Voit, K. V. Rao, J. Kehr, B. Bjelke, Scripta Mater., 44, 1713, (2001).
144 T. Ogawa, Y. Kanemitsu, *Optical Properties of Low-Dimensional Materials*, World Scientific, Singapore (1995).

145 Y. Kanemitsu, H. Tanaka, S. Mimura, S. Okamoto, T. Kushida, K. S. Min, H. A. Atwater, MRS Symp. Proc., 536, 223 (1999).

146 K. M. Hassan, A. K. Sharma, J. Narayan, J. F. Muth, C. W. Teng, R. M. Kolbas, Appl. Phys. Lett., 75, 1222 (1999).

147 H. Zhou, W. Cai, L. Zhang, Appl. Phys. Lett., 75, 495 (1999).

148 H. B. Liao, R. F. Xiao, H. Wang, K. S. Wong, G. K. L. Wong, Appl. Phys. Lett., 72, 1817 (1998).

149 Z. Li, H. Wang, H. Li, X. Wang, Appl. Phys. Lett., 72, 1823 (1998).

150 T. Kyprianidou-Leodidou, W. Caseri, U. W. Suter, J. Phys. Chem., 98, 8992 (1994).

151 F. Papadimitrakopoulos, P. Wisniecki, D. E. Bhagwagar, Chem. Mater., 9, 2928 (1997).

152 S. R. Panteny, R.Stevens, C. R.Bowen, Ferroelectrics 2000 UK (Editors N. McN. Alford & E. Yeatman), ISBN 1-86125-135-1, P75.

153 K. Tajima, H. J. Hwang, M. Sando, K. Niihara, J. Eur. Ceram. Soc., 19, 1179 (1999).

154 D. Hoppe, D. Ernst, H. Weiß, Proc. Eur. Conf.: Junior – Euromat '96, Lausanne, Switzerland, 26–30 August (1996), S. 18f.

155 S. Banerjee, D. Chakravorty, J. Appl. Phys., 85, 3623 (1999).

156 M. D. Jaeger, R. E. Soltis, J. A. Thomas, J. Hangas, A. E. Chen, E. M. Logothetis, MRS Symp. Proc., 549, 173 (1999).

157 A. Miller, E. Abrahams, Phys. Rev. B, 120, 745 (1960).

158 A. Aldea, M. Dulea, Phys. Rev. Lett., 60, 1672 (1988).

159 M. E. J. Newman, R. B. Stinchcombe, Phys. Rev. B, 43, 1183 (1991).

160 X.H.Yan, J.W.Ding, Q.B.Yang, Eur. Phys. J., B, 20, 157 (2001).

161 B. Abeles, Adv. Phys., 24, 407 (1975).

162 G. Deutscher, A. Kapitulnik, M. Rappaport, *Percolation Structures and Processes*, Adam Hilger, Bristol, UK (1983), Chapter 10.

163 G. Deutscher, O. Entin-Wohlman, S. Fishman, Y. Shapira, Phys. Rev. B, 21, 5041 (1980).

164 R. Zallen, *Physics of Amorphous Solids*, Wiley, New York, NY, USA (1983) Chapter 4.

165 H. Baessler, N. Herrmann, N. Riehl, G. Vaubel, J. Phys. Chem. Solids, 30, 1579 (1969).

166 H. Gleiter, Prog. Mater. Sci., 33, 223 (1989).

167 I. J. Youngs, J. Phys. D: Appl. Phys., 35, 3127 (2002).

168 J.N. Coleman, S. Curran, A.B. Dalton, A.P. Davey, B.M. McCarthy, W. Blau, R.C. Barklie, Phys Rev B, 58, 7492 (1998).

169 K. Beltsios, T. Steriotis, K. Stephanopoulos, N. Kanellopoulos, in *Handbook of Porous Solids*, ed. F. Schueth, K. Sing, J. Weitkamp, Wiley, New York, NY, USA, p. 2281 (2002).

170 E. Wilson, Chem. Eng. News, 77(5), 32 (1999).

171 K. Keizer, A.J. Burggraaf, J. Membr. Sci., 66, 259 (1992).

172 R.W. Baker, *Membrane Technology and Applications*, McGraw-Hill, New York, NY, USA, (1999).

173 S. Sircar, M. B. Rao, in *Recent Advances in Gas Separation by Microporous Membranes*, ed. N. Kanellopoulos, Membrane Science and Technology Series, vol. 6, Elsevier, Amsterdam, Netherlands, p. 473 (2000).

174 K. Kusakabe, Z.Y. Li, H. Maeda, S. Morooka, J. Membr. Sci., 103, 175 (1995).

175 Shekhar Garde, personal communication.

176 website: http://www.etis.net/balpyo/envi/envilz.pdf

177 B. Hamilton, Semicond. Sci. Technol., 10, 1187 (1995).

178 J.B. Hunter, US Patent 2, 773, 561 (1956).

179 J.P. O'Sullivan, G.C. Woods, Proc. R. Soc., London, A, 317, 511 (1970).

180 H. Masuda, H. Yamada, M. Satoh, H. Asoh, M. Nakao, T. Tamamura, Appl. Phys. Lett., 71, 2772 (1997).

181 C.N. Wu, K.J. Chao, T-G. Tsai Y.-H. Chiou, H.-C. Shih, Adv. Mater., 8, 1008 (1996).

182 P. Behrens, Adv. Mater., 5, 127 (1993).

183 H. Masuda, K. Nishio, N. Baba, Thin Solid Films, 223, 1 (1993).

184 T. Kyotani, T. Nagai, S. Inoue, A. Tomita, Chem. Mater., 9, 609 (1997).

185 R. Skomski, J. M. D. Coey, Phys. Rev. B, 48, 15812 (1993).

186 K. O'Donnel, C. Kuhrt, J. M. D. Coey, J. Appl. Phys., 76, 7068 (1994).

187 H. Fukunaga, J. Kuma, Y. Kanai, IEEE Trans. Magn., 35, 3235 (1999).

188 P. G. McCormick, W. F. Miao, A. A. I. Smith, J. Ding, R. Street, J. Appl. Phys., 83, 6256 (1998).
189 H. Zeng, J. Li, J. P. Liu, Z. L. Wang, S. Sun, Nature, 420, 395 (2002).
190 T. Hayashi, S. Hirono, M. Tomita, S. Umemura, Nature, 38, 772 (1996).
191 M. Yu, Y. Liu, D. J. Sellmyer, J. Appl. Phys., 85, 4319 (1999).
192 M. Yu, Y. Liu, A. Moser, D. Weller, D. J. Sellmyer, Appl. Phys. Lett., 75, 3992 (1999).
193 B. Cord, J. Scherer, IEEE Trans. Magn., 36, 67 (2000).
194 C.L. Chien, J. Appl. Phys., 69, 5267 (1991).
195 B. Abeles, P. Shoeing, M. D. Coutts, Y. Arie, Adv. Phys., 24, 407 (1975).
196 L. Maya, M. Paranthaman, J.R. Thompson, T. Thundat, R. J. Stevenson, J. Appl. Phys., 79, 7905 (1996).
197 M. F. Doerner, K. Tang, T. Arnoldussen, H. Zeng, M. F. Toney, D. Weller, IEEE Trans. Magn., 36, 43 (2000).
198 "Seagate announced 23.8 Gbits/in2, Read-Rite 26.5 and 36 Gbits/in2, and IBM 35.3 Gbits/in2 demos in recent press releases," Data Storage Magazine, p. 8, July 1999.
199 J. Li, M. Mirzamaani, X. Bian, M. Doerner, S. Duan, K. Tang, M. Toney, T. Arnoldussen, M. Madison, J. Appl. Phys., 85, 4286 (1999).
200 H. N. Bertram, H. Zhou, R. Gustafson, IEEE Trans. Magn., 34, 1845 (1998).
201 M. F. Doerner, K. Tang, T. Arnoldussen, H. Zeng, M. F. Toney, D. Weller, IEEE Trans. Magn., 36, (2000).
202 D. Weller, A. Moser, L. Folks, M. E. Best, W. Lee, M. F. Toney, M. Schwickert, J.-U. Thiele, M. F. Doerner, IEEE Trans. Magn., 36, 10 (2000).
203 S. Stavroyiannis, I. Panagiotopoulos, D. Niarchos, J. A. Christodoulides, Y. Zhang, G. C. Hadjipanayis, Appl. Phys. Lett., 73, 3453 (1998).
204 Q. Liu, Z. Xu, J. A. Finch, R. Egerton, Chem. Mater., 10, 3936 (1998).
205 B. Z. Tang, Chemtech, page 9, November 1999.
206 N. R. de Tacconi, J. Carmona, K. Rajeshwar, J. Phys. Chem. B, 101, 10151 (1997).
207 R. R. Oberle, M. R. Scanlon, R. C. Cammarata, P. C. Searson, Appl. Phys. Lett., 66, 19 (1995).
208 S. D. Burnside, V. Shklover, C. Barbe, P. Comte, F. Arendse, K. Brooks, M. Gratzel, Chem. Mater., 10, 2419 (1998).
209 G. Schmid, Chem. Rev., 92, 1709 (1992).
210 H. Yonezawa, T. Matsune, T. Kunitake, Chem. Mater., 11, 33 (1999).
211 I. Ichinose, H. Senzu, H. Kunitake, Chem. Mater., 9, 1296 (1997).
212 J. W. Haus, N. Kalyaniwalla, R. Inguava, C. W. Bowden, J. Appl. Phys., 65, 1420 (1989).
213 A. P. Alivisatos, Science, 271, 932 (1996).
214 M. J. Ko, Adv. Mater. Opt. Electron., 8, 173 (1998).
215 A. I. Belogorokov, L. I. Belogorokhova, A. Perez-Rodriguez, J. R. Morante, S. Gavrilov, Appl. Phys. Lett., 73, 2766 (1998).
216 G. L. Hornyak, C. J. Patrissi, C. R. Martin, J. Phys. Chem., 101, 1548 (1997).
217 K. Yasumoto, N. Koshizaki, J. Mater. Sci. Lett., 16, 215 (1997).
218 T. Nagai, H. J. Hwang, M. Yasuoka, M. Sando, K. Niihara, J. Am. Ceram. Soc., 81, 425 (1998).
219 T. Takagi, Bull. Ceram. Soc. Jpn., 28, 539 (1993).
220 B. Djuricac, S. Pickering, D. McGarry, P. Tambuyser, P. Thomas, J. Mater. Sci., 34, 1911 (1999).
221 K. Zhang, C. H. Chew, G. Q. Xu, J. Wang, L. M. Gan, Langmuir, 15, 3056 (1999).
222 A. Matsuda, Y. Kotani, T. Kogure, M. Tatsumisago, T. Minami, J. Am. Ceram. Soc., 83, 229 (2000).
223 N. A. Dhas, A. Gedanken, Chem. Mater., 9, 3144 (1997).
224 D. J. Aldrich, K. M. Jones, S. Govindarajan, J. J. Moore, T. R. Ohno, J. Am. Ceram. Soc., 81, 1471 (1998).
225 J. I. Lee, N. L. Hecht, T. Mah, J. Am. Ceram. Soc., 81, 421 (1998).
226 F. Ruiz, W. D. Sun, F. H. Pollak, C. Venkatraman, Appl. Phys. Lett., 73, 1802 (1998).
227 J. J. Senkavitch, D. E. Leber, M. J. Tutor, N. A. Heiks, G. A. Ten Eyck, D. W. Scherrer II, J. Vac. Sci. Technol., B, 17, 2129 (1999).
228 http://www.sanken.osaka-u.ac.jp/labs/scm/about_resE.html
229 http://aldix.mpi-stuttgart.mpg.de/E_head.html
230 http://www.msm.cam.ac.uk/phase-trans/2000/RMS.pdf

231 http://eande.lbl.gov/ECS/aerogels/sapho-to.htm

232 http://eande.lbl.gov/ECS/aerogels/sapho-to.htm

233 http://www.chem.ox.ac.uk/researchguide/jsloan.html

234 http://www.seas.upenn.edu/~luzzi/re-search.html

235 J. Lee, H. Kim, S.-J. Kahng, G. Kim, Y.-W. Son, J. Ihm, H. Kato, Z. W. Wang, T. Okazaki, H. Shinohara & Young Kuk, Nature, 415, 1005 (2002).

236 http://www.che.udel.edu/~strano/structu-re.html

2
Polymer-based and Polymer-filled Nanocomposites

Linda S. Schadler

2.1
Introduction

Polymer composites are important commercial materials with applications that include filled elastomers for damping, electrical insulators, thermal conductors, and high-performance composites for use in aircraft. Materials with synergistic properties are chosen to create composites with tailored properties; for example, high-modulus but brittle carbon fibers are added to low-modulus polymers to create a stiff, lightweight composite with some degree of toughness. In recent years, however, we have reached the limits of optimizing composite properties of traditional micrometer-scale composite fillers, because the properties achieved usually involve compromises. Stiffness is traded for toughness, or toughness is obtained at the cost of optical clarity. In addition, macroscopic defects due to regions of high or low volume fraction of filler often lead to breakdown or failure.

Recently, a large window of opportunity has opened to overcome the limitations of traditional micrometer-scale polymer composites – nanoscale filled polymer composites – in which the filler is <100 nm in at least one dimension (Figure 2.1). Although some nanofilled composites (carbon black [1] and fumed silica [2, 3] filled polymers) have been used for more than a century, research and development of nanofilled poly-

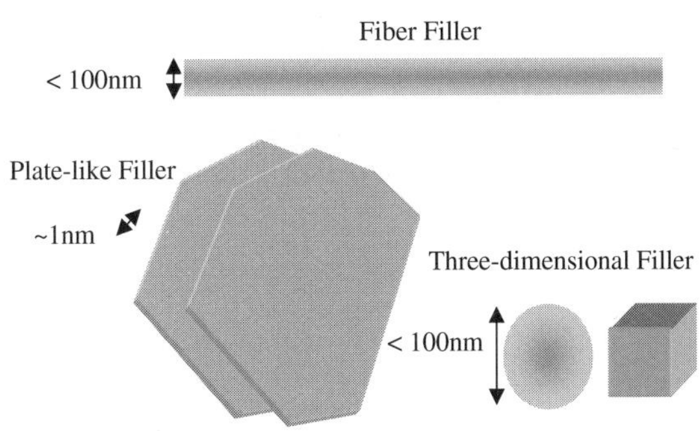

Fig. 2.1 Schematic of nanoscale fillers

mers has greatly increased in recent years, for several reasons. First, unprecedented combinations of properties have been observed in some polymer nanocomposites [4]. For example, the inclusion of equi-axed nanoparticles in thermoplastics, and particularly in semicrystalline thermoplastics, increases the yield stress, the tensile strength, and Young's modulus [5] compared to pure polymer. A volume fraction of only 0.04 mica-type silicates (MTS) in epoxy increases the modulus below the glass transition temperature by 58% and the modulus in the rubbery region by 450% [6]. In addition [7], the permeability of water in poly(e-caprolactone) decreases by an order of magnitude with the addition of 4.8% silicate by volume. Yano et al. [8] showed a 50% decrease in the permeability of polyimides at a 2% loading of MTS. Many of these nanocomposites are optically transparent and/or optically active.

A second reason for the large increase in research and development efforts was the 'discovery' of carbon nanotubes in the early 1990s [9]. Although more careful review has shown that nanotubes have been observed since the 1960s [10], it was only in the mid-1990s that they were made in the quantities required for property evaluation of composites. The properties of these carbon nanotubes, particularly strength and electrical properties, are significantly different from those of graphite and offer exciting possibilities for new composite materials.

Third, significant development in the chemical processing of nanoparticles and in the in situ processing of nanocomposites has led to unprecedented control over the morphology of such composites. It has also created an almost unlimited ability to control the interface between the matrix and the filler.

Thus, this is an exciting time to study nanocomposites, because of the unique combinations of properties that are achievable and because of the high potential for successful commercial development. Although the technical community has made advances in the processing of nanocomposites, we are just beginning to assemble the interdisciplinary teams required to understand, tailor, and optimize properties. We have at our fingertips, however, the ability to change the size, shape, volume fraction, interface, and degree of dispersion or aggregation. Thus, the opportunities may well become limitless when theory and experiment have assembled enough information to guide further development.

A relevant question addressed throughout this chapter is: What is unique to nanofillers compared to micrometer-scale traditional fillers, and how do the composites compare to their macroscopic counterparts? The most obvious difference is the small size of the fillers. For example, very small nanoparticles do not scatter light significantly, and thus it is possible to make composites with altered electrical or mechanical properties that retain their optical clarity. In addition, the small size means that the particles do not create large stress concentrations and thus do not compromise the ductility of the polymer. A similar concept applies for electrical breakdown strength.

The small size of nanofillers can also lead to unique properties of the particles themselves. For example, single-walled nanotubes are essentially molecules, free from defects, and have a modulus as high as 1 TPa and strengths that may be as high as 500 GPa. Single-crystal particles that are optically active, but are unmanageable on the macro scale can be combined in a polymer to achieve the optical gain of the material and the ease of processing afforded by the polymer.

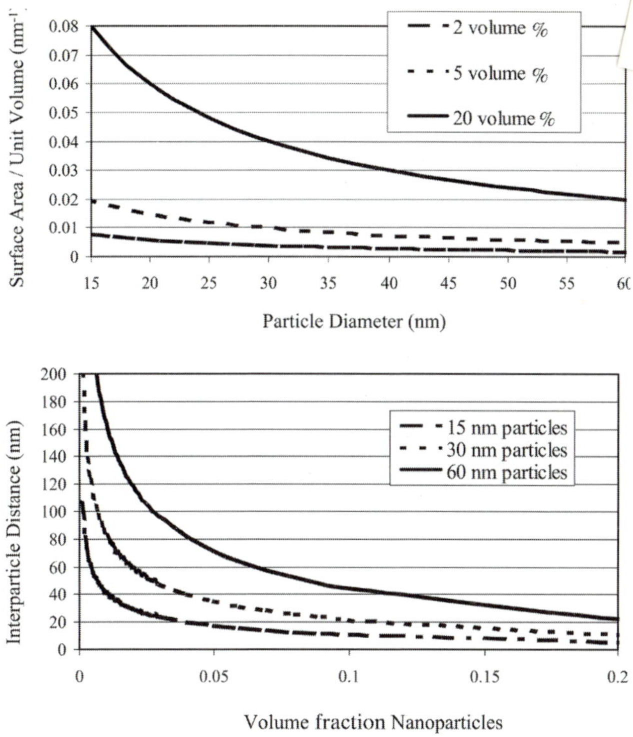

Fig. 2.2 (a) Surface area per unit volume vs. particle size for spherical particles that are ideally dispersed, and (b) interparticle distance for spherical particles that are ideally dispersed (Thanks to B.J. Ash for preparation of this figure)

In addition to the effect of size on particle properties, the small size of the fillers leads to an exceptionally large interfacial area in the composites. Figure 2.2a shows the surface area per unit volume as a function of particle size for spherical particles that are ideally dispersed. The increase in surface area below 100 nm is dramatic. The interface controls the degree of interaction between the filler and the polymer and thus controls the properties. Therefore, the greatest challenge in developing polymer nanocomposites may be learning to control the interface. Thus, it seems relevant to define the interfacial region and discuss its properties.

As defined in traditional composites, the interfacial region is the region beginning at the point in the fiber at which the properties differ from those of the bulk filler and ending at the point in the matrix at which the properties become equal to those of the bulk matrix [11]. It can be a region of altered chemistry, altered polymer chain mobility, altered degree of cure, and altered crystallinity. Interface size has been reported to be as small as 2 nm and as large as about 50 nm. Figure 2.2b shows interparticle spacing as a function of particle size for an ideally dispersed nanoparticle composite: at low volume fractions the entire matrix is essentially part of the interfacial region. For example, for 15-nm particles at a filler loading of 10 vol %, the interparticle spacing is only 10 nm. Even if the interfacial region is only a few nanometers, very quickly the entire polymer matrix has a different behavior than the bulk. If the interfacial region is more extended, then the polymer matrix behavior can be altered at much smaller loadings. Therefore, by controlling the degree of interaction between the polymer and the nanofiller, the properties of the entire matrix can be controlled.

To implement the novel properties of nanocomposites, processing methods that lead to controlled particle size distribution, dispersion, and interfacial interactions are critical. Processing technologies for nanocomposites are different from those for composites with micrometer-scale fillers, and new developments in nanocomposite processing are among the reasons for their recent success.

This chapter highlights what we feel are important concepts for understanding and developing polymer nanocomposites. We start with a description of nanotube or nanofiber fillers, plate-like fillers, and equi-axed fillers. We then introduce the interfacial region and attempt to give a broad understanding of how to control it, without delving too deeply into the chemistry involved. This is followed by a section on the processing of composites. Finally, we describe some of the properties that have been achieved in nanofilled polymer composites.

2.2 Nanoscale Fillers

Nanoscale fillers come in many shapes and sizes. For ease of discussion, we have grouped nanofillers into three categories (Figure 2.1). Fiber or tube fillers have a diameter <100 nm and an aspect ratio of at least 100. The aspect ratios can be as high as 10^6 (carbon nanotubes). Plate-like nanofillers (Figure 2.1) are layered materials typically with a thickness on the order of 1 nm, but with an aspect ratio in the other two dimensions of at least 25. Three dimensional (3D) nanofillers are relatively equi-axed particles <100 nm in their largest dimension. This is a convenient way to discuss polymer nanocomposites, because the processing methods used and the properties achieved depend strongly on the geometry of the fillers.

2.2.1 Nanofiber or Nanotube Fillers

2.2.1.1 Carbon Nanotubes

Micrometer-size carbon tubes, which are similar in structure (but not in dimensions) to the recently discovered multi-walled carbon nanotubes, were first found in 1960 by Roger Bacon [12]. These nanosized, near-perfect whiskers (termed nanotubes) were first noticed and fully characterized in 1991 by Sumio Iijima [9] of NEC Corporation in Japan. He was investigating the surface of carbon electrodes used in an electric arc-discharge apparatus that had been used to make fullerenes. Several exciting developments have taken place in this field since then, and several books document recent progress [13].

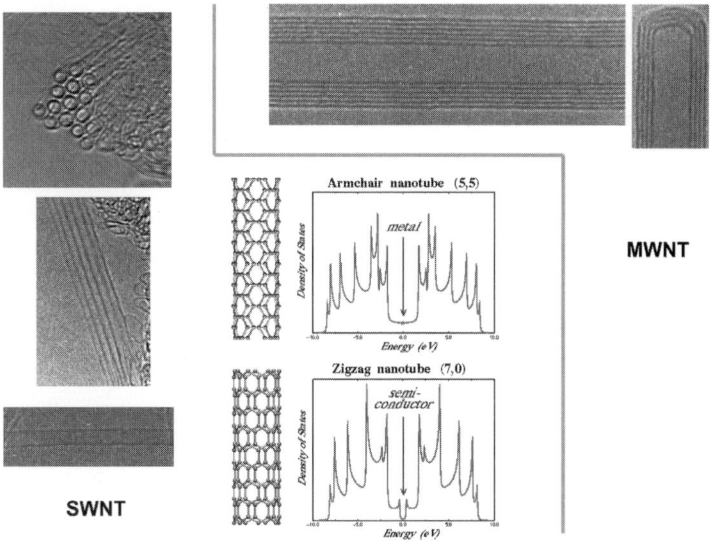

Fig. 2.3 (a) HRTEM image showing the SWNT in bundles, (b) HRTEM images of a MWNT along its length and at the end, (c) schematic showing two examples of the helicity that occurs, zigzag and armchair

The first nanotubes observed were multi-walled nanotubes (MWNT). MWNTs consist of two or more concentric cylindrical shells of graphene sheets coaxially arranged around a central hollow core with interlayer separation as in graphite (0.34 nm) [14]. In contrast, single-shell or single-walled nanotubes [15, 16] (SWNT) are made of single graphene (one layer of graphite) cylinders and have a very narrow size distribution (1–2 nm). Often many (tens) single-shell nanotubes pack into larger ropes. Figure 2.3 shows electron micrographs of SWNT and MWNT. Both types of nanotubes have the physical characteristics of solids and are microcrystals, although their diameters are close to molecular dimensions. In nanotubes, the hexagonal symmetry of the carbon atoms in planar graphene sheets is distorted, because the lattice is curved and must match along the edges (with dangling bonds) to make perfect cylinders. This leads to a helical arrangement of carbon atoms in the nanotube shells. Depending on the helicity and dimensions of the tubes, the electronic structure changes considerably [17, 18]. Hence, although graphite is a semi-metal, carbon nanotubes can be either metallic or semiconducting. Nanotubes are closed by fullerene-like end caps that contain topological defects (pentagons in a hexagonal lattice). The electronic character of the ends of these tubes differs from the cylindrical parts of the tubes and is more metallic, due to the presence of defects in these regions [19].

The discovery of nanotubes has complemented the excitement and activities associated with fullerenes [20]. Although fullerenes have fascinating physical properties, their relevance in the nanocomposite field is limited, so we shall restrict our discussion of carbon nanostructures to nanotubes.

Tab. 2.1 Theoretical and experimentally measured properties of carbon nanotubes.

Property	Nanotubes	Graphite
Lattice structure	(Cylindrical) hexagonal lattice helicity Nanotubes: ropes, tubes arranged in triangular lattice with lattice parameters of a = 1.7 nm, tube–tube distance = 0.315	Planar hexagonal, plane-to-plane distance c = 0.335
Specific gravity	0.8–1.8 g cc^{-1} (theoretical)	2.26 g cc^{-1}
Elastic modulus	~1 TPa for SWNT ~0.3–1 TPa for MWNT	1 Tpa (in-plane)
Strength	50–500 GPa for SWNT, 10–60 GPa for MWNT	
Resistivity	~5–50 micro-ohm-cm	50 (in-plane)
Thermal conductivity	3000 W $m^{-1}K^{-1}$ (theoretical)	3000 W $m^{-1}K^{-1}$ (in-plane) 6 W $m^{-1}K^{-1}$ (c axis)
Thermal expansion	Negligible (theoretical)	-1×10^{-6} K^{-1} (in-plane) 29×10^{-6} K^{-1} (c axis)
Oxidation in air	>700 °C	450–650 °C

The properties of carbon nanotubes are unique compared to other graphite fibers. Their structure remains distinctly different from that of traditional carbon fibers [21], which have been used industrially for several decades (e.g., as reinforcements in tennis rackets, airplane body parts, and batteries). Nanotubes represent the ideal, most perfect, ordered carbon fiber, the structure of which is entirely known at the atomic level. Table 2.1 shows measured and theoretical properties of both SWNT and MWNT. A nice review of the mechanics of nanotubes was recently published [22]. The mechanical properties reported in Table 2.1 were determined directly on nanotubes by using various methods of loading with an AFM. Salvetat et al. [23] were able to place SWNT ropes across a silicon bridge and bend them using an AFM tip. They found that the modulus decreased dramatically as the size of the bundle increased (Figure 2.4). Walters et al. [24] did a similar study and measured the complete stress–strain curve. The modulus of an SWNT rope, which, at this point for composites, is the relevant form of SWNT, was low at small strains and increased at higher strains. As further support of this, Wood et al. [25] found similar results with a more indirect measurement of the stress–strain curve. Because individual SWNTs are theorized to have a modulus as high as 1 TPa, this low modulus of larger ropes at low strains implies that slippage occurs between the individual nanotubes within the rope. The mechanics of this are not clear, but may be similar to that of rope mechanics [26]. If so, individual strands can slip until they are tightly packed, and then there is large friction between them; only then is the modulus of the individual fibers evident. When SWNT ropes were pulled directly by two AFM tips [27], a realistic modulus for the individual tubes within the bundle was achieved only if the SWNT in contact with the tips were

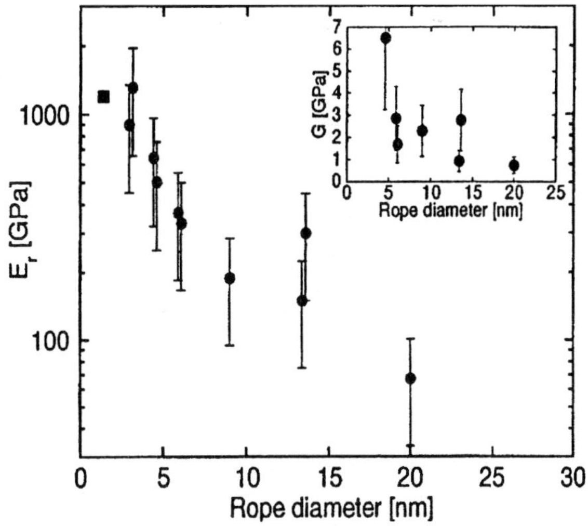

Fig. 2.4 A plot of modulus vs. SWNT rope diameter showing the decrease in the modulus of SWNT ropes as a function of rope diameter. This highlights the poor load transfer between the SWNT within the rope and the need for separating SWNT from the ropes for use in composites. Reprinted with permission from [23] copyright 1999, American Phyiscal Society

counted as the cross sectional area. If the whole rope was used to determine the cross sectional area, the modulus was much lower. The important implication of this finding is that the modulus of SWNT can be realized only when the SWNT are isolated from

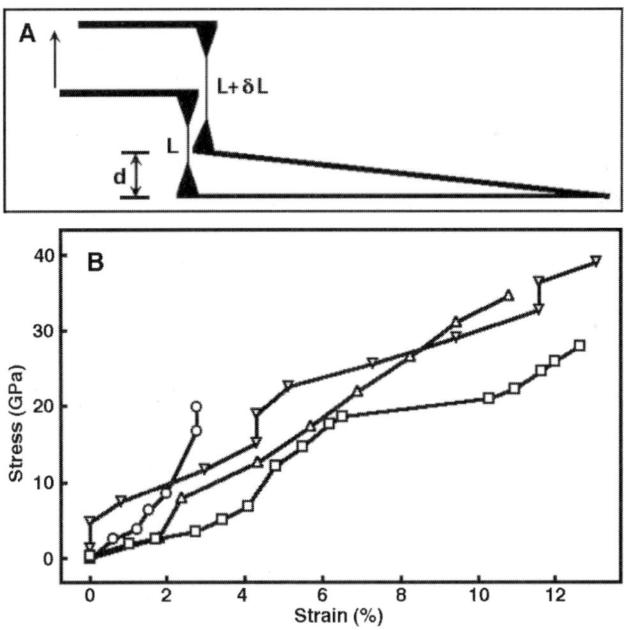

Fig. 2.5 Stress vs. strain curves for MWNT. The variation in the curves is partially a function of the difference in MWNT diameter. Reprinted with permission from [28]

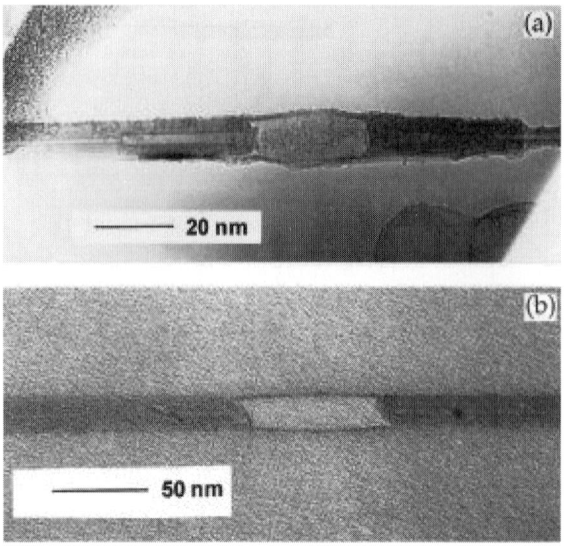

Fig. 2.6 Transmission electron micrographs showing the telescopic failure that can occur in MWNT. Reprinted with permission from [29] copyright 1998, American Institute of Physics

the bundles and/or the bundles are small enough that matrix is in contact with each one.

The most direct measurement of the mechanical properties of MWNT was conducted by Yu et al. [28] using two AFM tips (Figure 2.5). They found a range of modulus and strength. The range in modulus may again be due to difficulty in determining the cross sectional area. Failure occurred via a sword-and-sheath mechanism (telescoping) first observed by Wagner et al. [29] (Figure 2.6), implying that only the outermost nanotube is carrying the load and the rest are slipping inside. Therefore, a smaller cross sectional area may be required for appropriate calculations of the modulus. Thus, the variability in the number of layers in an MWNT probably accounts for the broad scatter in the modulus (Table 2.1). The important implication of this in composites, however, is that MWNT do not have the high modulus that is achievable with isolated SWNT. MWNT do have reversible deformation in compression, which may prove useful. Bower et al. [30] showed that MWNTs buckle under a compressive strain of about 4.7% and that this buckling is reversible under very large strains, probably due to the ability of the nanotubes to slide within each other.

The fracture and deformation behavior of nanotubes is intriguing. Simulations have shown that highly deformed nanotubes switch reversibly into different morphological patterns with an abrupt release of energy. Nanotubes become flattened, twisted, and buckled as they deform [31]. They sustain large strains (40%) in tension without showing signs of fracture. These changes in deformation modes, such as buckling, have been recorded by TEM. The flexibility is related to the in-plane flexibility of the planar graphene sheet and the ability of the carbon atoms to rehybridize, with the degree of sp^2-sp^3 rehybridization depending on the strain. Such nanotube flexibility under mechanical loading is important for their potential application as nanoprobes and in nanocomposites. This behavior also offers an advantage against breakdown that could occur during composite processing.

In addition to the deformation mechanics described, new modes of plastic behavior have been predicted in nanotubes at higher temperatures (relevant to nanotube/ceramic composites, for example). Pairs of a 5–7 pair defect, called a Stone–Wales defect in sp^2 carbon systems, are created at high strains in the nanotube lattice and become mobile. Such mobility leads to a step-wise size reduction (localized necking) of the nanotube. It also introduces changes in helicity in the region where the defects have moved (similar to a change in lattice orientation where a dislocation passes through a crystal) [32].

High-temperature fracture of individual nanotubes under tensile loading has been studied by molecular dynamics simulation. Elastic stretching elongates the hexagons until, at high strain, some bonds are broken. The local defect is then redistributed over the entire surface by bond saturation and surface reconstructions. The final result of this process is that instead of fracturing, the nanotube lattice unravels into a linear chain of carbenes (carbon atoms linked by double bonds into a chain) [31]. Such behavior is unusual in crystals. It could play a role in increasing the toughness of nanotube-filled ceramic composites during high-temperature loading by increasing the energy absorbed during deformation.

The electrical properties of nanotubes are also unique. Calculations have predicted that all the armchair tubes (Figure 2.3c) are metallic, whereas zigzag and helical tubes are either metallic or semiconducting [33]. The electronic conduction process in nanotubes is unique because the electrons are confined in the radial direction in the singular plane of the graphene sheet. Conduction in the armchair (metallic) tubes occurs through gapless modes, because the valence and conduction bands always cross each other at the Fermi energy [34]. In most helical tubes that contain a large number of atoms in the unit cell, the 1D band structure shows an opening of the gap at the Fermi energy, lending it semiconducting properties. This unique electronic behavior occurs only in small nanotubes. As the diameter of the tubes increases, the band gap (which varies inversely as the tube diameter) tends toward zero, yielding a zero-gap semiconductor that is equivalent to the planar graphene sheet [35]. In an MWNT, the electronic structure of the smallest inner tubes is overshadowed by the outer, larger, planar, graphene-like tubes. Experiments in which the band structure obtained from individual MWNT resembles that of graphite substantiate this phenomenon. In a later section, discussion of the electrical conduction and optical properties of nanotube-filled polymer composites demonstrates the electronic importance of nanotube properties.

2.2.1.2 **Nanotube Processing**

The best methods to date for producing ideal nanotubes are based on the electric-arc and laser-ablation processes [36]. The material prepared by these techniques has to be purified by tedious chemical separation methods. Not one of these techniques is suitable for producing the quantities needed for industrial applications (e.g., in composites). In addition, this barrier has hindered research and development. In recent years, work has focused on developing chemical vapor deposition techniques using catalyst particles and hydrocarbon precursors to grow nanotubes [37]. In the past, such techniques were used to produce large quantities of hollow nanofibers of carbon. The drawback of catalytic CVD-based nanotube production is the inferior quality of the structures, which contain gross defects (twists, tilt boundaries, etc.), particularly because the structures are created at much lower temperatures (600 – 1000 °C) compared to the arc or laser process (~2000 °C).

However, recently a major breakthrough has resulted in the efficient production of single-walled nanotubes by what is called a HiPCO process [38]. The technique, developed at Rice University, has become the main source for commercially available high-purity SWNT in gram quantities. The technique uses the high-pressure (several atmospheres) disproportionation of CO gas in the presence of iron carbonyl catalyst vapor [39]. The plan proposed by the Rice group is to produce nanotubes in pound quantities by the year 2005 at a reasonable price (a few dollars per gram). Nanotubes obtained by this method are greater than 80% pure by weight, and the 20% impurity consists principally of iron particles from the catalyst. These particles should not have much effect on the mechanical properties of nanotubes when used as filler but could certainly have effects on their electrical, magnetic, and optical properties.

In his first experimental report, Iijima [9] showed MWNT sticking to the ends of graphite electrodes that were used in the production of fullerenes. Fullerenes are formed in the vapor phase from the evaporation of graphite electrodes. MWNT are formed on the cathode surfaces used in such soot generation. A year after the discovery of MWNT, it was found that, if conditions are right, the evaporated carbon can be made to condense continuously on the cathode surface as a cigar-shaped deposit. This builds up into a boule a few centimeters long, made of a graphite shell packed with nanotubes and other forms of closed graphite nanoparticles [40]. The technique is similar to what Roger Bacon used almost 30 years ago to generate large (micrometer size) arc-grown carbon whiskers, although he used very different conditions in his experiment. The method is similar to the Kratschmer – Huffman method of generating fullerenes [41]. To generate MWNT, a dc arc is normally used (with a modest voltage of 20 V and currents of $<$100 amp) in an inert atmosphere of 500 torr helium. Surprisingly, such perfect structures as nanotubes self-assemble in the plasma created in the interelectrode region, where the temperature is close to 3500 °C. The time scale for formation of these nanostructures is extremely short; a MWNT 5 nm in diameter and 1000 μm long grows in about 10^{-4} s [42].

The inside of the deposit formed in the electric arc-discharge contains a highly porous network of randomly oriented MWNT structures that are organized on a macroscopic scale into pencil-like columns aligned in the axial direction of the deposit. Although MWNT grown by the electric-arc method has so far been the most perfect in terms of structure and properties, the technique suffers from drawbacks. Being a batch process limits the amount of material that can be produced, and the material in the deposit contains substantial amounts of nanoparticles that have polyhedral shapes and are low in aspect ratio. Several parallel attempts to make nanotubes through catalytic vapor deposition have overcome some of the problems of the arc process. In general, catalytic metal particles are exposed to a medium containing gaseous hydrocarbon species. Nanofibrils (twisted hollow fibers) are formed [43]. One good example

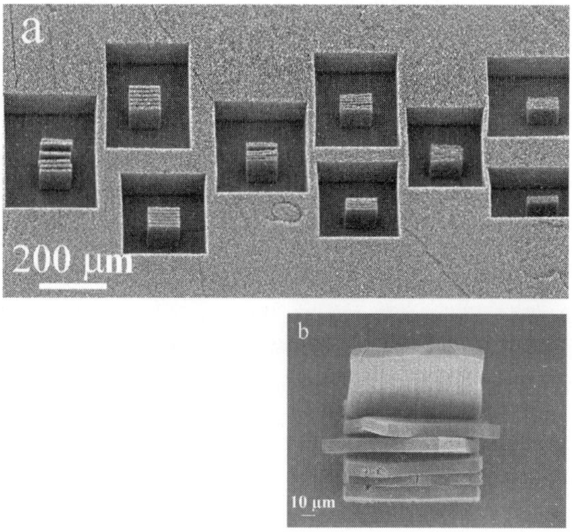

Fig. 2.7 Scanning electron micrographs showing MWNT grown by catalytic vapor deposition onto catalysts prefabricated into patterned arrays resulting in well aligned nanotube assemblies

is the Hyperion fibers that have been in commercial use for some time [44]. Controlling the size of the catalyst seed has enabled achieving uniformity in the size of the fibers. The process can be scaled to produce large amounts of material. In some cases, when the catalysts are prefabricated into patterned arrays, well-aligned nanotube assemblies are produced (Figure 2.7) [45]. Similarly, template-based approaches are also in use: the aligned pores of a nanoporous membrane (such as electrodeposited porous alumina) are filled with carbon species through vapor deposition and later graphitized to produce nanotubes. The template membrane is removed to obtain aligned nanotube arrays [46, 47].

Single-walled nanotubes are made with a combination of catalyst and dense carbon vapor. Both are simultaneously introduced into an inert atmosphere by an electric arc [48] or through laser ablation [49]. In the former, the setup is similar to that used for MWNT synthesis, but a hole is drilled in the anode and packed with a mixture of the metal catalyst and graphite powder. Several metals and combinations of metals have been used to obtain good yield of nanotubes. The best so far consists of a mixture of Ni and Y with graphite in a 15:5:80 weight ratio. When the electric arc is struck with such a modified electrode, spectacular growth occurs in the reaction vessel, which becomes decorated with a network of webs that contain SWNT. Upon closer look, the webs contain ropes of nanotubes that consist of tens of SWNT (Figure 2.8). The maximum density of nanotubes is seen in the product that grows around the cathode like a collar, which consists of >50 % wt. nanotubes. The rest of the carbon soot contains fullerenes, partially graphitized carbon (glassy carbon), amorphous carbon, and finely distributed catalyst particles. In the laser-ablation technique, a metal (Ni-Co) containing a small percentage (<1 % wt.) of graphite target placed in an oven is ablated with a strong laser pulse under an inert atmosphere. A felt of nanotube material is generated and is collected onto a water-cooled target by forced gas flow. Using two coordinated laser pulses breaks down the carbon formation better, producing a very high yield of nanotubes, with >90 % of the collected material being SWNT.

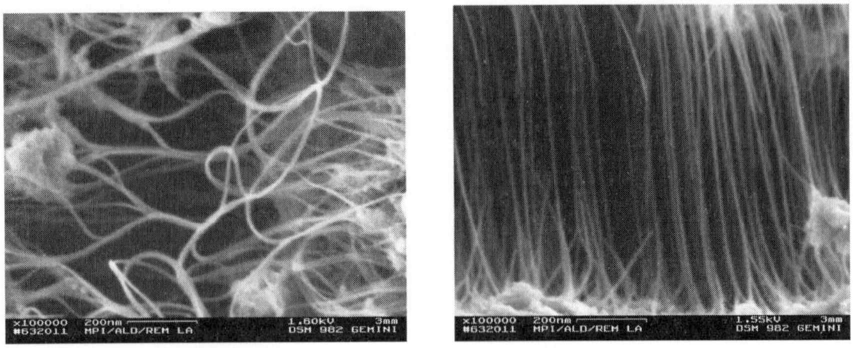

Fig. 2.8 Scanning electron micrograph showing SWNT grown by a catalyzed arc-discharge method

2.2.1.3 Purity

Nanotubes consisting of large impurity fractions are not useful for either laboratory studies or composite manufacture. For example, MWNT processed by the arc-discharge method have a significant volume fraction of other carbonaceous materials. In addition, they may contain a high content of scrolled layers instead of nanotubes. Figure 2.9a shows typical material as produced. Because the extra carbonaceous material and catalyst can play a significant role in determining the composite's properties, they must be removed. In addition, we still need better control in manipulating (slicing, joining, and making larger structures from nanotube building units) nanotubes, especially through generalized approaches such as chemistry.

Nanotubes are insoluble, and hence all the purification procedures must use filtration-based techniques. As a result, the nanotube products can never be as pure as, for example, fullerenes. All purification procedures follow certain essential steps: preliminary filtration to get rid of large graphite particles, dissolution to remove fullerenes (in organic solvents) and catalyst particles (in concentrated acids), microfiltration, settling, and chromatography to separate MWNT and nanoparticles or SWNT from the amorphous carbon impurities [50]. Nanotubes must be kept separated when in suspension and are typically dispersed with a surfactant (e.g., sodium dodecyl sulfate) prior to the last stage of separation. Some reports claim that SWNT samples with

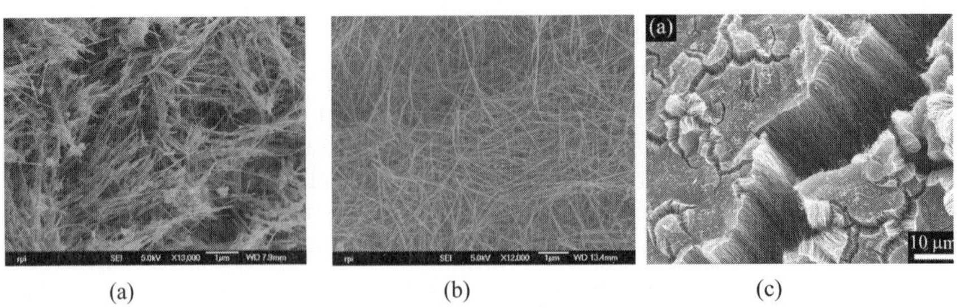

Fig. 2.9 Scanning electron Micrographs showing the purity of MWNT (a) as received MWNT produced by arc-discharge, (b) purified MWNT, (c) MWNT grown by CVD

>99 % purity can be made by repeating some of these steps [51]. Recent efforts [52] to use size-exclusion chromatography have yielded good separation between nanotubes and nanoparticles. Curran et al. [53] achieved a 20 % yield by using toluene. Oxidation of any remaining carbonaceous material in a 600 °C oven [54] can follow. Figure 2.9b shows the purity that can be obtained with this method, but the yield is very low [55]. Other oxidation methods, such as plasma or acid oxidation, work but damage the nanotubes [56]. Processing MWNT by chemical vapor deposition or pyrolysis [57] is an alternative, which yields MWNT contaminated with catalyst but with little else in the way of impurities. A higher concentration of defects may result (Figure 2.9c). MWNT processed by laser-deposition techniques, including the nanofibers produced by Hyperion [58], have a significant number of defects relative to pure MWNT but are produced purely. The effect of such defects on the mechanical properties is unclear, but they lead to early oxidation and probably reduce both the strength and the conductivity.

It is now possible to cut nanotubes (SWNT) into smaller segments by extended sonication in concentrated acid mixtures [59]. The resulting pieces of broken nanotubes (open pipes that are typically a few hundred nanometers long) form a colloidal suspension in solvents and can be deposited onto substrates or further manipulated in solution and functionalized at the ends. Such segments can perhaps be joined with appropriate chemical bridges to construct long nanotube chains resembling polymers. A whole gamut of possible chemistry based on nanotubes is just beginning to unfold. In the future, it may be possible to dissolve such functionalized nanotubes in organic solvents and later separate them to produce high-purity samples.

Because size and helicity are two important parameters that affect nanotube properties, selection on the basis of both these factors is important. This is mostly relevant to SWNT, because the MWNT shows average properties that tend to be semi-metallic only. Fortunately, the size distribution produced during the high-yield synthesis of SWNT is quite narrow, with an overwhelming majority of the nanotubes having diameters close to 1.4 nm. The nanotube diameter can be tuned in the range between 1.2 and 5 nm by altering the temperature (from 700 °C to 1200 °C) under which they are formed. It is fortuitous that, when SWNT are formed in high yield, a majority of the tubes have armchair (or close to armchair) arrangement, and this arrangement is independent of the conditions used in synthesis. This size and helicity have a strong bearing on the possible usability of nanotubes, especially in electronics.

2.2.1.4 **Other Nanotubes**

Other types of nanotubes require mention, and these could have some role in the fabrication of novel nanocomposite materials. The closest in structure and mechanical properties to carbon nanotubes are hexagonal boron/nitrogen nanotubes, which can be produced by arc-discharge, laser ablation, and CVD processes and can be fabricated into multiwalled and single-walled structures. The modulus and strength of BN nanotubes are very similar to those of their carbon counterparts [60]. The advantages of using them would be better oxidation resistance and their electrically insulating nature (for certain dielectric applications). The lattice of carbon nanotubes can be doped (to certain levels) with boron and nitrogen, providing a broad range of possible BCN nanotube structures [61]. Boron has an exceptionally interesting effect: insertion of boron into carbon nanotubes during growth allows selection of their helicity (zigzag nanotubes are stabilized) [62]. The average length of nanotubes is also much higher if boron is added during growth, because boron acts as a surfactant near the growing ends, making it difficult for the structure to close. Nitrogen also can be inserted into the carbon nanotube lattice, but this insertion produces corrugated nanotube structures [63]. The presence of unsaturated nitrogen atoms in the lattice of nanotubes makes them more easily dispersed in solutions, and such nanotubes make good interfaces with certain polymers. Thus, doping nanotubes improves their surface reactivity. This could ultimately aid in the design of strong nanotube matrix (particularly polymer) interfaces, which is necessary in high-strength composites. These modified nanotube lattices also lead to modified electrical and optical properties and hence novel properties for composites in which they are embedded.

Most layered materials can be fabricated into nanotubes. Examples other than those noted above are nanotubes made of dichalcogenides (MoS_2, WS_2, etc.) [64], nanotubes of several oxides (V_2O_5, MoO_3, etc.) [65], and organic nanotubes [66]. These materials (including the BCN types) are not currently available in bulk quantities, and hence very little data is available on the mechanical and electrical properties of composites that contain these nanostructures. However, research is moving quickly in synthesizing several such nanotubular structures, which could ultimately find novel applications in multifunctional nanocomposites.

2.2.2 Plate-like Nanofillers

The most common 2D fillers are layered silicates. The most well known layered silicate is mica. Mica is made up of large sheets of silicate (on the order of centimeters or more) with relatively strong bonds between the layers. Smectic clays or phyllosilicates, on the other hand, have relatively weak bonding between the layers, and the layers are small flakes. Each layer consists of two sheets of silica tetrahedra (corner shared) with an edge-shared octahedral sheet of either alumina (aluminosilicates) or magnesia (magnesium silicates) [67]. Due to isomorphic substitution of alumina into the silicate layers (Al^{3+} for Si^{4+}) or magnesium for aluminum (Mg^{2+} for Al^{3+}), each unit cell has a negative charge between 0.5 and 1.3. The layers are held together with a layer of charge-compensating cations such as Li^+, Na^+, K^+, and Ca^+. These charge-compensating cations provide a route to the rich intercalation chemistry and surface modification required to disperse clays at the nanoscale into polymers. The cation-exchange capacity (CEC) defines the number of exchangeable interlayer cations and is usually described as mEq/100 g. Values range from 60–120 for smectic clays. A typical structure is shown in Figure 2.10 [68]. As they occur in nature or synthetically, the layers are 20–200 nm in diameter laterally and come in aggregates known as tactoids, which can be ~1 nm or more thick. Examples of smectic clays include montmorillonite (CEC ~76.4–119 mEq/100 g), which is an aluminosilicate; sapo-

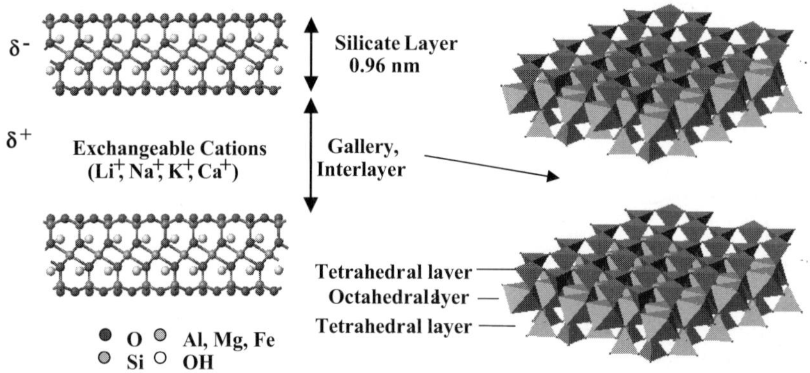

Fig. 2.10 Schematic of the crystal structure of 2:1 layered silicates (smectites). Van der Waals interlayer or gallery containing charge-compensating cations (M^+) separates covalently bonded oxide layers, 0.96 nm thick, formed by fusing two silica tetrahedral sheets with an edge-shared octahedral sheet of alumina or magnesia. Reprinted with permission from [68]

nite, a synthetic material that closely resembles montmorillonite; hectrite, which is a magnesium silicate (CEC ~55 mEq/100 g), and its synthetic equivalent Laponite. Montmorillonite tends to have sheets that are up to 200 nm wide, and Laponite sheets are 25 – 30 nm wide.

For these layered silicates to be useful as nanocomposites, the layers must be separated. Silicate clays are inherently hydrophilic, but polymers tend to be hydrophobic. This presents an interesting challenge in terms of being able to disperse the silicate layers in a polymer. Fukushima and Inagaki [69] demonstrated that an exchange reaction could be used to replace the inorganic exchange ions in the galleries between the layers with alkylammonium surfactants (dioctadecyldimethylammonium). This opens the galleries enough and makes them hydrophobic enough that a Nylon monomer could be intercalated between the layers, resulting in a clay/Nylon nanocomposite. Figure 2.11 [70] shows the increase in interlayer spacing, measured by x-ray diffraction, as a decrease in the peak position for montmorillonite swelled with alkylammonium surfactants of various lengths. The intercalation also modifies the polarity of the layers by lowering the interfacial free energy of the silicate [67]. The number of onium ions that can pack into the galleries depends on the charge density of the clay and the cation-exchange capacity. This, as well as the length of the surfactant chain, determines the distance between the layers. At lower charge densities, the surfactant packs in monolayers and, as the charge density increases, bilayers and trilayers can form (Figure 2.12) [4]. At very high CECs ($\geq$120 mEq/100 g) and long surfactants (>15 carbons), the packing can be ordered in a paraffin-type structure. The positively charged heads of the onium ions prefer to pack close to the clay layer, and the organic ends tend to fill the middle. Further processing of the layered silicates to form polymer composites is discussed in the processing section below.

In general, the properties of the clay layers are unknown, and estimates are made based on composite properties of similar materials in bulk. The properties are not unusual. For these fillers, it is the aspect ratio that is the most significant. Their

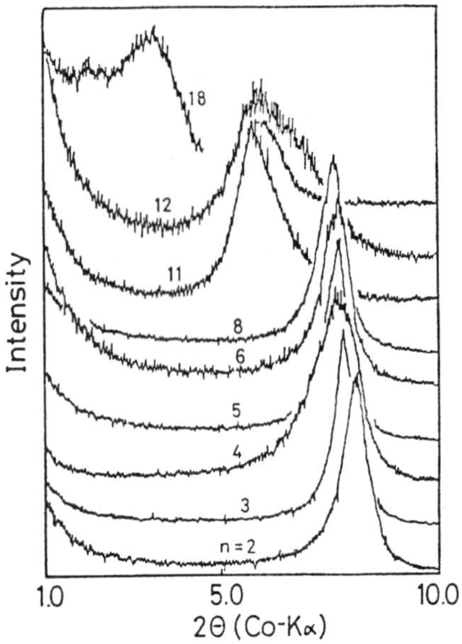

Fig. 2.11 X-ray diffraction patterns of n-montmorillonite as the length of the alkylammonium chain length increases. Reprinted with permission from [70]

high aspect ratio theoretically makes them efficient for carrying loads. Such clay layers are almost impermeable to gases and water through the layer thickness and thus provide an excellent barrier to gas and water in composites.

Layered silicic acids provide an alternative to smectic clays. The intercalation chemistry is similar to that of smectic clays, and they can be obtained with high purity and with structural properties that compliment those of clays [71]. Examples of layered silicic acids include kanemite ($NaHSi_2O_5$), makatite ($Na_2Si_4)_9 \cdot nH_2O$), octosilicate ($Na_2Si_8O_{17} \cdot nH_2O$), magadiite ($Na_2Si_{13}O_{29} \cdot nH_20$), and kenyaite ($Na_2Si_20O_{41} \cdot nH_2O$). The thickness of the layers varies from 0.5 nm for makatite to 1.77 nm for kenyaite [72]. The general structure of layered silicic acids involves layers of SiO_4 tetrahedra with an abundant hydroxyl siloxane surface. Interlayer alkali ions can be ex-

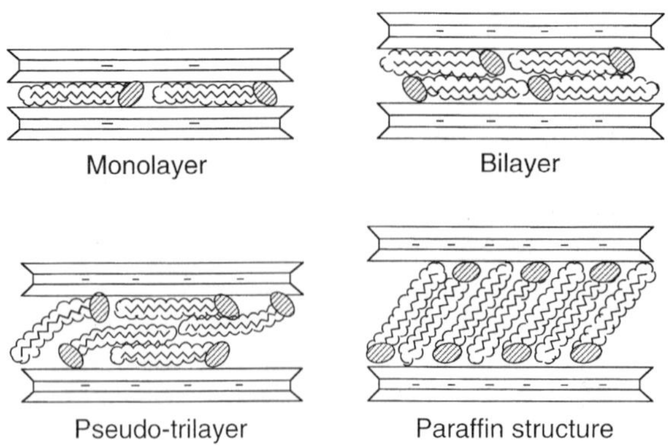

Fig. 2.12 Schematic of the orientation of alkylammonium ions in the galleries of layered silicates with different layer charge densities. Reprinted with permission from [4]

changed with the resulting active hydroxyl sites, leading to enhanced bonding with the intercalates.

Layered double hydroxides (LDH), for example, $Mg_6Al_2(OH)_{16}CO_3 \cdot 4H_2O$, have a positive (instead of negative) charge on the $Mg(OH)_2$ layers. They are produced synthetically [73] and provide an opportunity for intercalation with organic anions. They have been used successfully to make nanocomposites [74, 75].

Transition-metal dichalcogenides are layered materials that can be intercalated with lithium. After intercalation, MoS_2 and WS_2 can be placed in distilled water and exfoliation occurs [76, 77]. Titanium and tantalum dichalcogenides have also been used to make nanocomposites [78, 79]. Essentially any layered material, including graphite [80], with appropriate chemistry can be expanded, creating the potential for intercalation of a polymer.

2.2.3 Equi-axed Nanoparticle Fillers

Production of nanoparticles with controlled size and degree of aggregation is the goal of many research efforts. The primary driver for this interest is the effect of particle size on their properties. For example, the optical absorption spectrum of Au changes with the size of the Au particles. The electroluminescence of semiconducting nanoparticles is also size-dependent. One of the advantages of using nanoparticles in a composite structure is that the particle size and distribution can be stabilized. Materials that cannot be grown easily as single crystals can be used at the nanoscale and dispersed in a polymer to take advantage of the single-crystal properties. In addition, the particles can lend properties to the polymers that they cannot achieve alone or with traditional fillers. For example, nanoparticle-filled polymers can lead to an increase in modulus and strength, but also maintain the polymer's ductility. This is because the nanoparticles are such small defects relative to their micrometer-sized counterparts.

Nanoparticles have been available for more than a century. Carbon black made by pyrolysis is available in a range of surface areas and degrees of aggregation. Silica made by a variety of methods, including the wet chemical process developed by Stober [81], the commercial Ludox® process [82], as well as flame processes, have been available for years. Although recent developments have been made [83], these are well-established processes. These nanoparticles have found significant applications in the rubber industry, in catalysis, chemical mechanical polishing, water treatment, and other applications. In recent years, significant progress in the diversity of processes for making nanoparticles has led to a more diverse set of nanoparticles and better control of particle size, morphology, and surface properties. Recent developments in in-situ processing of nanoparticles within a polymer matrix are discussed in the composite processing section 2.4.3. In this section, we focus on the preparation of discrete particles that are subsequently incorporated into a polymer matrix.

Aerosol methods usually result in the formation of nanoparticles by condensation from a gas phase [references in 84–87]. One example of a highly successful commer-

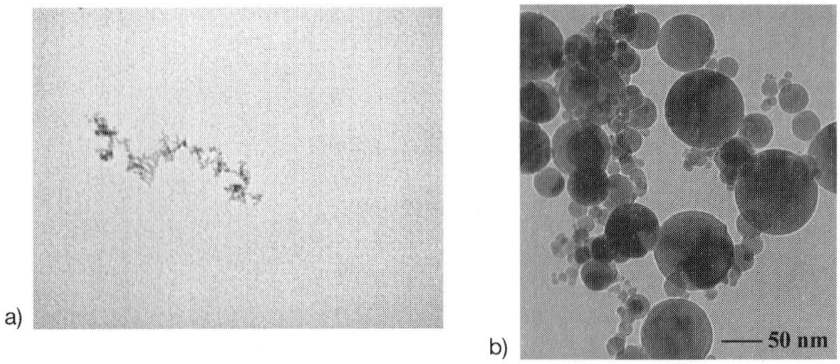

Fig. 2.13 Micrographs showing (a) typical silica nanoparticle agglomerates and (b) titania made via gas condensation. (Thanks to T. Li for these micrographs)

cial aerosol processes is flame hydrolysis. Here, a vapor precursor (such as silicon tetrachloride to make silica) is burned in a hydrogen/oxygen fuel mixture to produce the metal oxide. Titania, alumina, and zirconium oxide can all be produced this way. For silica, particle sizes range from 7 to 27 nm, and the surface areas range from 100 to 380 $m^2 g^{-1}$. The specific surface area and agglomerate structure can be controlled by adjusting the temperature and time of reaction. At higher temperatures, particle coalescence is fast, and low-surface-area materials result. At lower temperatures, coalescence is slow relative to the collision rate, and fractal agglomerates result [88]. Figure 2.13a shows a typical silica agglomerate.

Pyrolysis is another aerosol method and is commonly used to make carbon black, an amorphous form of carbon. The particles have a size range of 20 – 300 nm with surface areas of 20 – 500 $m^2 g^{-1}$.

In a third method, called gas condensation [89, 90], a metal vapor is formed in either an oxygen atmosphere or an inert gas. In an oxygen atmosphere, the metal reacts with the oxygen to form metal oxides, which are collected by cooling the gas and condensing the nanoparticles. A large range of metal oxides can be processed this way, including TiO_2, Al_2O_3, CuO, CeO_2, ZnO, ZrO_2, Fe_2O_3, and Y_2O_3 [91]. In an inert atmosphere, metal nanoparticles result. Advantages of this method are that the particle size distribution is narrow, the particles are crystalline, and the surfaces are clean. Figure 2.13b shows typical metal oxide particles that result from this process. They tend to be isolated, crystalline, and spherical. In the past dozen years or so, great progress has been made in understanding the processes by which such nanoparticles are made and the fundamental relationships between their sizes and structures and their special properties [92 – 95]. In the late 1990s these processes were improved and ton quantities of gas-condensed metal-oxide nanoparticles have been produced.

Laser ablation processes are another form of gas condensation. Here, targets (metal, metal oxide, or semiconductor) are ablated with a laser (often a pulsed laser). An inert or reactive gas is supplied to cool, condense, and sometimes react with the target material. By varying the pulse frequency, gas flow rate, and total pressure, the size of the particles and particle aggregates can be controlled. For example, increasing the gas flow rate decreases the primary particle size, but increasing the laser fluence and total gas pressure produces larger primary particles. Significant quantities of nanoparticles can be produced this way. Many types of materials can be processed, including SiO_2, MgO, Fe_3O_4, Mg_2SiO_4, $CaTiO_3$, $MgAl_2O_4$ [96], Al_2O_3 [96–98], TiO_2 [97], Fe_2O_3 [97], and many others [99, 100].

Other methods for creating a vapor that condenses into nanoparticles include plasma and chemical vapor condensation, spray pyrolysis, electrospray, and plasma spray [101]. These processes can form a wide range of nanoparticles. One interesting variation on the aerosol process developed recently combines sol–gel processing with the aerosol process. Hydrolysis and condensation of TEOS is carried out in a laminar flow chamber at temperatures ranging from room temperature to 100 °C [102]. The advantage of this low-temperature process is that a large specific area ($400\,m^2\,g^{-1}$) can be produced because of the low coalescence rate relative to the collision rate. In addition, the agglomerates tend to be linear instead of highly branched.

A host of wet-chemistry methods exist in which one or two precursors are placed in an appropriate solution and nanoparticles form. These can be stabilized with an emulsion, surfactant, or a macromolecule that surrounds and protects the nanoparticles from agglomeration. Metal [103], metal-oxide [104], semiconducting [105–107], superconducting, and magnetic particles [108] have all been made this way. Using ZnO as an example, zinc acetate $[(CH_3COO)_2Zn \cdot 2H_2O]$ can be reacted with NaOH [109] in the presence of poly(vinylpyrrolidone) to create very stable ZnO particles with an average diameter of 4 nm [110].

Metal nanoparticles can also be made by sonication chemistry [111]. This process takes advantage of the formation, growth, and implosive collapse of bubbles within a liquid to create localized spots with temperatures as high as 5000 K and pressures as high as 1800 atm. Under appropriate sonication conditions, precursors such as $Fe(CO)_5$ and $Co(CO)_3(NO)$ form nanoparticles. If sonication is done in the presence of an alkane, the particles can be separated, but sonication can also be done in the presence of a polymeric ligand to make stable colloids or in the presence of an inorganic support.

Metal oxides an also be formed by taking advantage of hydrodynamic cavitation [112]. Here, hydrodynamic cavitation is produced by using a high-pressure fluid system in the presence of a sol–gel solution. The local high temperatures and pressures due to collapse of bubbles result in the creation of nanoparticles. The pressure and the exposure time can be adjusted to control the particle size.

This section has clearly left out many details of nanoparticle processing, and many significant contributions have not been referenced. We introduce only the basics of nanoparticle processing here, because this topic is so broad and our focus is on nanocomposites.

2.3
Inorganic Filler–Polymer Interfaces

As mentioned in the Introduction, nanocomposites have at least an order of magnitude more interfacial area than traditional composites. This increase in interfacial area results in a 3D interfacial region within the polymer matrix that can have properties significantly different from those of the bulk polymer. The local chemistry, degree of cure, chain mobility, chain conformation, and degree of chain ordering or crystallinity can all vary continuously from the filler/matrix boundary to some point in the polymer bulk. To understand the significant effects nanoscale fillers have on polymer behavior, this interfacial region must be understood. The paragraphs below outline some of the literature describing the behavior of polymers near a surface.

Consider first a pure thermoplastic polymer film interacting with either air or a very flat inorganic surface. We know that surface/polymer interactions greatly influence the glass transition temperature (mobility, relaxation spectra) of thin polymer films. For example, the presence of a free surface, as observed in free-standing ultrathin films, reduces the glass transition temperature [113–116] (Figure 2.14), but an attractive surface increases the glass transition temperature [117, 118]. The mechanism causing these changes in Tg is currently under discussion in the literature. Molecular dynamics simulations [119] show that the decrease in Tg for free-standing thin films is caused by a decrease in the density, which in turn increases the mobility of the polymer chains. Monte Carlo simulations show that the presence of a neutral surface can increase the density and decrease the mobility of the polymer chains [120], although experimental data suggest that the surface must be attracting before Tg increases.

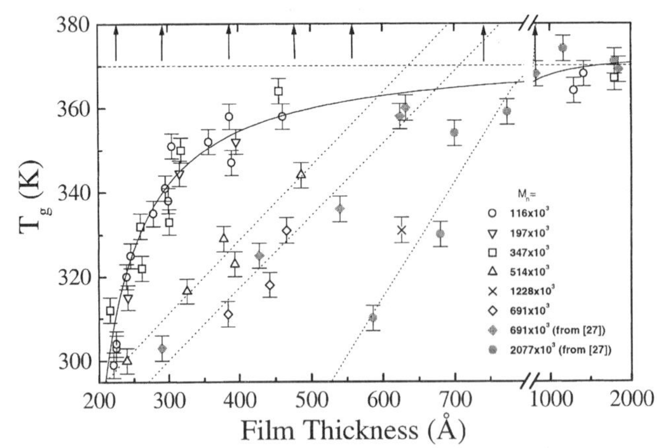

Fig. 2.14 The glass transition temperature as a function of film thickness for low and high molecular weight polystyrene free-standing films. Reprinted with permission from [113]

Composites, however, do not contain ideal 2D interfaces. In fact, the highly curved surfaces and the size scale close to that of a typical radius of gyration for a polymer chain make it difficult to apply 2D observations and measurement techniques directly to nanocomposites. Based on some early nanocomposite work on polymers filled with carbon black, one description of the interfacial region in nanofilled polymers is a bound polymer layer (where 'bound' implies a region of immobile polymer) with a distinct interface between the bound and unbound regions. The evidence for this is an insoluble fraction of polymer that results from the addition of nanoparticles [121]. In addition, the properties of the nanofilled polymers often scale with a parameter somewhat larger than the average particle diameter [122]. The argument has been that the stronger the polymer–particle interaction, the larger the bound polymer fraction, up to 2–9 nm thick [123].

The discrete bound polymer layer could result in a second glass transition temperature, according to some researchers [124]. Research on thin film, however, does not point to a discrete layer of affected material but to a far-field effect. Figure 2.14 [113, 118] shows the effect of film thickness on the glass transition temperature for free-standing polystyrene films of various molecular weights. The higher the molecular weight, the greater the film thickness at which the decrease in Tg begins. This suggests that some multiple of the radius of gyration may be a relevant parameter. Studies on diffusion using SIMS show that the effect can be up to 10 times the radius of gyration [125]. In addition, the thinner the film, the lower the Tg becomes. This indicates that the effect of the surface is a continuous function within the polymer. In addition, recent results [126] have shown that the nonlinear viscoelastic behavior of nanofilled polymer melts is strongly dependent on the surface treatment of the filler particles. The results suggest that the filler surface has a far-field influence on the chain motions. However, these effects are only apparent when finite shear–strain magnitudes are used to probe the behavior. Small-strain studies on the same composites in the vicinity of the glass transition temperature show that, although the glass transition temperature is increased by filler addition, the curves of normalized loss modulus vs. frequency superpose perfectly. These studies suggest that the distribution of relaxation times is shifted but otherwise unaltered. Taken together, these results indicate that the bound polymer layer is essentially a monolayer and that the far-field influence is due to entanglement trapping.

Further clarification reassures us that a discrete bound layer and a more diffuse region are not contradictory. First, a *bound polymer layer*, as originally defined, is simply the amount of polymer that is 'stuck' to a particle surface; the term was not intended to completely describe the mobility within that region or the effect of that region on the surrounding polymer. Thus, let us define a bound polymer layer as any polymer that is bonded to the particle surface. This sometimes occurs in nanocomposites, but not always. Let us define the interfacial region as the region of polymer surrounding the nanoparticle and having altered chain conformation and/or mobility due to the presence of the filler. Clearly, polymer chains strongly bound to the surface result in a different interfacial region than chains either repelled by the surface or weakly interacting. The magnitude of the effect this interfacial region has on the properties of a composite depends on the property being measured. For example, the effect

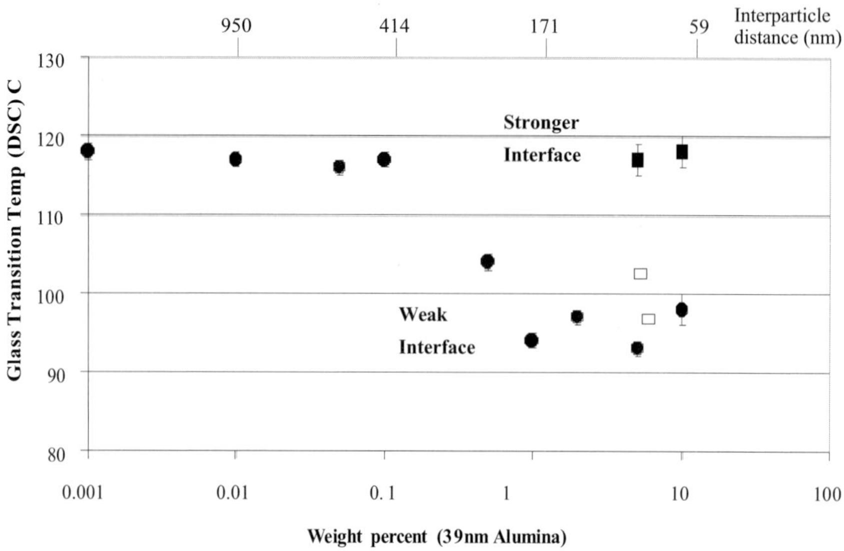

Fig. 2.15 Glass transition temperature vs. volume fraction of alumina particles for a PMMA matrix composite. The glass transition temperature does not decrease until a critical volume fraction is reached. In addition, when the nanoparticles interact strongly with the polymer, the glass transition temperature does not decrease. Reprinted with permission from [127]

of the interfacial region on the glass transition temperature, which is the result of cooperative segmental motions, may be different from its effect on diffusion, which depends on larger-scale mobility.

Consider for a moment, ignoring the specifics of their morphology and properties, the implications of both the bound polymer layer and the interfacial region on the behavior of nanofilled polymer composites. The film thickness at which a significant change in glass transition temperature occurs depends on the molecular weight, but can be as high as 80 nm. For a 30-nm particle dispersed in a polymer matrix, the average interparticle spacing is 80 nm at filler loadings as low as 1.5 vol. % (Figure 2.2). Thus large effects on bulk properties can be expected at low filler loading.

The glass transition temperature of a bulk part can be raised and lowered by the addition of nanoparticles [126–128]. Ash et al. [127] have shown that, for a non-wetting nanoparticle/polymer composite (alumina/PMMA), the glass transition temperature starts to decrease at a specific filler volume fraction corresponding to an interparticle distance of about 200 nm (Figure 2.15). On the other hand, if the PMMA is adsorbed on the particle surface, the glass transition temperature is stabilized. More commonly, an increase in Tg is observed [129, 130], which can be as large as 30° [80].

The situation becomes even more complicated for composites in which the polymers are confined between two flat layers (as in clay composites) [131]. For example, NMR studies on PEO-intercalated nanoclay showed that confinement of the chains increases segmental mobility at low temperatures, but at higher temperatures a less mobile fraction is revealed. This suggests a broadening in the relaxation spectra. In addition, the glass transition temperature is suppressed. In thin films this is attributed to a highly immobilized polymer layer. Krishnamoorti [131], armed with the knowledge that the rate of polymer intercalation into the clay layers can be as fast as the diffusion rate in the bulk, suggested that this phenomenon is due only to non-cooperative motion. This is clearly an area in need of further study and is discussed more below.

In addition to changes in the mobility of a polymer chain, the average chain conformation can be altered and/or the type and degree of crystallinity can change. Dhinojwala and coworkers recently reported measurements on a polystyrene (PS)/sapphire interface [132, 133], which revealed a nearly perpendicular orientation of the phenyl rings to the PS/sapphire interface. It is unclear how these conformations change in the presence of a nanostructured surface, particularly one with a high radius of curvature. Recent modeling, however, suggests that studies on flat surfaces may be relevant [134]; however, it is unclear how the many techniques used to study behavior on flat surfaces could be modified to measure this. Studies have focused on the effect of structural details of surfaces on protein structure and have found that the density of alkane chains on an alkylated surface directly affects the amount of protein adsorbed [135]. Other studies [136–138] showed that vitronectin (a mediator of osteoblast adhesion) remained folded on monolithic alumina with micrometer-sized grains, but showed enhanced adsorption as well as conformational changes when it was adsorbed on nanograined alumina. Vitronectin interactions with nanoceramics were correlated with significantly enhanced osteoblast adhesion. The change in polymer chain conformation, in contrast to that of proteins, is likely to be much more local, but could still affect bulk properties.

Crystallinity is also sensitive to surface interactions. This is seen dramatically in graphite fiber/polypropylene composites in which a transcrystalline layer develops perpendicular to the fiber [139]. Nanofillers can also change the degree of crystallinity, change the rate of crystallization, and/or suppress the formation of the thermodynamic crystal phase and stabilize the metastable phase [140, 141]. In silica-filled Nylon 11 that was thermally sprayed, the crystallinity increased due to the presence of the nanoparticles [142]. In montmorillonite/HDPE nanocomposites, the polymer crystals decreased in size and the crystallization rate decreased [143], whereas in polyamide (Nylon 6) composites the metastable hexagonal phase formed instead of the typical monoclinic phase [144]. Another example in Nylon 6 occurs under high pressure: by annealing montmorillonite/ Nylon 6 at high pressures, a higher-temperature monoclinic phase forms (α), and the orthorhombic (γ) phase is suppressed [145].

Finally, changes in the chemistry can occur in the interfacial region. A dramatic example of this is the preferential adsorption of a curing agent onto the nanofiller. When this occurs, the region surrounding the nanoparticle consists of a 'layer' of stoichiometrically crosslinked thermoset with excess curing agent, surrounded by a 'layer' of depleted curing agent and thus less-than-stoichiometric crosslinking. This phenomenon has been observed in fiber-filled polymers [146] and is presumed to exist in nanoscale titania-filled epoxy [147]. More subtle changes could also occur, such as preferential adsorption of low-Mw material to the surface.

In summary, the interfacial region is complex, and when the interfacial area is very large, the whole polymer matrix may essentially be interfacial region. This presents one of the essential challenges in polymer nanocomposites: to develop technology to control the interface, to describe the interface mathematically, and to be able to predict properties taking into account the interfacial region. Traditional composite theory, although very far advanced in describing properties that are relatively independent of the interface, is still in its infancy in taking into account the role of the interfacial region. Significant progress has been made in understanding the interaction of polymers with flat surfaces, but the interaction of polymers with highly curved surfaces and at scales similar to that of the radius of gyration is not well understood. The nanocomposites community has their work cut out for them.

2.4 Processing of Polymer Nanocomposites

One of the key limitations in the commercialization of nanocomposites is processing. Early attempts at clay-filled polymers required processing that was not commercially feasible, but this situation has changed. Similarly, processing of other nanocomposites is becoming easier and more commercially viable as our understanding improves. A primary difficulty is proper dispersion of the fillers. Without proper dispersion and distribution of the fillers, the high surface area is compromised and the aggregates can act as defects, which limit properties. To facilitate discussion, we will define the state of aggregation in those nanocomposites. Distribution of a nanofiller describes the homogeneity throughout the sample, and the dispersion describes the level of agglomeration. Figure 2.16 schematically illustrates good distribution but poor dispersion (a), poor distribution and poor dispersion (b), poor distribution but good dispersion (c), and good distribution and good dispersion (d). The state of aggregation is further defined in layered silicate/polymer nanocomposites later in this chapter.

2.4.1 Nanotube/Polymer Composites

The processing of nanotube/polymer composites is still in its infancy. Although nanotubes have been incorporated into composites commercially, the literature describing the processes is limited. Significant issues remain to be solved about purification, dispersion, and bulk processing.

The ability to disperse SWNT and MWNT into a polymer may be the most critical processing parameter for controlling properties. Nanotubes that are in clumps or are agglomerated with other carbonaceous materials create defect sites that will initiate failure. In addition, they limit the efficiency with which the nanotubes carry load. This limitation has been illustrated explicitly in both polymer and ceramic matrix composites [148, 149]. CVD-grown MWNT, which are easily dispersed and less agglomerated, increased the modulus and strength of polystyrene without compromising the

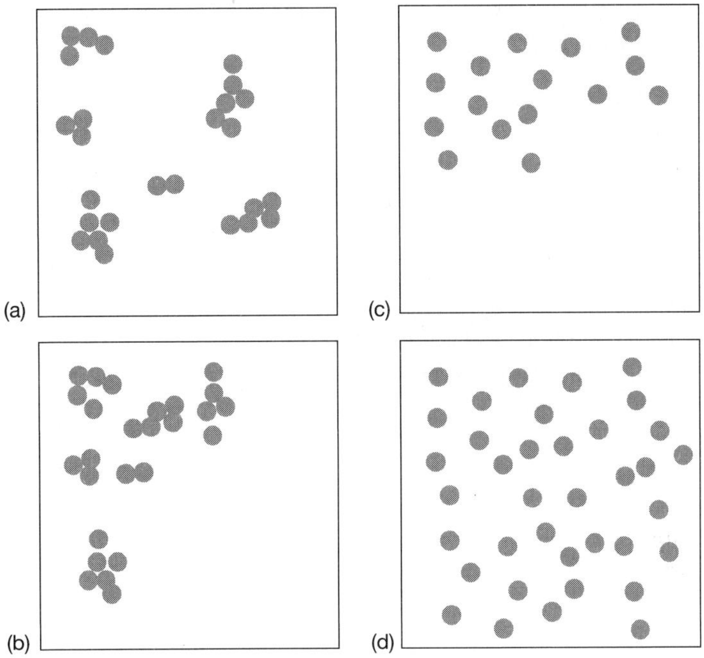

Fig. 2.16 A schematic illustrating the difference between dispersion and distribution and giving examples of good and poor for each

strain-to-failure factor significantly. Other work on arc-discharge-grown MWNT, which were not fully purified and not as well dispersed, did not show the increase in toughness observed for well-dispersed MWNT. Similarly, the toughness of MWNT/alumina composites with excellent dispersion increased significantly compared to composites with somewhat worse dispersion.

Dispersion has been achieved primarily by sonication of nanotubes in a solvent. The most focused efforts have been on the dispersion of SWNT. Chemical modification of the surface with the aid of surfactants [150], by functionalization of the end-caps with long aliphatic amines [151], or by functionalization of the sidewalls with fluorine [152] or alkanes [153] has resulted in stable suspensions of nanotubes. None of these methods is ideal for composite processing. The use of surfactant results in an impurity in the composite. Functionalizing the ends limits further chemical modifications for controlling bonding with the matrix, and modification of the sidewalls can affect the mechanical properties [154]. As an alternative, the best solvents for direct dispersion of SWNT were identified as NMP, DMF, hexamethylphosphoramide, cyclopentane, tetramethylene sulfoxide, and ε-caprolactone [154]. These solvents are all strong Lewis bases without hydrogen donors, although not all solvents with these characteristics were good solvents for SWNT.

After dispersion, drying the dispersion on a glass slide and placing resin directly onto a thin film of nanotubes [29, 155, 156] can produce small-scale composites. Mix-

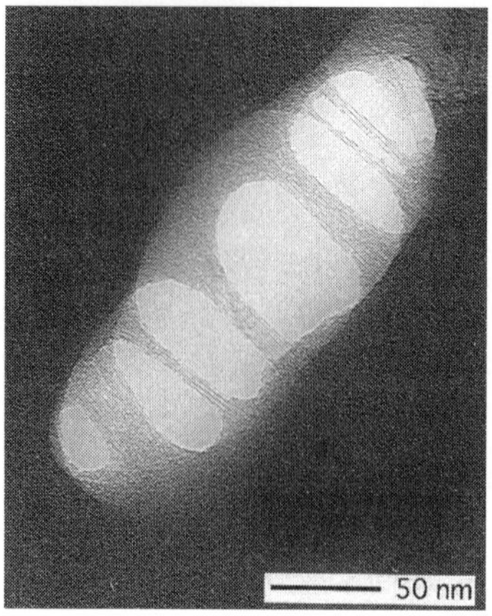

Fig. 2.17a Transmission electron micrograph of aligned singlewall carbon nanotube ropes bridging an elliptical hole in a polymer film. Reprinted with permission from [157]

ing both the nanotubes and the polymer in the presence of a solvent [25, 30, 157, 158], often with the help of a surfactant [159], can also produce composites. Figure 2.17a shows the wetting and the dispersion of SWNT dispersed in ethanol and then mixed with an epoxy resin [157]. Figure 2.17b [148] shows the dispersion achieved with CVD-grown MWNT, dispersed in toluene with dissolved polystyrene and casting films. Excellent dispersion was achieved. Nanotubes have also been dispersed directly into liquid urethane acrylate polymer [160], methylmethacrylate monomer [161], and epoxy resin [162], followed by curing or polymerization. Damage to the nanotubes is a tradeoff that must be considered.

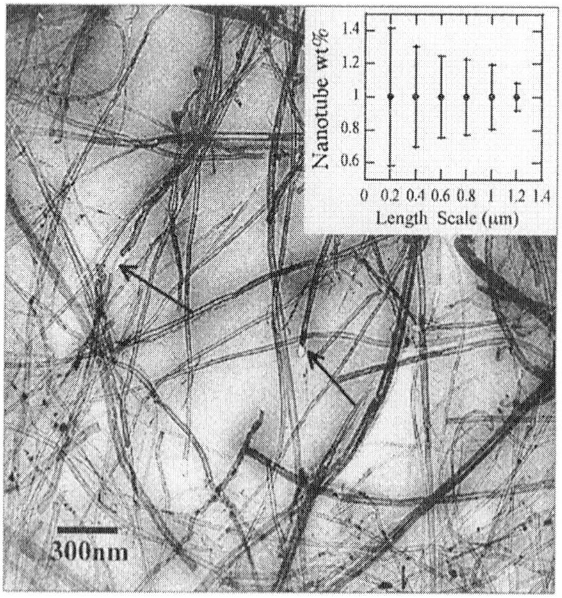

Fig. 2.17b Treansmission electron micrograph of MWNT-PS film in which the nanotubes are homogeneously distributed in the polystyrene matrix at a ~1 μm length scale. The inset is the MWNT weight fraction as a function of length scale, which was determined by measuring the weight fraction of nanotubes at areas in the image. The error bars represent the standard deviation of the experimental data, which reflects the homogeneity of the distribution. Reprinted with permission from [148]

Nanofibers have been successfully melt-mixed with polyphenylene ether/polyamide matrices in a twin screw extruder. This process has led to a commercial product in conductive thermoplastics for electrostatic painting without the loss of mechanical properties [163].

An interesting composite processed from nanotubes is a macroscopic fiber that is a mixture of nanotubes and a traditional material used for carbon fibers, such as pitch or PAN. This was achieved with a pitch-based matrix and led to interesting improvements in properties, because the pitch fibers were not subjected to the typical high temperature processing [164].

2.4.2 Layered Filler–Polymer Composite Processing

Scientists have known for about 40 years that polymers interact strongly with montmorillonite and that the clay surface can act as an initiator for polymerization [165, 166]. Patents for clay/Nylon 6 composites were not issued until the 1980s, at which point the clay/polymer nanocomposites were commercialized [167]. The improvement that led to commercialization was the appropriate dispersion of the clays at the nanometer scale.

The first step in achieving nanoscale dispersion of clays in polymers is to open the galleries and to match the polarity of the polymer or monomer so that it will intercalate between the layers. This is done by exchanging an organic cation for an inorganic cation (Figure 2.18). The larger organic cations swell the layers and increase the hydrophobic properties of the clay [168] (Figure 2.18), resulting in an organically modified clay. The organically modified clay can then be intercalated with polymer by several routes. Solution processing involves dispersion of both the organically modified clay and the polymer in a common solution. Variations on this process include emulsion or suspension polymerization [169, 170]. Highly polar polymers such as Nylon [171] and polyimides [8] are more easily intercalated than nonpolar polymers such as polypropylene, because polar polymers have a higher affinity for the polar clay galleries. In situ polymerization intercalates monomer directly into the organically modified clay galleries, and the monomer can either adsorb onto the layer surface [172] or be anchored by free radical techniques [173]. Melt intercalation involves mixing the clay and a polymer melt, with or without shear. The success of melt intercalation is surprising, given that the gallery spacing is only about 2 nm and the radius of gyration of the polymer is significantly larger than this. Even more surprising is that the speed of melt intercalation is faster than that of self diffusion of polymers and scales with the inverse of the molecular weight [174]. Coarse-grained molecular dynamics studies predict a rate twice that of self-diffusion, with the rate scaling inversely with the molecular weight. The results of molecular dynamics and experimental studies agree that the stronger the clay/polymer interaction, the slower the intercalation rate [175]. In addition, layer flexibility seems to control the mechanism of intercalation [176], perhaps due to a recently proposed mechanism called the 'kink' model of melt intercalation [177], in which sufficient shear force causes a kink to form in the clay sheet (a form

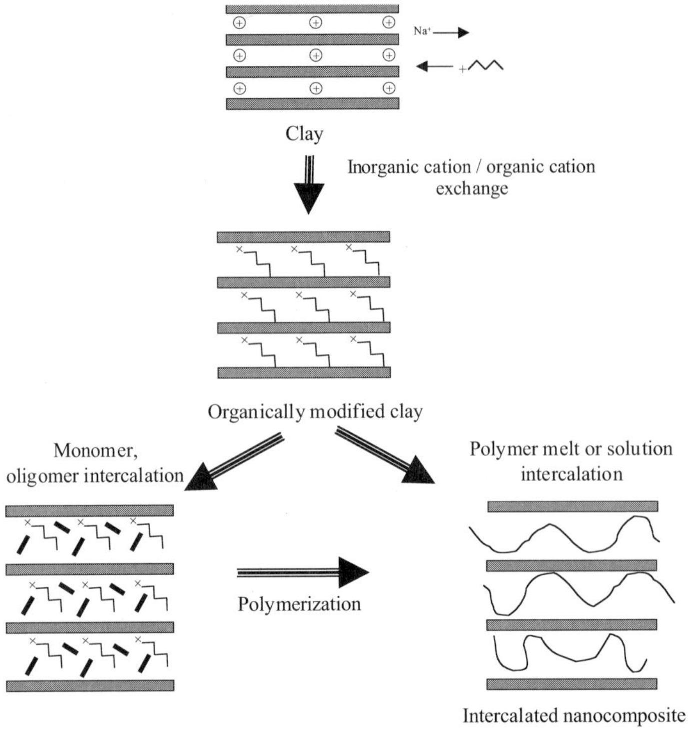

Fig. 2.18 Schematic of the basic steps in processing clay-filled polymers

of compression failure) (Figure 2.19). Polymer can then penetrate the new space between the layers. This kink can propagate along the layer, and more polymer can intercalate. This mechanism can explain both the fast intercalation rate (which is enhanced by the space created by kinking) and the dependence on layer flexibility. Layers with a lower modulus should kink more easily. A rigorous test of this model is still required. Models have also been developed to test the tendency for chains to melt-intercalate in the absence of shear [178, 179]. These studies showed that there is an optimal interlayer separation (slightly larger than the pseudobilayer for intercalation of polystyrene) and an optimal chemical interaction (the greater the hydrophilicity of the polymer, the shorter the organic modifier should be). This minimizes unfavorable interactions between the polymer and the aliphatic chain. Polar polymers are also more likely to intercalate.

As the layer spacing increases, the process can be monitored by x-ray diffraction (XRD). Intense peaks between 3° and 9° indicate an intercalated composite, but if the peaks are extremely broad or disappear completely, this indicates complete exfoliation. Figure 2.20 [180] shows the XRD pattern of organically modified montmorillonite (d), an intercalated montmorillonite (c), and two exfoliated montmorillonite nanocomposites (a, b).

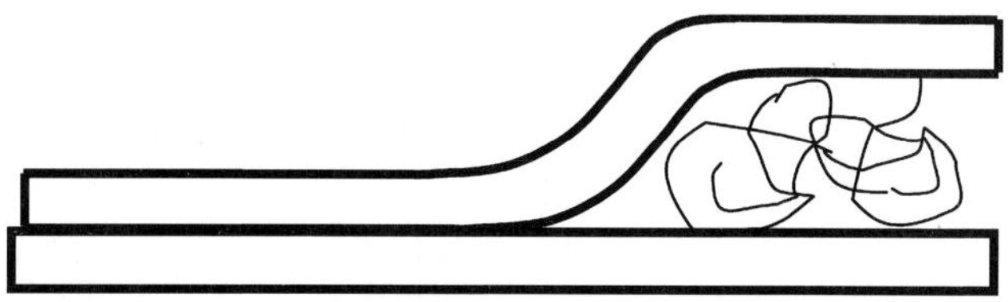

Fig. 2.19 Schematic of the kink model of melt intercalation. Ag denotes the agglomerated (immiscible) portion of the clay, and Int. denotes the intercalate state. Redrawn from Fig. 1(a) of [177]

Once the clays have been intercalated with polymer, the nanocomposites can be processed by traditional melt-processing methods. This final processing of the composite is important in determining the final properties. Mixing facilitates nanoscale dispersion and can lead to clay and/or polymer chain alignment. The degree of shear during molding determines, not only the degree of clay layer alignment, but also the degree of crystallite alignment. For example, extruded Nylon sheet with a draw ratio of 4:1 had a higher modulus than a sheet processed by injection molding [181]. This may be due to a higher degree of platelet and crystallite alignment and is a common phenomenon in the drawing of Nylon. In some cases, the crystallinity increases

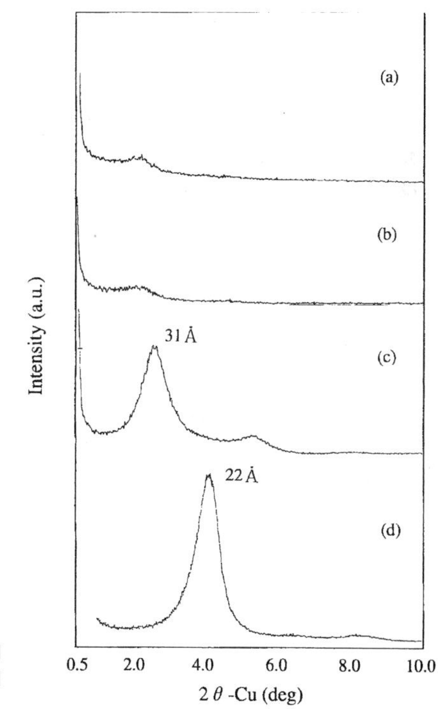

Fig. 2.20 X-ray diffraction data showing the diffraction patterns that result from (a), (b) exfoliated clays, (c) intercalated clays, (d) organically modified clays. Reprinted with permission from [180]

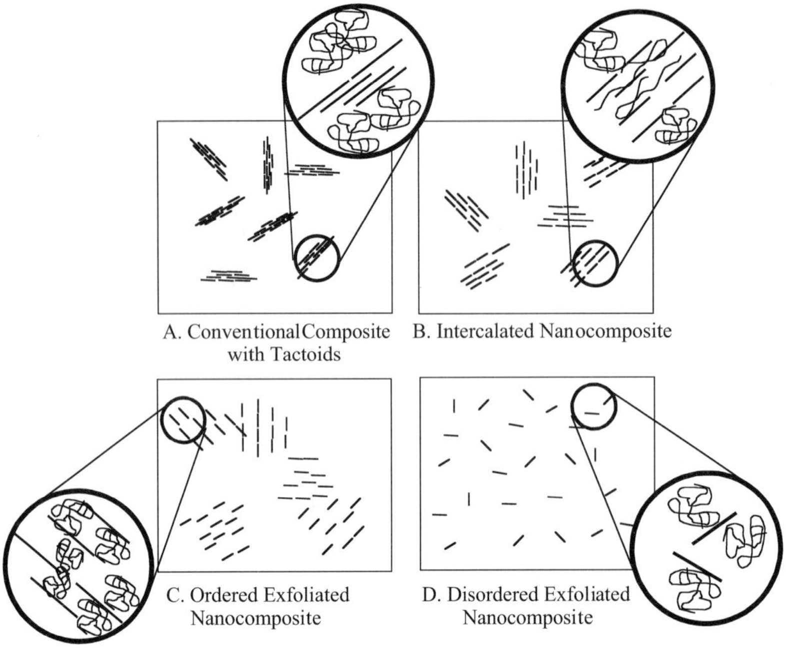

Fig. 2.21 Schematic of the microstructures that can develop in clay-filled polymer composites: (a) a conventional composite with tactoids, (b) an intercalated nanocomposite, (c) an ordered exfoliated nanocomposites, (d) a disordered exfoliated nanocomposite

(36%–38%, compared to 31% for unfilled Nylon) and remains constant with filler content [140]. Other studies, however, have shown decreases in crystallinity [182]. Increasing the pressure during processing to 0.1–0.6 GPa favors the α phase [145]. The clay platelets also enhance alignment of the Nylon 6 chains and the crystallites [140], though the extent is governed by the degree of shear and can vary across the thickness of an injection-molded part [183]. Understanding and optimizing the mixing and taking advantage of interesting rheology and other phenomenon such as strain hardening and rheopexy is in its infancy, and significant fundamental work still needs to be done [184].

The resulting nanocomposites can have several structures (Figure 2.21). The structure of an intercalated nanocomposite is a tactoid with expanded interlayer spacing, but the clay galleries have a fixed interlayer spacing. Exfoliated nanocomposites are formed when the individual clay layers break off the tactoid and are either randomly dispersed in the polymer (a disordered nanocomposite) or left in an ordered array.

The following sections give some details on processing clay/polymer nanocomposites. The discussion is organized by matrix type.

2.4.2.1 **Polyamide Matrices**

Nylon-6/Nylon-12/clay hybrid composites were the first exfoliated smectic clay composites made [187]. Montmorillonite, with a CEC of 119 mEq/100 g, was intercalated with 12-aminolauric acid, which increased the intergallery spacing from 1.0 to 1.7 nm. This '12-montmorillonite' was then mixed with ε-caprolactam, which increased the intergallery spacing even further, to 4.0 nm, indicating that the ε-caprolactam had intercalated into the galleries. Heating to 250 °C led to polymerization, forming a clay/Nylon-6 nanocomposite. Further research [185] determined that ε-caprolactam could intercalate directly into the galleries of montmorillonite in a hydrochloric acid solution and, upon intercalation, becomes oriented vertically in the galleries. The modified montmorillonite then mixed easily with additional molten ε-caprolactam and 6-aminocaproic acid, yielding a Nylon-6 homopolymer/clay nanocomposite. The montmorillonite was completely exfoliated. Recently [186], montmorillonite/Nylon 6 nanocomposites were processed by melt intercalation. Although the degree of exfoliation was not as high as in nanocomposites produced by the above methods, at weight fractions less than 0.1 the composites were primarily exfoliated.

2.4.2.2 **Polyimide Matrices**

The preparation of polyimide matrix clay nanocomposites involves several steps [8, 187] (Figure 2.22). By intercalating montmorillonite with the ammonium salt of dodecylamine, it becomes soluble in dimethylacetamide (DMAC). DMAC is also a solvent for 4,4'-diaminodiphenylether and pyrometllitic dianhydride, the precursors for polyamic acid and, as such, polyimides. After intercalation of the ammonium salt of

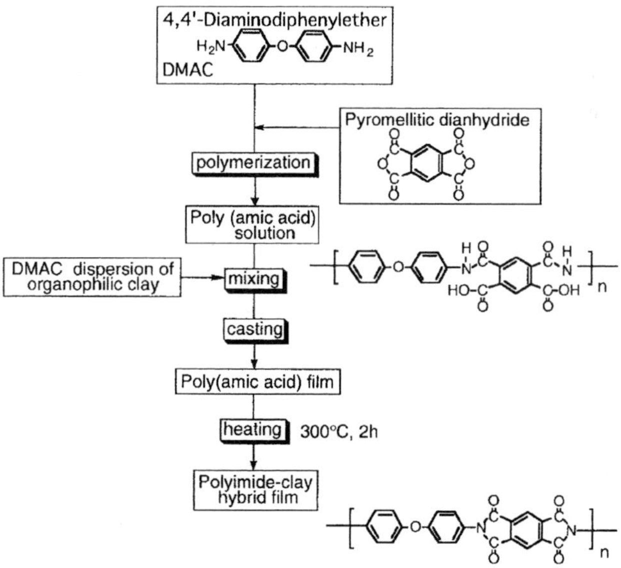

Fig. 2.22 Schematic of the synthesis of polyimide-clay hybrid film. Reprinted with permission from [187]

dodecylamine, x-ray studies [188] showed that hectrite (CEC = 55 mEq/100 g) has one monolayer of organic material between the layers, whereas saponite, montmorillonite, and synthetic mica (all with CEC > 100 mEq/100 g) have two. After composite formation, however, only the montmorillonite and the synthetic mica have exfoliated completely, but the hectrite and saponite remain in a somewhat aggregated state. Lan et al. [189] found aggregates of montmorillonite after using a similar procedure. More recently, P-phylenediamine in an HCl solution was also found to form organic-modified montmorillonite that dissolves in DMAC [190]. This same study showed that the presence of a small amount of nanoscale organoclay can decrease the imidization temperature by 50 °C (from 300 °C to 250 °C), and at 250 °C the imidization time decreased by 15 min. The activation energy decreased by 20 %. Clearly, the organoclay surface is acting as a catalyst.

2.4.2.3 Polypropylene and Polyethylene Matrices

Nonpolar polymers are very difficult to intercalate into smectic clays, because the clays are strongly polar. This challenge has been met [191, 192] by first intercalating stearylamine into montmorillonite and synthetic mica. Melt-mixing the organoclays with maleic anhydride-modified polypropylene oligomers results in PP-MA intercalation. The modified organoclay is then melt-mixed with a polypropylene matrix. There is a balance between creating a polar oligomer with enough maleic anhydride to intercalate well, but nonpolar enough to mix with the polypropylene. Unfortunately, the oligomer limits the extent of property improvement achieved to date. Polyethylene has also been successfully melt-mixed with modified montmorillonite and saponite after ion exchange with dioctadecyldimethylammonium bromide. The degree of dispersion is not excellent, and the layers are certainly not exfoliated; yet, significant modification of both the crystal structure and properties has been observed [143].

2.4.2.4 Liquid-Crystal Matrices

Smectic clay/liquid crystalline nanocomposites exhibit interesting optical properties [193, 194]. Upon initial formation, the composites are opaque, but when an electric field is applied they became transparent. To process these composites, montmorillonite was intercalated with lauryl ammonium or 4-(4'-cyanobiphenyl-4-oxy) butyl ammonium (CBAM), or 4-cyano-(4'-biphenyloxy) undecyl ammonium cation and solution-mixed with a nematic crystal matrix such as 4-pentyl-4'-cyanobiphenyl (5CB) and nematic TFALC (a mixture of low-molar-mass liquid crystals).

2.4.2.5 Polymethylmethacrylate/Polystyrene Matrices

The processing of clay/PMMA or clay/PS composites was first done by directly intercalating the monomer into the clay, followed by polymerization [195]. This method was not successful in exfoliating the clays. At issue again is the compatibility between the clay and the monomer. One solution for PMMA has been to use appropriate ammonium salts [196, 197], which may be reactive [198]. Another solution is to use a comonomer as a compatibilizer [199]. A similar solution was found for polystyrene by using the reactive cationic surfactant vinylbenzyldimethyldodecylammonium as the intercalant [200]. Exfoliated graphite/polystyrene composites have been made by similar processing methods [80]. Recently, a commercially viable process was developed [180] for polystyrene in which montmorillonite intercalated with octadecyl trimethyl ammonium chloride was melt-mixed with a styrene methylvinyloxazoline copolymer. This process resulted in complete exfoliation, which could not be achieved with pure polystyrene [201]. The hypothesis is that the hybridization is due to strong hydrogen bonding between the oxazoline groups and oxygen groups in the silicate clays.

2.4.2.6 **Epoxy and Polyurethane Matrices**

Epoxy is a widely used thermoset, with applications ranging from household glues to high-performance composites. To improve performance, increasing the Tg of epoxy and improving its properties above the Tg are desirable. Adding clays and layered silicic acids to epoxy [6, 71, 202–204] can greatly improve its mechanical performance, particularly at temperatures above Tg. The processing has been studied in detail [71, 205]. In the smectic clay/epoxy composites, the length of the intercalated organic amine determines the ease of exfoliation, and only clays with primary and secondary onium ions form exfoliated nanocomposites [71, 206]. After intercalation of the organic amines, the epoxy resin or a combination of resin and curing agent can be intercalated into the smectic clays or layered silicic acids. If enough resin and curing agent are intercalated and the curing process is controlled, exfoliated nanocomposites result. Figure 2.23 shows a diagram of the process [207]. The acidic onium ions catalyze the intragallery polymerization or curing of the resin. If this reaction occurs more rapidly than extragallery curing, then the clay exfoliates. Otherwise, an intercalated nanocomposite results. Therefore, careful control of temperature and time is required [71], or the ratio of resin to curing agent must be significantly less than the stoichiometric ratio [205] in order to achieve exfoliation. An approach similar to that used for epoxy composites was used to make intercalated montmorillonite/polyurethane composites [208].

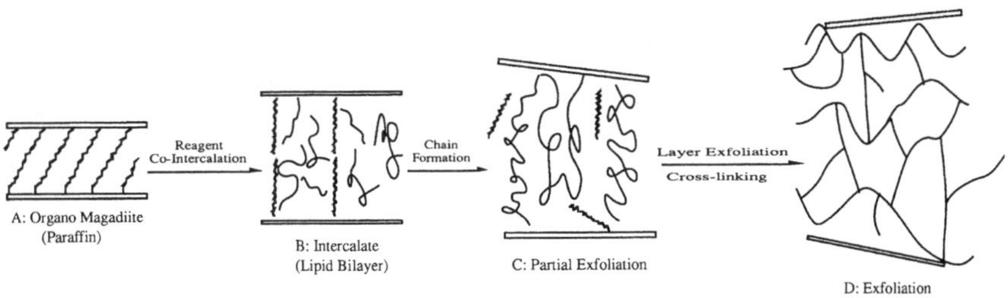

Fig. 2.23 Proposed pathway for formation of an epoxy-exfoliated magadiite nanocomposite: (A) Initial organomagadiite with a paraffin-like gallery structure of onium ions and neutral amine. (B) Reorientation of the alkylammonium ions into a lipid-like bilayer structure to accommodate the cointercalation of epoxide and curing agent. (C) Rapid intragallery formation of a polymer gel and expansion of the gallery height beyond a lipid-like layer. (D) Silicate nanolayers are completely exfoliated in a fully crosslinked epoxy polymer network. Reprinted with permission from [207]

2.4.2.7 **Polyelectrolyte Matrices**

Polyelectrolytes can be used in electrochemical devices such as solid-state batteries, electrochromic devices, and sensors [209]. The addition of layered silicates to polyelectrolytes increases the conductivity, improves the mechanical stability, and improves the interfacial stability with electrode materials. Polyelectrolytes are characterized by a large number of ionizable groups and thus are highly polar. This makes them excellent candidates for intercalation into smectic clays. Polyvinylpyridines are of particular interest because of the variety of processing methods available [210]. Intercalated nanocomposites can be formed easily from the water-soluble hydrobromide salt of the 1,2 or 1,6 polyelectrolyte (1,2 or 2,6 polyvinylpyridinium cations). However, only a single layer of polymer intercalates, and exfoliation does not occur. A slower, but ultimately more effective process, uses neutral poly-4-vinylpyridine and results in an exfoliated composite. A second method involves intercalation of 4-vinylpyridinium salts, followed by polymerization.

Poly(ethylene oxide) (PEO) matrix composites have also been processed both by intercalating PEO in solution into organically modified smectic clays [209] and by melt- mixing clay with PEO and PEO/PMMA mixtures [211,212]. In neither case does an exfoliated composite result. Aranda and Ruiz-Hitzky [209] dissolved PEO in acrylonitrile and found that the structure of the PEO changed when the interlayer cation was changed. Use of Na^+ montmorillonite or NH_4^+ montmorillonite resulted in either a helical PEO or a bilayer zigzag PEO structure in the galleries. The PEO arrangement was reversible with exchange of the interlayer cations.

2.4.2.8 **Rubber Matrices**

Several applications of rubbers might benefit from inclusion of exfoliated clays. Their greatly reduced permeability [213] would be useful for the inner liners of tires and inner tubes [214]. In addition, modification of the glass transition temperature and/or the loss modulus might be useful in a variety of damping applications. Montmorillonite has been ion-exchanged with a protonated form of butadiene and acrylonitrile copolymer. This was subsequently mixed with nitrile butadiene rubber in the presence of crosslinking agents and resulted in highly dispersed nanocomposites. Nanocomposites have also been prepared from dioctadecyldimethyl ammonium-exchanged montmorillonite in poly(styrene-*b*-butadiene) matrices [215].

Silicone rubbers have also been successfully intercalated into smectic clays. Burnside and Giannelis [216] intercalated Na^+ montmorillonite with dimethylditallow ammonium bromide and then sonicated in a silanol-terminated poly(dimethylsiloxane) (Mw = 18,000) with a tetraethyl orthosilicate and tin 2-ethylhexanoate crosslinking agent. Hexadecyltrimethylammonium bromide has also been used successfully as the intercalating agent [217]. Recently, a latex method was developed in which unmodified clays were mixed with a rubber latex, leading to improved properties over solution intercalation [218].

2.4.2.9 Others

Clay/polymer nanocomposites that include poly(ε-caprolactone) have been made via in-situ polymerization [7]. Composites that include poly(p-pheylenevinylene) have been made via intercalation of poly(xylylenedimethylsulfonium bromide) and subsequent elimination of the dimethylsulfide and HBR [219]. Those including cyclic polycarbonate [220] or polyethyleneterephthalate have been made via monomer intercalation and subsequent polymerization [221]; and those including polyaniline via in-situ polymerization of aniline monomer [222].

2.4.3 Nanoparticle/Polymer Composite Processing

There are three general ways of dispersing nanofillers in polymers. The first is direct mixing of the polymer and the nanoparticles either as discrete phases or in solution. The second is in-situ polymerization in the presence of the nanoparticles, and the third is both in-situ formation of the nanoparticles and in-situ polymerization. The latter can result in composites called hybrid nanocomposites because of the intimate mixing of the two phases.

2.4.3.1 Direct Mixing

Direct mixing takes advantage of well established polymer processing techniques. For example, polypropylene and nanoscale silica have been mixed successfully in a two-roll mill [5], but samples with more than 20 wt. % filler could not be drawn. This is typical and is a limitation of this kind of processing method. Nanoscale silica/PP composites have been processed in a twin-screw extruder, but the dispersion was successful only after modification of the silica interface to make it compatible with the matrix [223]. A Brabender high-shear mixer has been successfully used to mix nanoscale alumina with PET, LDPE [224]. Thermal spraying has also been successful in processing nanoparticle-filled Nylon [142]. When these traditional melt-mixing or elastomeric mixing methods are feasible, they are the fastest method for introducing new products to market, because the composites can be produced by traditional methods. This has been successful in many cases, but for some polymers, the viscosity increases rapidly with the addition of significant volume fractions of nanofiller, which in turn can limit the viability of this processing method.

In addition to viscosity effects, nanoparticles can either enhance or inhibit polymer degradation. One method for measuring degradation is to place the polymer in a high-shear mixer and measure the torque as a function of time and temperature. As the material crosslinks, the torque begins to increase (at a constant speed), and when chain scission begins, the torque decreases. This leads to a peak in the torque, whose position is often used as a measure of the degradation time. A recent study on ZnO/LDPE composites showed that, for nanoparticle/matrix mixtures, the time at the peak increased by a factor of almost 2 [224]; however, micrometer-size particles decreased the degradation time. In other instances, the catalytic nature of the nanoparticles greatly decreases the degradation time [225]. Whether degradation is inhibited or enhanced depends on the particle surface activity and the increased interfacial area (particle size).

2.4.3.2 Solution Mixing

Some of the limitations of melt-mixing can be overcome if both the polymer and the nanoparticles are dissolved or dispersed in solution. This allows modification of the particle surface without drying, which reduces particle agglomeration [226]. The nanoparticle/polymer solution can then be cast into a solid, or the nanoparticle/polymer can be isolated from solution by solvent evaporation or precipitation. Further processing can be done by conventional techniques.

2.4.3.3 In-Situ Polymerization

Another method is in-situ polymerization. Here, nanoscale particles are dispersed in the monomer or monomer solution, and the resulting mixture is polymerized by standard polymerization methods. One fortunate aspect of this method is the potential to graft the polymer onto the particle surface. Many different types of nanocomposites have been processed by in-situ polymerization. A few examples are silica/Nylon6 [227, 228], silica/poly 2-hydroxyethylmethacrylate [229], alumina/polymethylmethacrylate [230], titania/PMMA [231], and $CaCO_3$/PMMA [232]. The key to in-situ polymerization is appropriate dispersion of the filler in the monomer. This often requires modification of the particle surface because, although dispersion is easier in a liquid than in a viscous melt, the settling process is also more rapid.

2.4.3.4 In-Situ Particle Processing – Ceramic/Polymer Composites

An interesting method for producing nanoparticle-filled polymers is in-situ sol–gel processing of the particles inside the polymer. The process has been used successfully to produce polymer nanocomposites with silica [233] and titania [234] in a range of matrices [235]. The overall reaction for silica from tetrethylorthosilicate (TEOS) is shown below.

$$Si(OC_2H_5)_4 + \text{excess } H_2O \rightarrow SiO_2 + 4C_2H_5OH$$

The composites can be formed in one of several ways. In the first method, a copolymer of the matrix polymer and silica precursor are mixed, and the sol–gel reaction is allowed to progress. Figure 2.24 shows an example of the process for silica/polystyrene composites [233]. During drying, the polymer blocks phase-separate and the silica regions coalesce. A second approach mixes a silica precursor such as TEOS [236] with a polymer such as polyvinylacetate [237], polyetherimide [238], or polymethylmethacrylate [239]. For example, a recent paper on the processing of TiO2/poly(styrene maleic anhydride) resulted in excellent dispersion of TiO_2 in PSMA [234]. Direct addition of TiO_2 to a PSMA solution resulted in serious aggregation. Therefore, PSMA was dissolved in THF, and then tetrabutyl titanate was added under the appropriate conditions. Because the uncondensed TiOH and maleic anhydride could react, the polymer

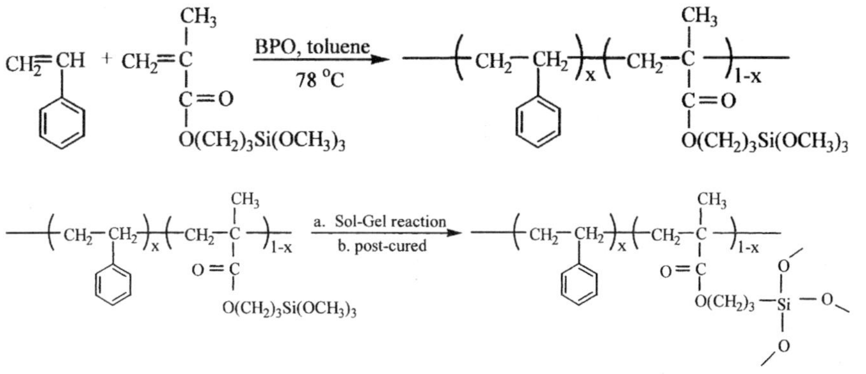

Fig. 2.24 Example of in-situ synthesis of silica particles in polystyrene. Reprinted with permission from [233]

coated the titania particles as they formed and prevented agglomeration. A combination of these first two methods has been used for silica/polydimethylsiloxane (PDMS) composites in a two step process. First, an unfilled PDMS network was formed, with TEOS as the end-linking agent. The network was then swelled with TEOS and the sol–gel reaction was catalyzed [240]. Significant work on determining the parameters affecting particle size has been accomplished with this method [241].

Finally, the polymer and ceramic can be polymerized simultaneously, and composite materials ranging from a few percent to 100% inorganic phase can be formed [242]. For example, silica/polyacrylate nanocomposites were formed with the simultaneous polymerization of HEMA (2-hydroxyethyl methacrylate) and tetraethoxysilane (TEOS) [229]. This resulted in a continuous interpenetrating network of silica and polyacrylate. There is a rich chemistry here that has been reviewed elsewhere on the processing of materials that range from glasses to polymers [243, 244] in which there is little if any phase separation between the organic and inorganic. We focus here on systems with discrete, isolated phases.

An interesting variation on in-situ formation of nanoparticles and polymerization is CVD deposition of nanocomposite films [245]. Here, the chemical precursors are modified separately and are mixed only just before deposition. Dichloro-*p*-xylylene was used as a precursor for poly(dichloro-*p*-xylyene), and diacetoxydi-tert-butoxysilane was used a precursor for silica. The resulting microstructure has discrete polymer and silica phases. This method is promising for the creation of interlayers with a low dielectric constant.

Barium titanate/polymer nanocomposites have also been prepared in situ by mixing a titanium alkoxide with a polymer and then casting and drying films. The films are then reacted with an aqueous solution containing barium, resulting in nanoscale barium titanate particles within a polymer matrix [246, 247]. Making the precursor and the polymer compatible improves the degree of dispersion, and using compounds that tend to form micelles can introduce controlled heterogeneity [246].

Nanosized calcium phosphate particles have been processed by an in-situ deposition technique in the presence of polyethylene oxide (PEO) or PEO/PVAc mixtures [248].

In this method, a polymer-calcium chloride complex is blended with trisodium phosphate. As the phosphate and chloride ions diffuse through the polymer, a gel-like precipitate forms. This can be separated; the size and morphology can be varied by changing the polymer and/or the polymer concentration.

2.4.3.5 In-Situ Particle Processing – Metal/Polymer Nanocomposites

Metal/polymer nanocomposites have also been processed via in-situ formation of metal particles from suitable metal precursors. The reaction occurs in the presence of a protective polymer, which limits the size of the particles. Figure 2.25 [249] shows a reduction scheme that uses a gold precursor and polypyrrole. Once a stable suspension of metal particles is prepared in the presence of a polymer, the composite can be cast, or additional monomers, of the same or a different polymer type, can be added to form a nanocomposite.

Mayer [250] wrote an excellent review paper on the processing of metal/polymer nanocomposites via in-situ methods in 1998. She reviewed the parameters that affect particle size, stability, and morphology. Several primary parameters controlled the particle size, including the choice of metal precursor and the metal/polymer interaction. For example, if $PdCl_2$ is compared with $(NH_4)_2PdCl_4$, the former tends to form halogen-bridged complexes and thus tends to form agglomerates of nanoparticles, but the latter does not. The interaction of the metal precursor with the polymer is also important. If the polymer has a stronger interaction with the precursor, then the particle size tends to be reduced [251], because the metal precursors are prevented from phase separating. In general, because most polymers are hydrophobic, precursors with a more hydrophobic character result in stronger interactions and thus smaller particle size. Complicating this situation is the ability of the polymer to form a stable suspension in solution. The rate of reduction affects the particle size, with faster reduction methods resulting in smaller particles [252, 253].

Another method giving more specific control over particle size and morphology is the use of either micelles formed from amphiphilic block copolymers (ABC) [254] or crosslinked/gelled matrices. For example, through the use of crosslinking and gelation

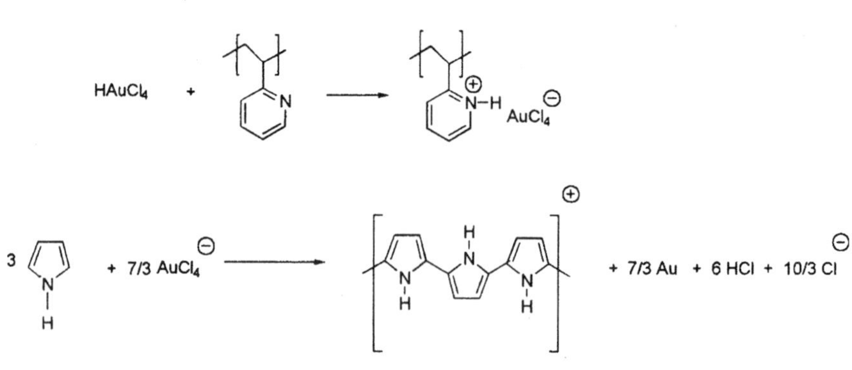

Fig. 2.25 Schematic of reaction leading to in-situ gold particles in a polymer. Reprinted with permission from [249]

(a) (b)

(c) (d)

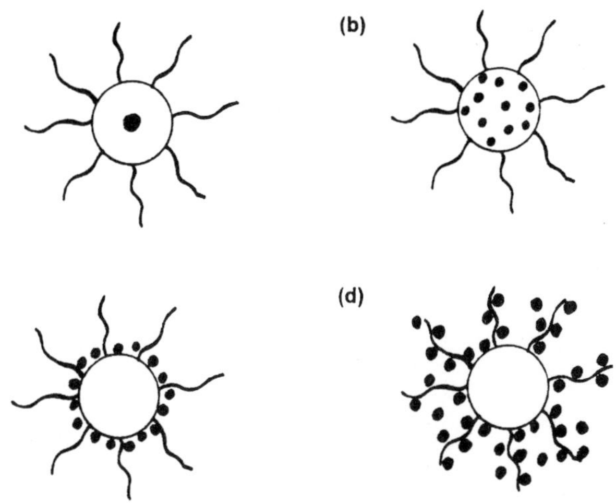

Fig. 2.26a Overall morphologies of block copolymer-metal systems involving spherical micelle formation of the amphiphilic block copolymers: (a) cherry morphology, (b) raspberry morphology, (c) strawberry morphology, (d) proposed red currant morphology. Reprinted with permission from [250]

reactions, dendritic gold particles can be formed [255]. Amphiphilic block copolymers provide another avenue of control. Here, block copolymers are used to form micelles. Metal salts then either penetrate into the micelles or are stable in the micelle corona. The reducing agent can be added, and metal particles form either within the micelles or in the corona, resulting in several morphologies shown schematically in Figure 2.26a [250]. Figure 2.26b [294] shows an example of the raspberry morphology for Pd colloids in a polystyrene-b-poly (4-vinylpyridine) block copolymer matrix that was reduced with hydrazine. Changing the block length can control the particle size.

This same principle is applicable to other block copolymer morphologies, such as rods and layers, and leads to interesting morphologies. One thing that is not clear, and needs to be addressed by theory, is which of these morphologies will lead to the best properties. Because of the range of metal particles that can be formed, including gold, silver, palladium [256], platinum, semiconductors [257], and metal oxides [234], tremendous opportunity exists to tune the properties of these systems.

Other methods are available for producing metal nanoparticle-filled polymers, such as deposition of a metal film onto the polymer [258, 259] and subsequent annealing to form particles, or electrochemical methods [260], both of which lead to highly controlled thin-film composite structures.

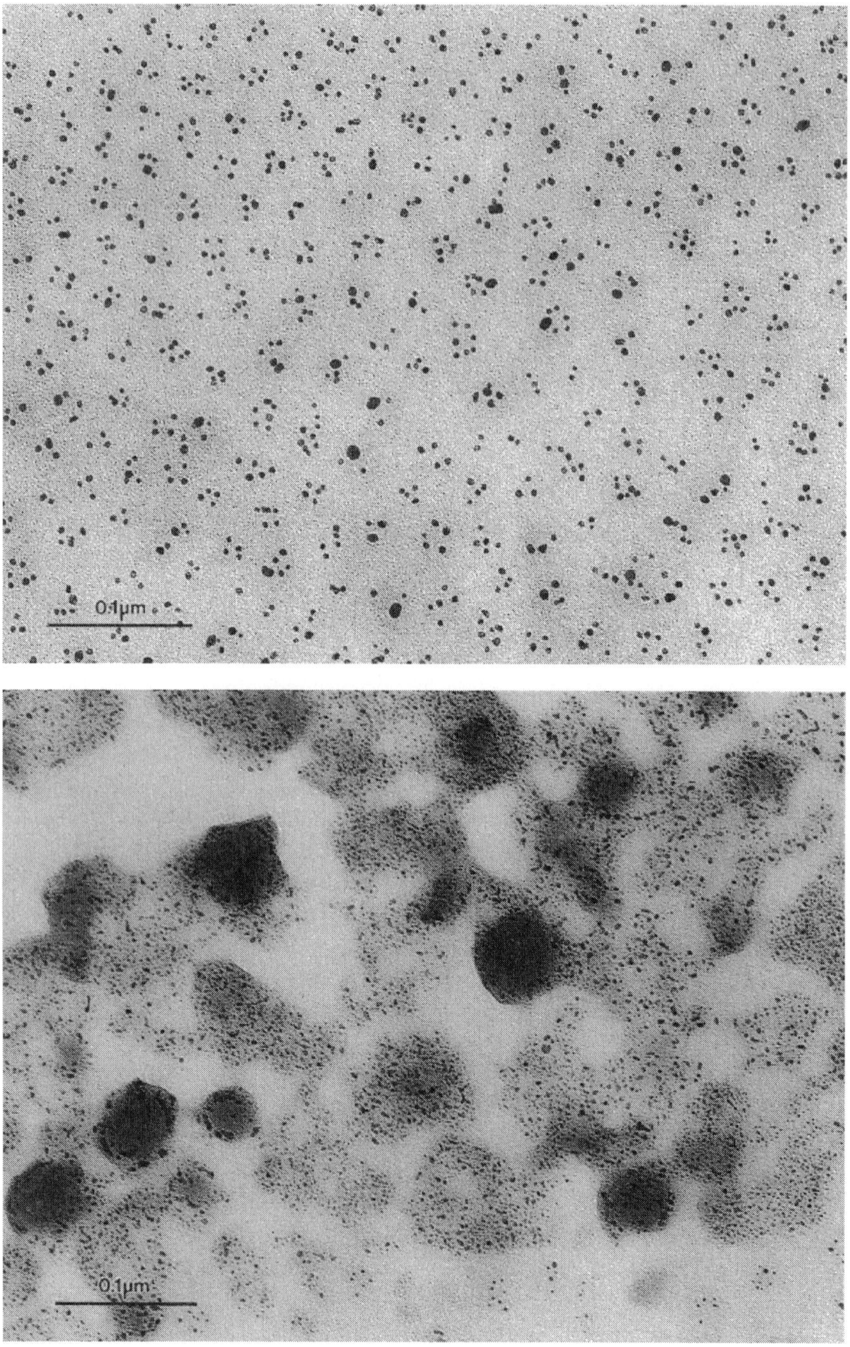

Fig. 2.26b Electron micrographs of Pd colloids synthesized in Ps-*b*P4VP block copolymers via reduction with hydrazine. Reprinted with permission from [253]

2.4.4 Modification of Interfaces

Literature on the modification of interfaces of traditional composite fillers is extensive. Many of the methods, however, are not directly applicable to nanocomposites for several reasons. Carbon and glass fibers are usually coated as long fibers on a spool, which is not practical for nanofibers. In addition, methods that require drying the particles are not appropriate for metal oxide and metal nanoparticles, because during drying (even at only 100 °C), they may agglomerate significantly. Freeze-drying methods are more appropriate. Although describing the relevant chemical processes is important for understanding nanofiller interface modification, this section focuses primarily on the results of modification, instead of on the detailed chemistry.

2.4.4.1 Modification of Nanotubes

Several reasons exist for modifying nanotubes. The first is to render them more dispersible in solution, thus providing a route toward further functionalization or use in applications. The second is to improve the interaction of the nanotubes with polymer matrices. Modifying the surface activity of MWNT has focused primarily on doping them with other elements. Doping MWNT with boron has resulted in changes in the electronic states [261, 262] and formation of a p-type semiconductor. In addition, boron substitution, as well as other electron-rich substitutions, may very well lead to out-of-plane bonding configurations that increase the reactivity of the surface with a polymer matrix. Nitrogen doping of MWNT creates an n-type semiconductor and leads to a bamboo-like structure having ordered regions connected via disordered regions [263]. The nitrogen creates nitrogen-rich cavities in the otherwise graphitic nanotube. Preliminary work has shown that nitrogen doping changes the interface enough to improve the dispersion of nanotubes and results in enough bonding with epoxy matrices to increase the glass transition temperature by 20° as well as in increased modulus and strength [264]. In addition, acid treatments and subsequent chemical reactions can lead to attachment of epoxy groups or even particles to MWNT [265, 266].

SWNT are assembled as ropes or bundles. To maximize their interaction in a polymer matrix, they need to be chemically modified and separated from the bundles. One way to increase the chemical activity is to use an SWNT with a smaller radius of curvature [267]. A second method is to add functional groups to the ends of SWNT. Carboxylic groups have been linked to SWNT ends, and then the SWNT have been tethered to nanoparticles via thiol linkages [268–270]. SWNT have been placed in solvent through the addition of octadecylamine groups on the ends and a small amount of dichlorocarbenes on the sidewalls [271]. The first successful method for adding functional groups to the sidewalls involved fluorination at elevated temperatures. The tubes could then be solvated in alcohols and reacted with other species, particularly strong nucleophiles such as alkylithium reagents [272]. A simpler route based on a process used on other carbon species [273, 274] was developed, in which an aryl diazonium salt was reduced electrochemically, resulting in a free radical that can attach to the carbon surface of small diameter tubes [271] (Figure 2.27).

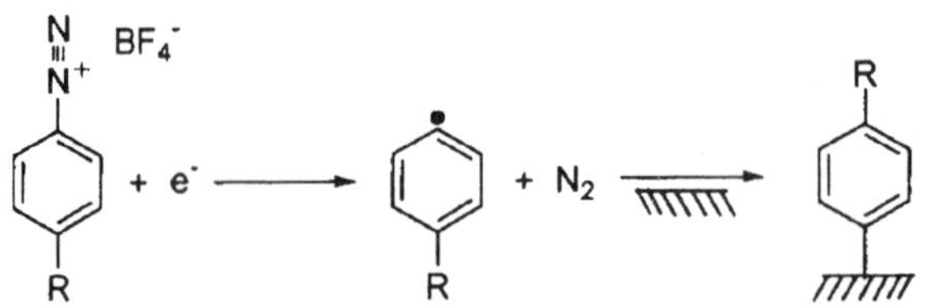

Fig. 2.27 Electrochemical reduction of an aryl diazonium salt to yield, giving a reactive radical that covalently attaches to a carbon surface. Reprinted with permission from [271]

The most exciting development in terms of composites, however, is the dissolution of full-length SWNT that have been separated from the bundles [275]. Given the issues of slipping of nanotubes within a bundle discussed earlier, their use in composites requires separation from the bundles. The process involves, first, partially breaking up the SWNT bundles in an acidic solution. Carboxylic acid groups are then added to the ends of the tubes. Finally, the SWNT are placed in an octadecylammonium (ODA) melt for several days. This results in exfoliated nanotubes. This important result should eventually lead to SWNT/polymer composites with the SWNT separated in the polymer, which will provide an avenue for taking advantage of the extraordinary properties of SWNT.

The development of methods to attach polymers to nanotubes has progressed significantly. First, separation of the nanotubes from their aggregates is required. This separation can be achieved by dispersion in the appropriate solvents with or without the aid of a surfactant [276] and/or by choosing an appropriate synthesis method. A polymer chain can then be attached to a bonding site by sonication [277], plasma activation [278], chemical etching of the tube ends [279], or chemical adsorption [280]. These advances should lead in the near future to some very interesting copolymers with SWNT as one block in the polymer and a rich interfacial chemistry. Some evidence suggests that polymers can be patterned by the nanotube surface, providing a method for templating polymer structures near the surface [281]. It is not clear yet what role this interface control will have in terms of properties of composites, but to develop technology that fully exploits the properties of nanotubes in composites, control over the interface is required.

2.4.4.2 Modification of Equi-axed Nanoparticles

A recent review by Frank Caruso [282] provides extensive background on the modification of nanoparticle surfaces. He cites two primary methods for modifying an inorganic nanoparticle surface with organic molecules. The first requires connecting a short-chain molecule such as a siloxane onto the surface through grafting or strong hydrogen bonding. The second involves application of a coating by polymerizing a polymer onto the nanoparticle. In addition, inorganic coatings can be applied.

2.4.4.3 Small-Molecule Attachment

Both metal and metal-oxide nanoparticles are readily modified with small molecules. They tend to have hydroxyl groups on the surface, although the number of hydroxyl groups and the strength of the metal-OH bond varies. The most common method

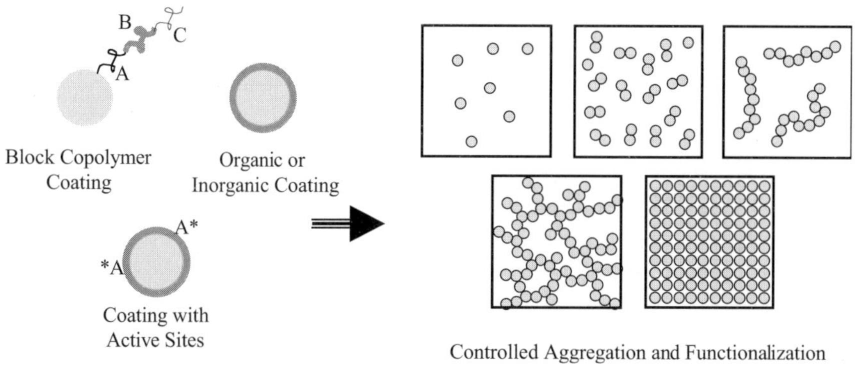

Fig. 2.28 Schematic summarizing many of the possibilities for modifying equi-axed nanoparticles

takes advantage of a long history of modifying metal-oxide glass fiber or filler surfaces. The hydroxyl group on the nanoparticle surface reacts with a silane coupling agent. The relative ease of the reaction depends on the nature of the nanoparticle surface (basicity vs. acidity). The silane coupling agent can have a large variety of functionalities attached to the silanol to modify the surface. The chains can be long or short, hydrophilic or hydrophobic, and linear or bulky. This provides tremendous flexibility in choosing a coupling agent. Changing the coupling agent can lead to control over the strength of the interaction between the filler and the matrix, from covalent bonding to repulsion. The change in bonding leads to increases or decreases in both glass transition temperature and the modulus, but large effects occur only for particles much less than 100 nm [128].

Of course, reacting with molecules other than silanes is also common. Nanoscale titania has been coupled with chelating agents such as lauryl sulfate [283] and with carotenoids (natural pigments) containing a terminal carboxylate group that reacts with OH groups on the surface [284].

One creative study underway provides a method for placing one, two, or several active functional groups on the surface of nanoparticles, followed by passivation of the rest of the particle. This will result in controlled aggregation of the particles into dimers, chains, or branched chains [285] (Figure 2.28).

2.4.4.4 Polymer Coatings

Historically, polymer coatings have provided control over the compatibility of the polymer with the matrix and the strength of the interaction. The coating can be grafted onto the surface of the particle or just highly adsorbed. After the particles are coated they can be dispersed easily in a matrix of similar chemistry to make low-volume-fraction composites. The exciting news is that controlling interfacial interactions has moved well beyond single-component coatings. The possibility for assembling coated nanoparticles directly into useful structures, the core-shell approach, has driven significant development in multicomponent organic and inorganic coatings. The possibilities of this approach seem almost endless. We currently know how to graft or strongly adsorb several types of polymers onto a surface. In addition, we know how to polymerize polymers with very monodisperse molecular weight and we know how to make block (di and tri) copolymers [286]. Progress has also been made in placing inorganic coatings onto particles. This provides flexibility in our ability to engineer the surface of the particles being placed in composites. No longer will the coating be only for controlling the extent of the interaction, for it may contain several layers. The role of the first layer may be to bond with the surface. The second layer may provide electrical, mechanical, or optical functionality, and the third layer may make the particle compatible with the matrix and help in controlling the properties. The final layer may have some reactive sites that could be used to bond particles together and lead to controlled aggregation of the nanoparticles (Figure 2.28).

In principle, polymer coatings can be produced on organic and inorganic particles by a variety of polymerization methods. Monomer adsorption and subsequent polymerization, heterocoagulation–polymerization [287], and emulsion polymerization [288] have been demonstrated on micrometer-scale filler surfaces [287], although not all have been applied to nanoparticles. Monomer adsorption has been used successfully in several studies, and initiation has occurred both chemically [289] and via irradiation [290]. One recently reported method can be used to place nanoparticles (in this case fumed silica) into a monomer such as styrene or methylmethacrylate. The mixture is

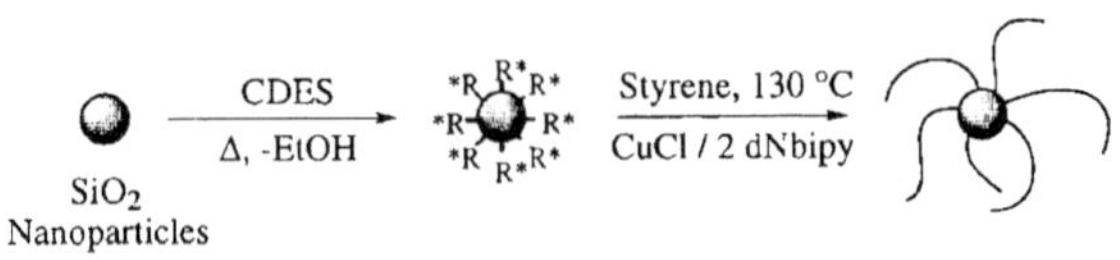

Fig. 2.29a Synthetic scheme for structurally well defined polymer-nanoparticle hybrids. Reprinted with permission from [292]

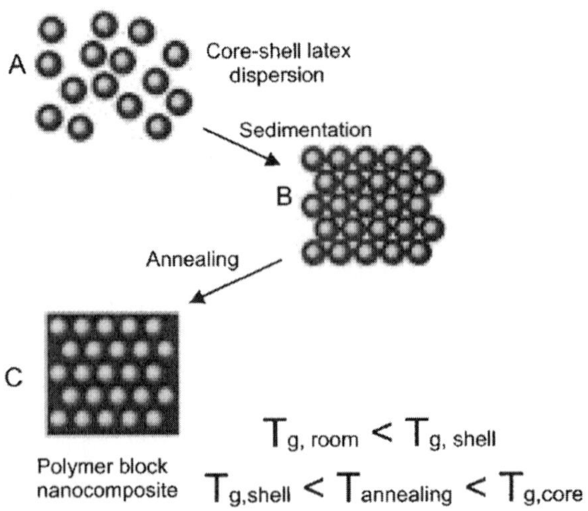

Fig. 2.29b Proposed approach to formation of 3-D polymer nanocomposite material. Stage A: synthesis of the core-shell latex particles with hard functionalized cores and soft inert shells. Stage B: assembly of latex particles in a 3-D close-packed structure. Stage C: heat treatment of the 3-D compact structure, which leads to flow of soft shells and formation of a nanocomposite polymers. Reprinted with permission from [294]

then irradiated [290]. The monomer penetrates the silica aggregates, and after grafting, the surfaces are more hydrophobic and blend more easily with polymers such as polypropylene. This treatment leads to more uniform distribution of the particles and greatly enhanced composite ductility. In some cases, the polymer is not directly grafted onto but interacts strongly with the surface via hydrogen bonding. For example, carboxylic acids are strongly adsorbed to alumina. If the acid contains a polymerizable group, then after adsorption, a polymer can be grown from the molecule attached to the surface. This has been shown for maleic acid adsorbed onto a surface and was used to attach AIBN-initiated 1-octadecene polymers to the surface [291].

Another method that leads eventually to a composite with an ordered array of nanoparticles involves grafting an initiator onto the surface of the nanoparticles and then polymerizing a grafted polymer onto the surface [292]. If living radical polymerization is used, then the particles can be processed into a composite without additional matrix, to make high-volume-fraction composites with particles in an ordered array. Figure 2.29a shows this for silica particles with grafted polystyrene packed into an ordered matrix [292]. This process lends itself nicely to block copolymer coatings and even triblock copolymers. Another method that can lead to multilayer structures is the use of self-assembled polymer layers in a 'layer-by-layer colloid templating strategy' [293]. Here, charged polymers are adsorbed onto charged particles, resulting in a coated particle with surface charge. A second charged polymer can then be exposed to the previously charged particles. When this polymer is adsorbed, the surface charge of the particle is reversed, and the particles have a bilayer coating. This method can be repeated and enables careful control of the coating thickness and functionality.

A similar core–shell approach has been used to coat a low-Tg polymer onto a high-Tg latex particle. The coated polymer particles were then taken out of solution and annealed to create an ordered array of polymer in polymer [294] (Figure 2.29b). The concept applies to coated nanoparticles other than latex as well.

2.4.4.5 **Inorganic Coatings**

Inorganic coatings have also been applied to submicrometer particles via either precipitation of the inorganic coating onto the particles or deposition via a sol–gel type process. With these methods, silica [295], yttria [296], titania [297], titanium nitride [298], and zirconia [299] have all been applied to support particles, usually other metal oxides. The advantage of such coatings is that the nanoparticles (both organic and inorganic) can be coated with inorganics for further functionalization. For example, conducting coatings can be applied, and the optical properties can be tuned.

Sonochemistry can also be used to apply inorganic coatings in much the same way as described above for the processing of nanoparticles. For example, ZnS coatings (1–5 nm) on colloidal silica have been prepared in this way [300].

2.5
Properties of Composites

2.5.1
Mechanical Properties

One of the primary reasons for adding fillers to polymers is to improve their mechanical performance. For example, the addition of high-modulus fillers increases the modulus and the strength of a polymer. In traditional composites, unfortunately, this often comes at the cost of a substantial reduction in ductility, and sometimes in impact strength, because of stress concentrations caused by the fillers. Well-dispersed nanofillers, on the other hand, can improve the modulus and strength and maintain or even improve ductility because their small size does not create large stress concentrations. In addition, the large interfacial area of nanocomposites provides an opportunity for altering the matrix properties in unique ways. This section highlights the potential effect of nanofillers on the modulus, failure stress/strain, toughness, glass transition, and wear behavior of polymers.

2.5.1.1 Modulus and the Load-Carrying Capability of Nanofillers

In traditional composites, the modulus can be predicted, within bounds, and the mechanism that leads to a change in modulus is load transfer. For example, if a high-modulus filler is added to a lower-modulus polymer, then load transfers from the matrix to the filler, leading to an increase in modulus. One important question for nanocomposites is how load is transferred from the matrix to the filler. This is a complicated question, because in some nanocomposites the polymer chain and filler are almost the same size, and in others, the fillers are atomically smooth, with few functional groups to which the polymer can bond. The importance of load transfer is easily illustrated with the shear lag discussion commonly applied to graphite fiber/polymer composites. Figure 2.30a shows a stress – strain curve for a typical high-modulus fiber, a thermoplastic matrix, and the resulting composite. At a given applied composite strain, the fiber carries more load than the matrix; therefore stress must be transferred from the matrix to the fiber. Figure 2.30b shows the stress-transfer behavior on the micromechanical level. At the fiber end (or at a fiber break) the fiber does not carry stress. Stress is transferred via a shear stress at the filler/matrix interface, and the rate of load transfer depends on the shear stress. This highlights the role of the interface in controlling the modulus of filled polymers. For example, the higher the interfacial shear stress and the shorter the load transfer length, the more efficient the filler is in carrying load and the higher the composite modulus. In addition, the higher the aspect ratio of the filler, the longer the length that carries the maximum load, and the higher the composite modulus.

To fully understand the modulus of nanocomposites, a second mechanism in addition to load transfer must be considered. For example, the filler can constrain the mobility of the polymer chains as well as their relaxation spectra [120], which can change the glass transition temperature [120, 121] and modulus of the matrix. Nano-

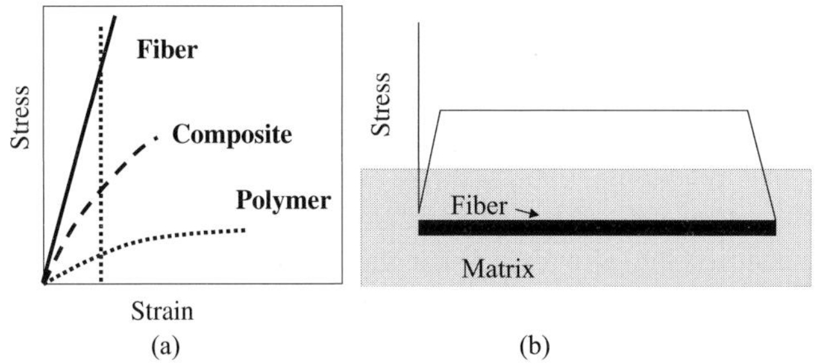

Fig. 2.30 Schematics of (a) stress vs. strain curve for a typical high modulus fiber, a thermosetting polymer, and the resulting composite, (b) transfer of strain from the matrix to the fiber near the fiber end

fillers can also change the degree or type of crystallinity in a polymer, which also changes the modulus [140, 141]. These two mechanisms of reinforcement are addressed for each geometry of nanofiller below.

Nanotubes

Consider nanotube-filled polymer composites. The modulus of nanotubes may be significantly higher than that of any graphite fiber, and therefore, they have the potential to yield higher composite moduli at the same filler volume fraction than a graphite fiber composite. In addition, because of their small size, if properly dispersed, nanotubes may not decrease the strain-to-failure. But, how is load transferred to a nanotube? Qian et al. [148] and Yu et al. [28] have shown, as first described by Wagner [29], that MWNTs fail via a sword-and-sheath mechanism (see section 2.2.1). This implies that, even if load is transferred to the outermost nanotube in a MWNT, the inner tubes slide within the concentric cylinder and do not carry much load. This situation limits the efficiency of MWNT/polymer composites, because only a small portion of the volume fraction of the MWNT carries load. Evidence for this was shown for MWNT/polystyrene composites, in which the effective modulus of the MWNT in the composite was only 500 GPa [148] but the measured modulus of the MWNT is close to 1TPa. Therefore, to minimize the number of layers not carrying load, 2–3 layers are preferable.

For SWNT composites, the SWNT are in a bundle and, as was suggested earlier, individual SWNTs may slip within the bundle. Work by Yu et al. [27] showed that, if only the nanotubes on the outer edge of a SWNT in a bundle are used to calculate the modulus, it is close to the predicted 1 TPa. However, if the whole area of the bundle is used, the calculated modulus is considerably lower. This suggests that, until the SWNT are isolated from the bundles or the bundles are crosslinked, the modulus of composites made from these materials will be limited. Fortunately, progress is being made in obtaining significant volume fractions of exfoliated nanotubes [279].

A final issue of relevance is the waviness of the nanotubes. If they are not straight when placed in the composite, then, as recently reported [301], the modulus of the composite is significantly decreased.

Clays

Plate-like fillers can also increase the modulus of a polymer significantly. A comparison of the upper bound modulus prediction (assuming incompressibility) shows that the modulus for well-aligned platelets can be three times that for well-aligned fibers [302]. In addition, there is direct evidence that nanoclay carries load; for example, as the aspect ratio of the clay layers increases, the modulus increases. Some indirect evidence [303] also shows that the stiffness of the clay layers affects the modulus (e.g., a stiffer filler leads to a higher composite modulus). Neither of these observations is surprising if the clay platelets are carrying load. Other studies, however, suggest that the modulus increase is not entirely due to the load-carrying ability of the platelets, but is caused by the volume of polymer constrained by the platelets [174]. This suggests that, to optimize the increase in modulus, the degree of dispersion must be optimized to maximize the degree of matrix/filler interaction. Work on PP nanocomposites [304], in which adding maleic anhydride (MA) to the matrix changed the degree of filler dispersion, supports this suggestion. Despite the plasticizing effect of MA, the modulus improved, due to enhanced dispersion of the clay. Additional evidence [305] (Figure 2.31) relates the interlayer spacing to the modulus. Lan and Pinnavaia [202] found that, as the degree of exfoliation increased by changing the length of the alkylammonium intercalating chain, the modulus and strength improved. As the polymer intercalates and swells, the layers and the area of interaction between the polymer and the filler increase, and the modulus increases significantly.

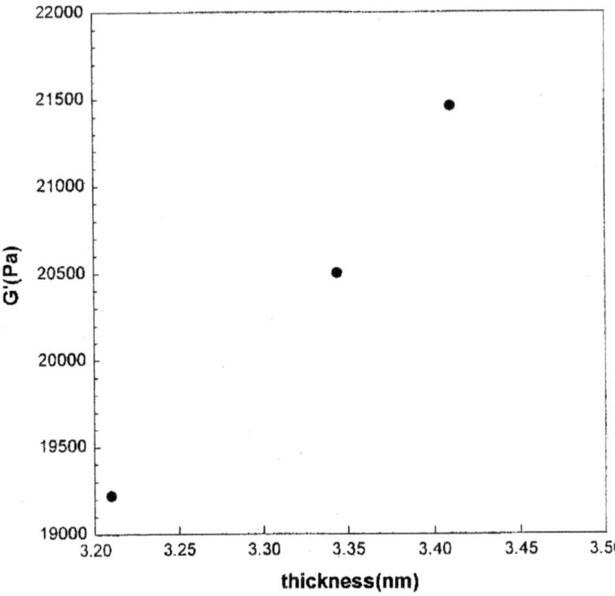

Fig. 2.31 Relationship between storage modulus and interlayer spacing during the melt intercalation process. Reprinted with permission from [305]

The specific role of the clay/polymer interface in controlling the modulus is also becoming clearer. An interesting study on the modulus of clay/Nylon 6 composites examined the modulus as a function of the ionic strength (as measured via ^{15}N-NMR chemical shifts) of the clay [182] (Figure 2.32). Another study determined the types of matrix/filler interactions responsible for modulus changes [306] (Figure 2.33). Using short-chain alkylammonium ions minimized the type-B interactions, allowing more of the polymer (epoxy) to interact directly with the clay surface (type-A interactions). Direct interaction of the matrix with the clay basal planes led to a larger increase in the modulus. It is not clear if the modulus increase caused by the type of interaction and/or the ionic strength is due to an increase in the interfacial shear stress (the load-bearing efficiency of the clays), the ability of the clay to constrain the polymer, or increases in the degree of crystallinity. These effects are all related, and the exact mechanism is still an open question.

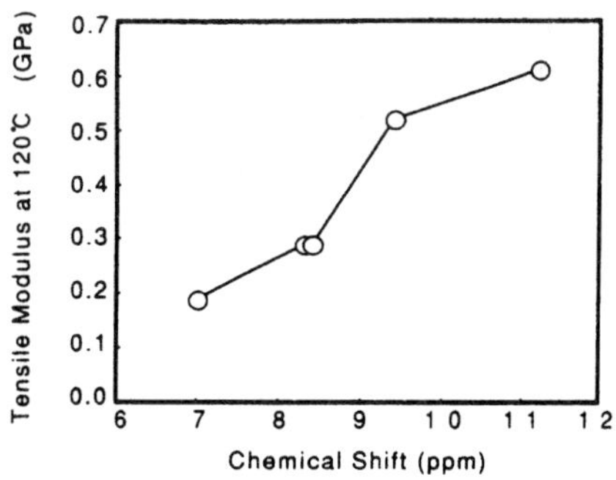

Fig. 2.32 Relationship between strength of polymer/filler interaction as measured by ^{15}N-NMR shift and the tensile modulus at 120 °C of clay/nylon hybrids. Reprinted with permission from [182]

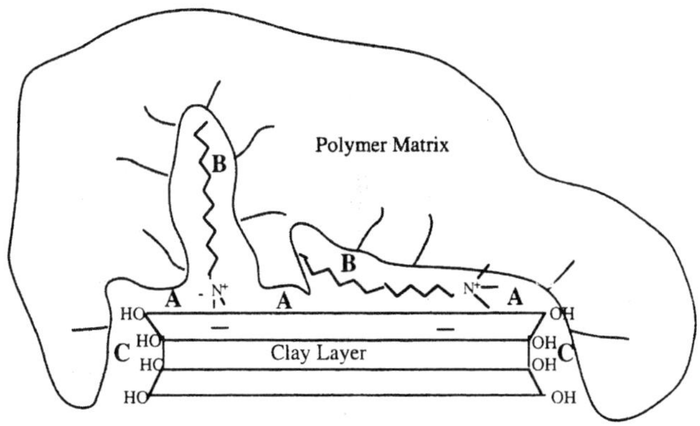

Fig. 2.33 Schematic of the types of interfacial interactions occurring in polymer-organoclay nanocomposites including direct binding (adsorption) of the polymer to the basal siloxane oxygens (type A), "dissolving" of the onium ion chains in the polymer matrix (type B), and polymer binding to hydroxylated edge sites (type C). Reprinted with permission from [206]

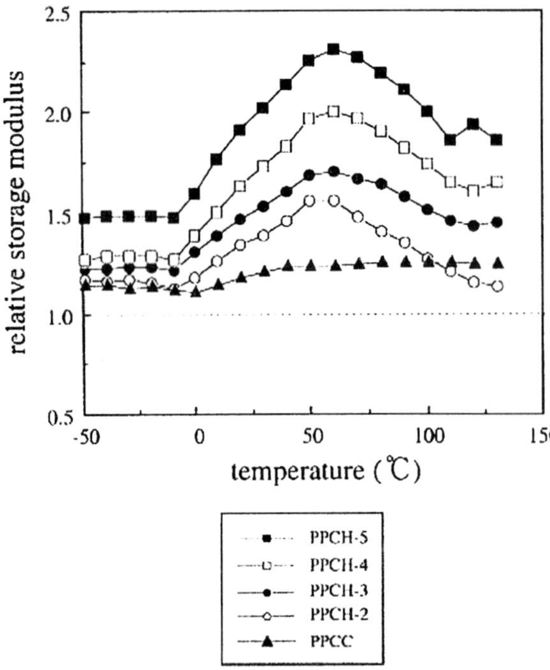

Fig. 2.34 Relative dynamic storage modulus of polypropylene clay hybrids as a function of temperature at filler loadings of 2–5 wt. percent. Reprinted with permission from [304]

Further evidence that the polymer/clay interaction plays a significant role in controlling mechanical behavior is that the improvement in properties tends to be higher above the glass transition temperature than below it [307]. Figure 2.34 shows this explicitly for montmorillonite/modified PP matrix composites. Below Tg, the modulus increases by a factor of 2, and above Tg it increases by a factor of 2.5. This behavior is also seen for epoxy matrices. When a brittle epoxy with a Tg well above room temperature was reinforced with organically modified montmorillonite (OMMT), the increase in modulus at room temperature was modest. In contrast, when an epoxy with a Tg below room temperature was used as the matrix, the modulus (and strength) improved by a factor of 10 at 15wt. % of filler [202]. A strong clay/polymer interaction affords more opportunity for immobilization above Tg than below Tg. In addition, if the surface area for interaction decreases, the modulus decreases. This result was also observed in OMMT/polyimide composites. The modulus increased by 1.4 over that of unfilled polyimide, but at higher volume fractions it started to decrease, presumably due to aggregation of the filler [308].

Equi-axed nanoparticles

The size scale for equi-axed nanoparticle/polymer composites ranges from hybrid nanocomposites, in which the matrix and filler are so intimately mixed that they are no longer truly distinct, to discrete particles in a continuous matrix. We focus here on discrete fillers more than approximately 5 nm in diameter. The matrix significantly influences changes that occur in modulus and strength. For example, in alumina/PMMA composites, the modulus decreased for very weak particle/polymer interactions [127]. In silica/polystyrene composites in which the particle was grafted to a polymer chain, however, the modulus increased [129]. In addition to the reinforcing effect that the particles can contribute, the increase or decrease in modulus is also related to the effect of the particles on the polymer's mobility. Changes in the Tg are strong evidence that polymer mobility is significantly altered. Tg increases for strong polymer/filler interactions and decreases for weak interactions [129]. Given the sensitivity of the polymer modulus to the temperature relative to Tg, changes in Tg can lead directly to changes in modulus. A third mechanism, namely localized yielding near the particles, may also decrease the modulus at very low and seemingly macroscopic elastic strains.

2.5.1.2 **Failure Stress and Strain – Toughness**

The addition of rigid micrometer-scale fillers to a polymer often increases its strength, but decreases the toughness. This tradeoff is a significant technical problem for commercial applications of filled polymers. The reason for the decrease in ductility is straightforward: the fillers or agglomerates act as stress concentrators, and the defects initiated at the filler quickly become larger than the critical crack size that causes failure.

Well-dispersed nanoparticles are much smaller than the critical crack size for polymers and need not initiate failure. Thus, they provide an avenue for simultaneously toughening and strengthening polymers. Proper dispersion is critical for achieving this. For example, Hasegawa et al. [307] studied the dispersion of clays in polypropylene. When they achieved exfoliation, the strain-to-failure ratio remained high (>200 %) at loadings as high as 3 %, but even a small amount of aggregation decreased the strain-to-failure ratio to 5 %–8 %. Another example is polyimide matrices, in which the exfoliation is not complete and the strain-to-failure ratio decreases by 72 % with just 1 % of OMMT [308]. Further addition of OMMT causes further aggregation and a precipitous drop in the strain-to-failure ratio. Many authors report an optimum volume percent of filler and claim that the decrease in strain-to-failure ratio above the optimum is due to agglomeration [147].

Although the small size of nanoparticles, based on the explanation above, should not decrease the strain-to-failure, it also should not cause the observed increases in the strain-to-failure [127]. Clearly, there is a second mechanism operating. The lack of discussion on toughening polymers with rigid particles means we must turn to the literature on rubber toughening of polymers to obtain hints about the second mechanism. An excellent review on rubber toughening [309] provides the background required for this discussion. First, the toughness of a material can be defined as the energy to cause failure. This is related to the deformation in the material, but also to the volume of material undergoing deformation. Even brittle amorphous polymers such as polystyrene and polymethylmethacrylate have the potential for large deformation, but the volume of material undergoing deformation is small. Material must first strain-soften after yielding and eventually strain-harden. If the material does not strain-soften, then defects in the material lead to stress concentrations that cause

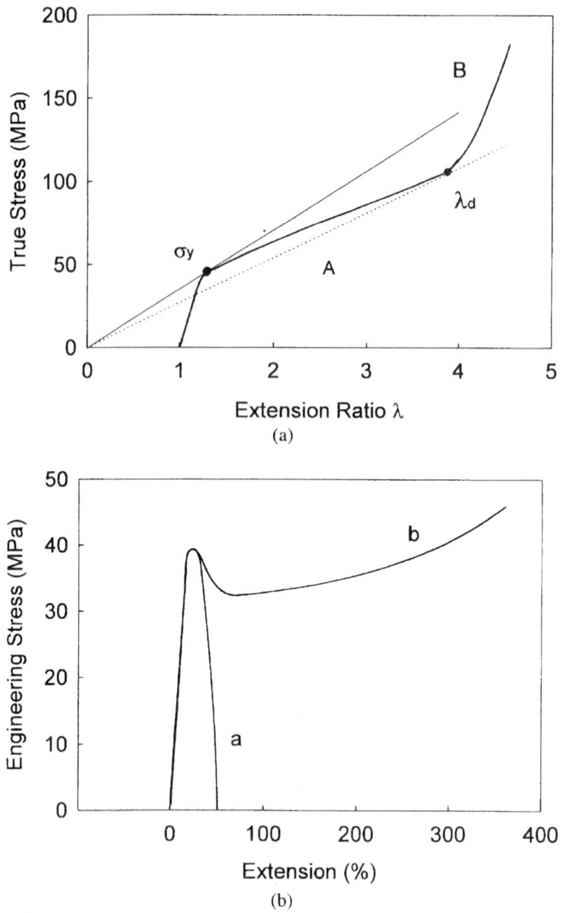

Fig. 2.35 Schematic of Considere's construction. True stress is plotted against extension ratio, and tangents are drawn through the origin. The first tangent defines the strain at which the tensile load reaches a maximum (the engineering yield stress, σ_y). The second tangent marks the point at which strain hardenening stabilizes the neck (the natural draw ratio λ_d). Failure at point A between the two tangents gives a low extension to break (curve a). Failure beyond the second tangent gives curve b. Reprinted with permission from [309]

catastrophic brittle failure. If strain softening occurs prior to strain hardening, then the strain-hardened region draws a larger volume of material into the deformation zone and a large strain-to-failure results. If strain softening occurs, but not strain hardening, the initially strain-softened material propagates the crack. Figure 2.35 [309] illustrates all three scenarios with Considere's construction.

Plastic deformation and strain hardening can occur via two different mechanisms: crazing and shear yielding. Crazing is the result of dilatational strains that cause an increase in volume and subsequent drawing of the material between the voids (fibrils). Thus, strain softening occurs due to dilatational strains in front of a craze tip and at the craze/matrix boundary, and strain hardening occurs, which allows drawing and strain hardening of the material in the craze. In principle, this allows the crazes to thicken and the volume of deformed material to increase significantly. Unfortunately, in many brittle amorphous polymers, failure initiates at the craze/bulk polymer interface and the volume of deformed material is small.

Shear yielding also results in strain softening and strain hardening and can result in stable neck formation and significant toughness. Due to the large change in shape that shear stress causes, a triaxial stress state forms in the center of the sample, which can lead to void formation and crack growth.

In rubber particles having a shear modulus <0.1 that of the matrix, the yield stress decreases due to the stress concentration caused by the soft rubber (which is essentially a void in terms of stiffness). If the particle does not cavitate, then the only effect of the addition is to delocalize the stress and allow a significant volume of material to yield and to limit craze formation. This effect leads to modest improvements in strain-to-failure and toughness and may occur at a critical interparticle spacing [310]. If, on the other hand, the rubber particle cavitates, then the stress state in the matrix between particles is biaxial or uniaxial and can lead to significant drawing (strain hardening) of the matrix. By this mechanism, large volumes of matrix are drawn into the deforma-

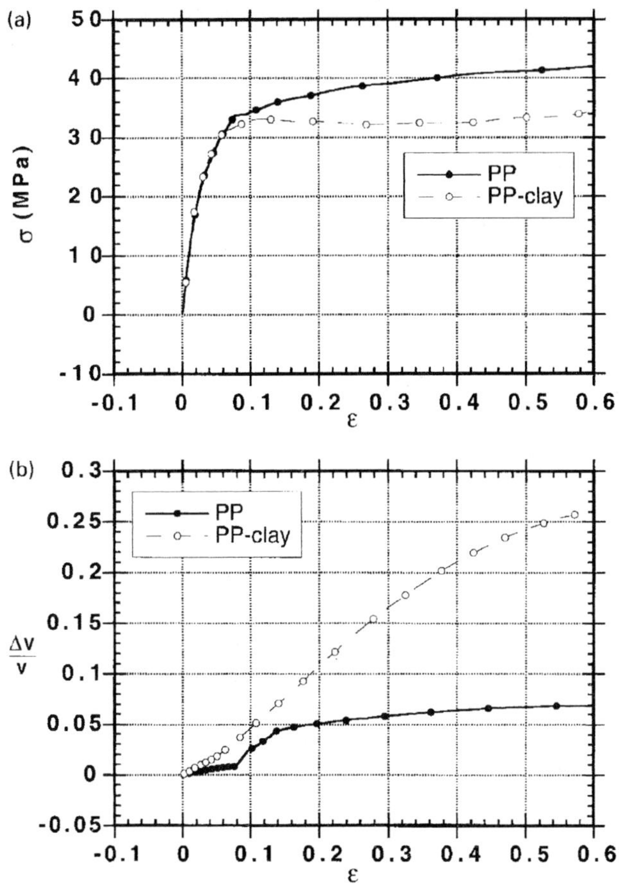

Fig. 2.36 (a) True stress–true strain curves at room temperature for PP-based materials. (b) Corresponding volume strain versus elongation for the PP-based materials. Reprinted with permission from [311]

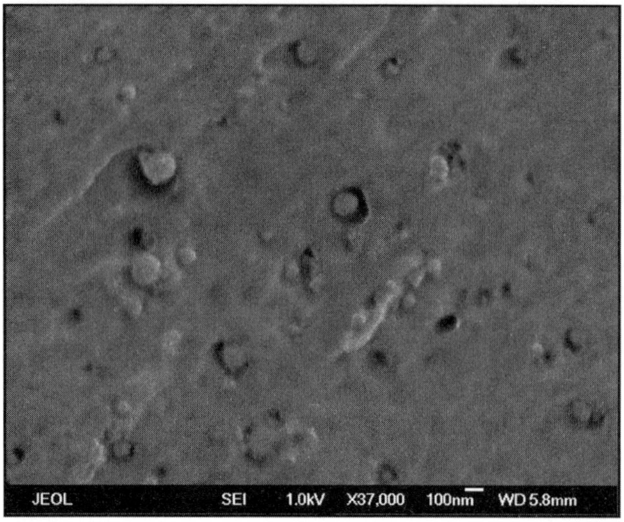

Fig. 2.37 Field-emission electron micrograph showing cavitation surrounding weakly bound 39 nm alumina particles in PMMA. (Thanks to B. Ash for this micrograph)

tion zone and toughening occurs. In addition, the rubber particles strain-harden and thereby strengthen the most highly deformed regions, allowing a large volume of material to participate in the deformation process.

How does this apply to nanoparticle-filled polymers? First of all, if the particle/polymer interaction is weak, then the nanoparticles can essentially act as voids, which initiate yield and allow for shear bands to form and then thicken. Evidence of this mechanism is provided by several studies. Video extensometry was used to monitor the volume strain as well as the true stress-vs.-strain behavior of clay-filled PP and nylon 6 [311]. The results are fascinating. The typical failure mechanism for nylon is shear banding, but the addition of well-dispersed montmorillonite encouraged cavitation. This added another mechanism of deformation and increased the volume of material involved in deformation. Figure 2.36 shows a graph of the volumetric strain, as well as a scanning electronmicrograph of the fibrillation. Interestingly, this result occurred only for weak polymer/clay interactions or for strong interfacial interactions at temperatures above Tg. In addition, crystalline alumina-filled PMMA [127] with weak particle/polymer interactions exhibited an order of magnitude increase in the strain-to-failure. Evidence for cavitation was seen via fractography (Figure 2.37).

Another example in which nanoparticle inclusions lead to significant improvements in strain-to-failure and/or toughness is rubber matrices. Here, even if the interaction between the particle and polymer is strong, the soft material around the particle can easily lead to interface failure. In rubbery epoxy, the strain-to-failure is maintained for exfoliated nanocomposites [202] and can be improved by using magadiite to reinforce relatively ductile epoxy [207]. The addition of OMMT to polyurethane improved the strain-to-failure by a factor of two [208]. This may be due to a plasticizing effect of

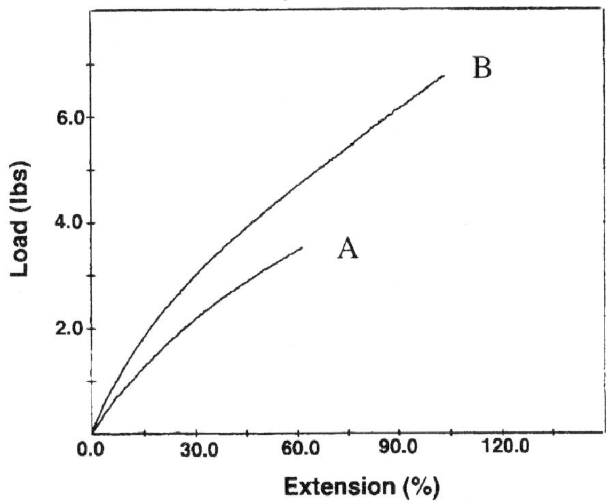

Fig. 2.38 Stress–strain curves for (a) a pristine polyurethane elastomer, and (b) a polyurethane-clay nanocomposite at a strain rate of 1.0 in/min. Reprinted with permission from [208]

the alkylammonium ions, but did not result in a decrease in strength or modulus. Figure 2.38 shows the stress–strain curve for a representative sample. Other investigators have found similar results [312], which can be maximized when the intercalating agent acts as a crosslinking agent.

2.5.1.3 **Glass Transition and Relaxation Behavior**

As discussed in section 2.1, the interfacial region is extremely large in nanocomposites. The interaction of the polymer with the nanoparticles gives significant opportunity for changing the polymer mobility and relaxation dynamics. For example, polystyrene chains intercalated between the layers of a smectic clay have more mobility locally than in the bulk polymer [313]. This greater mobility may be due to an ordering that occurs between the layers, which creates low- and high-density regions, thus providing the opportunity for mobility in the low-density regions [314]. If, however, the polymer is tethered to the clay, the relaxation spectra broaden to include slower relaxation times, suggesting that the clay can reduce the mobility locally. Recent work on silica/polyvinylacetate nanocomposites indicates that, for equi-axed particles, the overall effect of the filler is to create a weak network structure that contains both physical entanglements in the matrix phase and 'trapped' entanglements arising from temporary bonds to the filler particles. This suggests a greater far-field effect of altering chain dynamics than simply a local immobilization of the polymer chain [318]. Another example of altered mobility was found for gold particles dispersed in poly(*tert*-butylacrylate). For particle sizes <10 nm, the viscosity of the low molecular weight polymer at 30 vol. % filler increased by a factor of 4 [315]. The authors claim this result is due to bridging of the polymer chains between the particles. The bridges have relatively long lifetimes and decrease segmental motion at the interface compared to the bulk.

Although the specific mechanics of chain dynamics discussed above are not yet completely understood, it is very clear that the rheology/glass transition temperature of a polymer can be controlled by changing the polymer mobility with nanocomposite interfaces. For example, the glass transition temperature can be eliminated for clay nanocomposites with intercalated polymer chains. This indicates a limited ability for cooperative chain motion when the polymers are confined between the layers [131]. If the clays are exfoliated, and thus the polymer is not confined between layers, the Tg does not change significantly.

Without the help of confinement, the Tg can be increased if the polymer is tightly bound to the nanofiller. Increases of 10° have been observed for clay-filled Nylon, PMMA, and PS [196], as well as for polyimide composites [308]. The behavior is not limited to clays. A Tg increase of almost 30° was observed for exfoliated graphite/polystyrene composites [80], as was a 20° increase for nitrogen-doped carbon nanotube/epoxy composites [264]. The Tg also changes in nanoscale equi-axed particle-filled composites. The addition of $CaCO_3$ to PMMA resulted in a 35° increase Tg at 6 wt. % filler when the composites were prepared via an in-situ process [232]. Smaller increases were observed for silica/PMMA [128] nanocomposites.

The Tg can decrease if the interaction between the filler and the matrix is weak, as has been observed with silica [128] and alumina-filled PMMA [316]. This decrease in Tg was recently given more attention in nanoalumina/PMMA composites, where it was found that the Tg did not decrease until a critical volume fraction was reached. Furthermore, changing the particle surface and making it compatible with the matrix eliminated the Tg depression.

The explanation for the increase or decrease in Tg in the thermoplastics may be related to the thin film results discussed in section 2.3.

Of course, chain dynamics affect much more than the glass transition temperature. Recent studies have attempted to understand the role of filler/polymer interaction, to begin to understand the role of nanofillers in controlling damping, melt rheology, and other dynamic processes [131, 317, 318].

2.5.1.4 **Abrasion and Wear Resistance**

Although it is well known that the abrasion resistance of filled polymers depends on particle size, the incorporation of nanoscale fillers has led to unexpected results. For filler particles that are larger than the abrasive particles, most of the filler particles are stable and increase the abrasion resistance of the composite. As the filler size is decreased to a size similar to that of the abrading particles, filler particles are removed, and the abrasion resistance is compromised. This does not happen on the nanoscale, however. For example, the addition of nanoscale $CaCO_3$ to PMMA resulted in a significant decrease (factor of 2) in material loss due to abrasion at only 3 wt. % of filler [232]. In addition, nanoparticles can simultaneously improve wear resistance and decrease the coefficient of friction. With larger-scale fillers, increased wear resistance is accompanied by an increased coefficient of friction, which has been observed for silica/Nylon [319] as well as for polyetheretherketone matrix composites filled with ZrO_2, SiC, SiO_2, and Si_3N_4 [320–323] and for sheet silicates in polyimide [324]. The coefficient of friction measured as dry sliding against steel decreased monotonically with increasing weight percent of fillers. The authors attributed this finding to an improved and tenacious transfer film that was generated on the counterface, an area that needs further exploration.

2.5.2 Permeability

The reduced gas and liquid permeability of nanofilled polymers makes them attractive membrane materials. The large change in permeability of liquid or gas through a composite material can be explained from simple predictions [325, 326]. Given a plate-like material with an aspect ratio of L/W dispersed parallel in a matrix, Nielson calculated the tortuosity factor τ as

$$\tau = 1 + (L/2W)V_f \tag{1}$$

where V_f is the volume fraction of plate-like filler. The relative permeability coefficient Pc/Pp, where Pc and Pp are the permeability coefficients of the composite and polymer matrix, respectively, is given by

$$Pc/Pp = 1 \ / \ (1 + (L/2W)V_f) \tag{2}$$

A slightly more rigorous analysis for a random in-plane arrangement of plates yields

$$Pc/Pp = 1 \ / \ (1 + \mu(L/W)^2(V_f^2/(1 - V_f) \tag{3}$$

where μ depends on the distribution of the plate-like material. In either case, the higher the aspect ratio of the filler, the larger is the decrease in permeability. Yano et al. [188] compared Eq. 2 to data on clay-filled polyimide (Figure 2.39) and found a good fit. The large reductions in permeability (Figure 2.39) occur at only 2 wt. % for exfoliated composites (Figure 2.40) [8]. The barrier, however, is sensitive to the degree of dispersion and the alignment of the plates. Significant reductions in permeability were also reported for exfoliated clay/PET [327], Nylon [328], and polycaprolactone [7] composites. The reduction is linear with the wt. % of filler for polycaprolactone [7], but for polyimide [8] (Figure 2.40), the decrease in permeability with wt. % of filler is nonlinear.

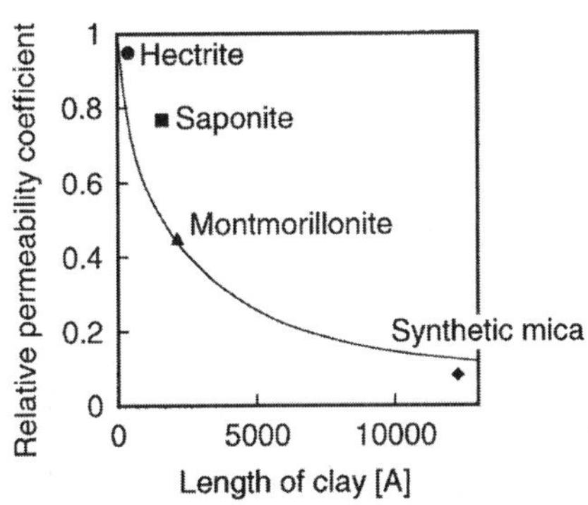

Fig. 2.39 Effect of clay layer size on relative permeability compared with the prediction from Eq. 4.2.2. for clay/polyimide nanocomposites. Reprinted with permission from [188]

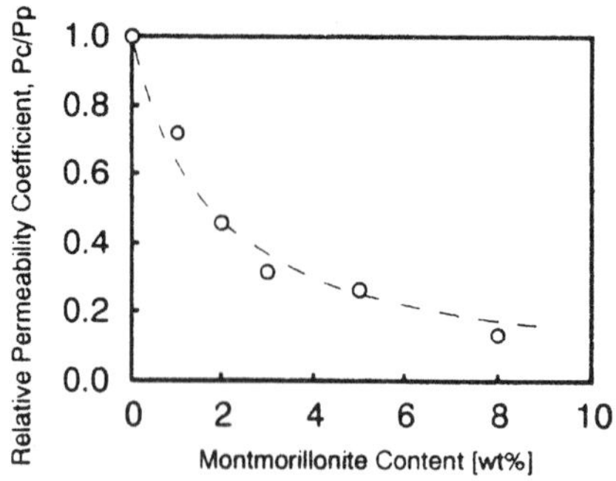

Fig. 2.40 Effect of montmorillonite content on relative permeability coefficient of water vapor in polyimide nanocomposites. Reprinted with permission from [8]

Several nanocomposites are being produced commercially for barrier applications [327]. They include a vermiculite clay/PET composite with oxygen permeation values of ≤ 1 cc m^{-2} d^{-2}, produced by ICI Corporation, which is significantly less than that of pure PET. Ube Industries produces a Nylon/clay hybrid whose permeability is less than that of Nylon by a factor of 2 at 2 wt. % filler.

Apparently, significant improvements in barrier properties are also achievable with nonplate-like nanoparticles [329]. Nano Material Inc. reports that a PVA/EVOH matrix composite with 7 nm silica and titania nanoparticles exhibits a gas permeability of 1 cc m^{-2} d^{-1} atm^{-1} and moisture permeability of less than 1 g m^{-2} d^{-1}. Although this is achieved at very high loadings, the material is melt processable.

The absorption of water into composites is significant. For example, one of the limitations of Nylon is the reduction in mechanical properties that accompanies the absorption of water. The addition of exfoliated montmorillonite increases the resistance to water permeation after 30 min from 2% to 1% at 5 wt. % of filler [330]. The mechanism of the reduction is attributed to the constrained region of the Nylon. If the constrained region is taken into account, the diffusion coefficient follows a rule of mixtures. Figure 2.41 shows the change in diffusion coefficient of water in Nylon in response to clay content.

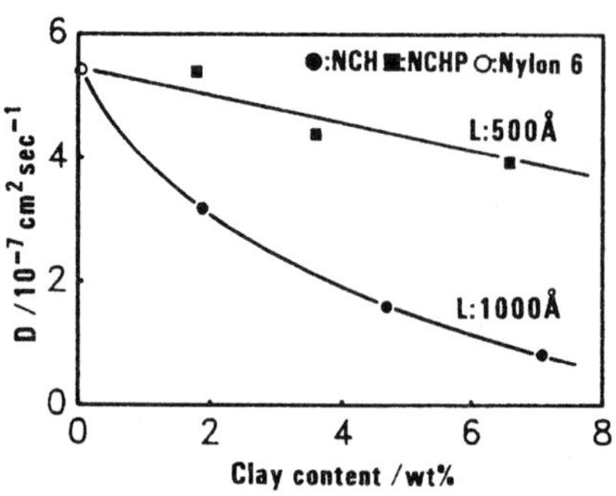

Fig. 2.41 Dependence of diffusion coefficient of water on clay content for montmorillonite with a layer width of 100 nm, and saponite with a layer width of 50 nm. Reprinted with permission from [330]

2.5.3 Dimensional Stability

Dimensional stability is critical in many applications. For example, if the layers of a microelectronic chip have different thermal or environmental dimensional stabilities, then residual stresses can develop that cause premature failure. Poor dimensional stability can also cause warping or other changes in shape that affect the function of a material.

Nanocomposites provide methods for improving both thermal and environmental dimensional stability. The first mechanism by which nanofillers can affect the coefficient of thermal expansion (CTE) of a polymer is also observed in traditional fillers. The composite CTE changes according to the volume average of the two CTEs. For traditional fillers, at low volume fractions of filler, this change in CTE is almost linear. Figure 2.42 shows the CTE for polyimide filled with exfoliated montmorillonite [8]. In contrast to traditional fillers, the dependence is nonlinear. This leads to the second mechanism effecting CTE – the interfacial region. It may be that the nonlinear effect is due to the increase in volume of polymer that is behaving similar to a thin film and has a lower CTE than the bulk matrix. Observations that, in thin films that interact with a substrate, the CTE decreases as the film thickness decreases below 10 nm [118] support this possibility.

Nonlinear dimensional changes are also observed for nanofilled rubbers that swell in the presence of a liquid. This is a well-known phenomenon [331] and is again due to the interfacial region around the particles, as discussed in section 2.3. The effect of this layer increases as the strength of the particle/polymer interaction increases [332]. The behavior is also observed in recently developed nanofilled rubbers. In comparisons of the behavior of montmorillonite/PDMS composites with other conventional aluminosilicates or carbon black [216], the relative swelling of the montmorillonite/PDMS composites was about half that of both unfilled PDMS and carbon black/PDMS. Meanwhile, kaolinite increased the swelling. The authors argued that the decrease in swelling is due to a bound polymer layer and that the large surface area and very strong polymer/filler interaction enhance swelling resistance. Reduced swelling was observed recently in nanoparticle titania/epoxy composites [147].

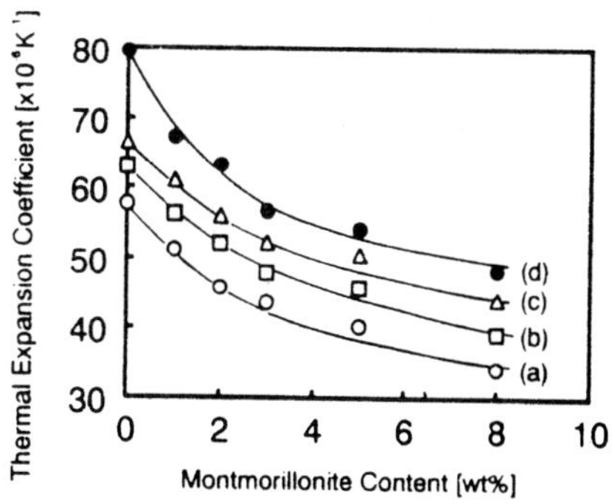

Fig. 2.42 Effect of montmorillonite content on thermal expansion coefficient of polyimide clay hybrids at (a) 150 °C, (b) 200 °C, (c) 250 °C, (d) 295 °C. Reprinted with permission from [8]

A recent paper described the addition of OMMT to the type of PMMA used in dental work [333]. Here, PMMA is added to MMA monomer, and the resulting mixture is polymerized under ultraviolet light. The shrinkage occurs largely during polymerization, which causes internal stresses and reduces the lifetime of the filling. The addition of OMMT to the PMMA/MMA mixture, however, reduces shrinkage considerably and thus lowers the stresses that form. This may provide an excellent method for increasing longevity.

2.5.4
Thermal Stability and Flammability

The first indications that nanoclays could increase the thermal stability of polymers was demonstrated in 1965 for montmorillonite/PMMA composites. The degradation temperature (defined as 50% mass loss) increased about 50 °C for a composite with only 10 wt. % PMMA (90 wt. % filler) [334]. More recently, the degradation temperature of PDMS was found to increase about 140 °C with 10 wt. % montmorillonite for a delaminated nanocomposite [216]. A 50 °C increase was found for intercalated clay/PE composites [143]. The dispersion of the clays is critical to increasing the degradation temperature. Delaminated composites have significantly higher degradation temperatures than intercalated nanocomposites or traditional clay composites [335]. Some speculate that this increase in stability is due to the improved barrier properties of the composites. If oxygen cannot penetrate, then it cannot cause oxidation of the resin [216]. In addition, the inorganic phase can act as a radical sink to prevent polymer

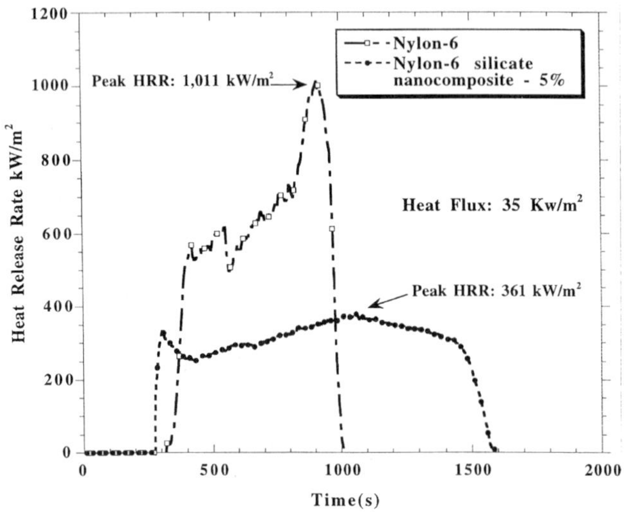

Fig. 2.43 Comparison of the heat release rate (HRR) plot for Nylon-6, Nylon-6 silicate-nanocomposite (mass fraction 5 percent) at 35 kW/m^2 heat flux, showing a 63 percent reduction in HRR for the nanocomposite. Reprinted with permission from [338]

chains from decomposing. The improved thermal stability of some composites may be limited by the lower thermal stability of alkylammonium ions. For example, in intercalated clay/polystyrene composites, the intercalating agent decomposes at about 250 °C. Bonding the intercalating ion to the polystyrene matrix noticeably improved the thermal stability. Polyimide [308] and polymethylmethacrylate [333] also become more thermally stable with the addition of OMMT.

In addition to thermal stability, the flammability properties of many clay/polymer nanocomposites are also improved [336, 337, 338]. Figure 2.43 shows typical plot of the heat release rate (HRR) vs. time for Nylon-6 and a clay/Nylon-6 nanocomposite with only a few wt. % of filler. The HRR increases at a lower temperature for the nanocomposite, but levels off quickly. Clay nanocomposites commonly show a 60 % reduction in peak HRR relative to unfilled polymer (Table 2.2). In addition, the mass loss rate is the same for the two materials until a char forms, and then the nanocomposite exhibits a significantly lower mass loss rate. Observation of the char suggests that the layered silicate collapses during combustion and forms a uniform layered structure (interlayer spacing about 1.3 nm). The layers act to reinforce the char and reduce the permeability of the char, reducing the rate of volatile product release. The structure of the char was fairly uniform for several different matrices and independent of the initial microstructure of the composite (intercalated or delaminated).

Combining traditional flame retardants with intercalated or exfoliated clays can result in further improvements in flame retardance [339, 340]. An additional advantage

Tab 2.2 Cone calorimeter data. Heat flux: 35 kW/m^2, Hc: heat of combustion, peak heat release rate, mass loss rate, and specific extinction area (SEA) data are reproducible to within ±10 %. The carbon monoxide and heat of combustion data are reproducible to within ±15 %. Reprinted with permission from [338].

Sample *(structure)*	**Residue yield (%) ± 0.5**	**Peak HRR (Δ %) (kW/m^2)**	**Mean HRR (Δ %) (kW/m^2)**	**Mean H_c (MJ/kg)**	**Mean SEA (m^2/kg)**	**Mean CO yield (kg/kg)**
Nylon-6	1	1010	603	27	197	0.01
Nylon-6 silicate-nanocomposite 2 % *delaminated*	3	686 (32 %)	390 (35 %)	27	271	0.01
Nylon-6 silicate-nanocomposite 5 % *delaminated*	6	378 (63 %)	304 (50 %)	27	296	0.02
Nylon-12	0	1710	846	40	387	0.02
Nylon-12 silicate-nanocomposite 2 % *delaminated*	2	1060 (38 %)	719 (15 %)	40	435	0.02
PS	0	1120	703	29	1460	0.09
PS silicate-mix 3 % *immiscible*	3	1080	715	29	1840	0.09
PS silicate-nanocomposite 3 % *intercalated*	4	567 (48 %)	444 (38 %)	27	1730	0.08
PS w/DBDPO/Sb_2O_3 30 %	3	491 (56 %)	318 (54 %)	11	2580	0.14
PP	0	1525	536	39	704	0.02
PP silicate-nanocomposite 2 % *intercalated*	5	450 (70 %)	322 (40 %)	44	1028	0.02

of nanoclay composites relative to traditional flame retardants is the improvement in other properties, such as the heat distortion temperature and the bending modulus.

The flammability resistance of clay-filled polymers indicates that their ablation resistance might also be excellent. As a material is heated during ablation, the surface of the material reacts and forms a tough char. If the char is not reinforced, it fails and is removed from the surface, exposing more material [341]. Traditional composites require a significant weight fraction of filler (more than 30 wt. %) to achieve significant ablation resistance. On the other hand, 2–5 wt. % nanoclay-filled Nylon 6 exposed to a mock solid rocket motor firing rig formed a layer of char on the surface that was tough and significantly retarded further erosion [341]. In addition, oxygen plasma forms a passivation layer on Nylon 6/layered silicate nanocomposites, which significantly retards further erosion of the composite surface [342]. This behavior is not a strong function of the organic molecules used to modify the clay or the strength of the clay/polymer interaction [342], but is a function of the degree of exfoliation [341].

2.5.5 Electrical and Optical Properties

The electrical and optical properties of nanofilled polymers are exciting areas of research. This is particularly true because of the possibility of creating composites with unique combinations of functionalities, such as electrically conducting composites with good wear properties that are optically clear. Such properties can result because nanoparticles, with diameters distinctly below the Rayleigh scattering limit, still display their solid-state physical properties when embedded in transparent matrices.

Optical composites have been defined as composites consisting of optically active nanoparticles embedded in a transparent host material, often a polymer [343]. Optical composites take advantage of the optical properties of materials that are hard to grow in single-crystal form or that require protection from the environment and give them the ease of processing afforded many polymers. In addition, sometimes the material must be used at the nanoscale to achieve specific optical properties, and the matrix is used just to hold the particles together and provide processability. For example, high-grade optical composites, with properties otherwise obtainable only in optical glasses, become accessible through the use of polymer molding techniques.

The following sections briefly review some of the literature in this area.

2.5.5.1 Resistivity, Permittivity, and Breakdown Strength

Electrical properties are expected to be different when the fillers get to the nanoscale for several reasons. First, quantum effects begin to become important, because the electrical properties of nanoparticles can change compared to the bulk. Second, as the particle size decreases, the interparticle spacing decreases for the same volume fraction. Therefore, percolation can occur at lower volume fractions. For example, the high aspect ratio of carbon nanotubes or exfoliated graphite leads to percolation at a lower volume fraction and thus increases in electrical conductivity of polymers at a

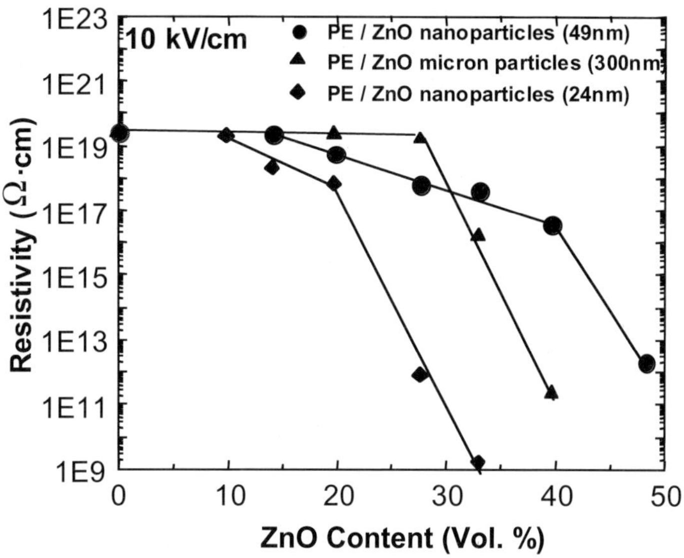

Fig. 2.44 Resistivity vs. volume fraction for nanoscale and micrometer scale ZnO filler in low density polyethylene. (Thanks to J. Hong for this figure)

lower volume fraction than in nonexfoliated graphite [80]. This can be understood from geometric considerations. Figure 2.44 [344] shows an example of this behavior for ZnO-filled low-density polyethylene. In addition, the rate of resistivity decrease is lower than in micrometer-scale fillers. This is probably due to the large interfacial area and high interfacial resistance.

The permittivity of polymers can also be increased with the addition of many metal oxide fillers, on both the micro- and nanoscale. The use of micrometer-scale fillers for this application, however, results in a significant decrease in breakdown strength due the field concentration created by the particles. One of the significant advantages of using nanoscale fillers instead of micrometer-scale fillers may be a reduction in the loss in breakdown strength. The breakdown strength decreases for micrometer-scale fillers because of the field concentration the particles create. Nanoscale fillers do not lead to as large a reduction in breakdown strength. Preliminary results using ZnO/low-density polyethylene composites show that at 50 wt. % of filler, the breakdown strength of the nanofilled LDPE is about 10% higher than that of LDPE filled with submicrometer ZnO [344]. In addition, fundamental to controlling the breakdown strength of insulating polymers is the cohesive energy density [345] and the free volume or mobility of the polymer [346].

Nanotubes can also be used to change the resistivity. Recently, such a behavior was demonstrated in a conjugated luminescent polymer, poly(*m*-phenylenevinylene-co-2,5-dioctoxy-*p*-phenylenevinylene) (PPV), filled with MWNT and SWNT. Nanotube/PPV composites have shown large increases in electrical conductivity compared to the pristine polymer, of nearly 8 orders of magnitude, with little loss in photoluminescence/electroluminescence yield. In addition, the composite is far more robust than the pure polymer with respect to mechanical strength and photo-bleaching properties (breakdown of polymer structure due to thermal buildup). Preliminary studies indicate that the host polymer interacts weakly with the embedded nanotubes but that the nanotubes act as nanometric heat sinks. Recent experimentation with composites of conjugated polymers such as PPV and nanotubes, show a promising new phenomenon. Apparently, the polymer chains wrap around nanotubes suspended in dilute solutions of the polymer. Microscopy suggests that coiled polymer chains can interact strongly with nanotubes and assemble on the nanotubes, leading to the creation of unique interfaces affecting the electronic and optical properties of the polymer [347].

An interesting example of how nanocomposites may affect electronics is the development of photo-patternable hybrid inorganic–organic polymers with negative resist behavior. They are composed of inorganic oxidic structures that are cross-linked or substituted by organic groups. They are synthesized from organosilane precursors reacted by sol–gel processing in combination with organic cross-linking of monomers. The processing is often integrated with micro-patterning technologies to fabricate dielectric and passivation layers in microelectronic systems and devices as well as cladding for optical applications [348].

2.5.5.2 Optical Clarity

A limitation in the development of optical composites or transparent nanocomposites with improved mechanical or electrical performance is minimizing light scattering due to the presence of the particles. The scattering power for light propagation through a collection of scattering particles can be predicted by Rayleigh scattering:

$$P_{scat} - 24\,\pi^4\,P_o\,\rho\,(\,(n' - n)/n^2)\,(V^2/\lambda^4)$$

Where P_o is the incident power, ρ is the concentration of particles, n' is the refractive index of the particles, n is the refractive index of the matrix, V is the volume of a single particle, and λ is the wavelength of light. Therefore, to minimize scattering, the particles must be as small as possible with an index of refraction as close as possible to that of the matrix material. This equation assumes, however, that the particles are much smaller than the wavelength of light, and if this is not true, Rayleigh–Debye calculations are required [349].

Relatively good optical clarity has been obtained in many nanocomposites, particularly at low volume fractions. Studies on PMMA modified with n-dodecylmethacrylate during polymerization were shown to maintain optical clarity up to 10 wt. % of bentonite [199]. In addition, when exfoliation was achieved in polyimide matrix composites, the 2 wt. % hybrid was relatively transparent [8] compared to nonexfoliated clay. A comparison between smectic clays and magadiite (a layered silicic acid) shows that magadiite results in much better optical clarity than smectic clay, probably due to index of refraction matching [203]. Optically clear nanofilled polyurethanes have also been prepared [208]. The addition of UV-absorbing nanoparticles have been used to create a clear coating on polymers to prevent degradation Figure 2.45 shows a quantitative

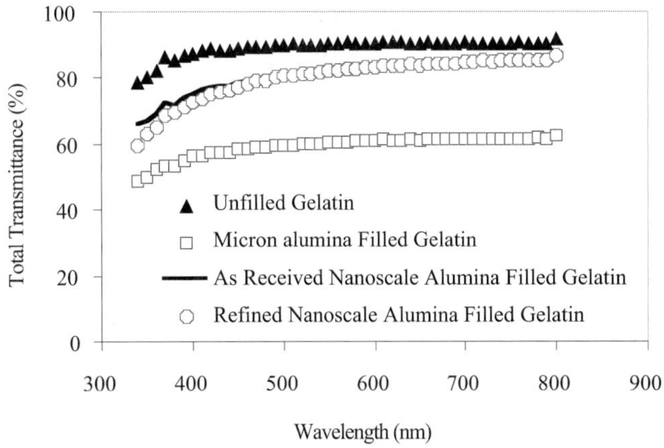

Fig. 2.45 Optical transmittance vs. wavelength for a nanoscale alumina-filled gelatin at 17 wt. percent filler. (Thanks to G. Irvin, Q. Chen for this figure [350])

example of the transmission achieved in nanoscale alumina/gelatin nanocomposites as a function of wavelength [350].

2.5.5.3 Refractive Index Control

Preparation of clear nanocomposites with refractive indices over the entire range of <1 to >3.9 is possible, and these values are by far the lowest and highest ever achieved for any polymer system. An excellent example of the use of nanoparticles to obtain a specific index of refraction was presented by Zimmerman et al. [351]. They found that the index of refraction scaled closely with the volume fraction of lead sulfide nanoparticles in a gelatin matrix. Indices of refraction as high as 3.9 were obtained [352]. The addition of nanoscale iron sulfide to polyethylene increased the index of refraction to between 2.5 and 2.8 [353].

2.5.5.4 Light-Emitting Devices

After the discovery of electroluminescence from conjugated polymer materials (such as PPV), polymer-based light emitting diodes (LEDs) have attracted much attention. The practical advantages for polymer-based LEDs are low cost, low operating voltage, ease of fabrication, and flexibility. Organic light emitting diodes (OLEDs), however, suffer from low quantum efficiency and low environmental stability (against oxidation, etc.).

Functionalization of nanotubes and doping of chemically modified nanotubes in low concentration into photoactive polymers such as PPV can affect the hole-transport mechanism and hence the optical emission of the polymer. Small loadings of nanotubes are used in these polymer systems to tune the color of emitted light from organic LEDs. Although there have been some initial attempts at quantification of nanotube/conducting polymer interactions, little has been done to understand how excitonic transport in a polymer might be modified by the presence of a high-aspect 1D metal such as a carbon nanotube. Recently, Ago et al. [354] proposed 'hole collecting' properties of multi-wall carbon nanotubes from a conjugated polymer at the composite interfacial region. Such hole collecting properties seem to be common in some conjugated polymers. Specifically, SWNTs in a polymer matrix trap holes injected from the anode in OLEDs [355].

The addition of layered silicates may combat the low quantum efficiency and poor stability to oxygen and moisture of OLEDs [356]. Devices made from organically modified clay and (poly[2-methoxy-5-(2'-ethyl-hexyloxy)-1,4-pheneylenevinylene]) (MEH-PEV) posses an external quantum efficiency 100 times that of the pure polymer (~0.38 % photons/electron) [357]. The composite also shows increased photoluminescence efficiency, due to increased excitation within the two-phase composite.

The addition of equi-axed nanoparticles to electroluminescent polymers also increases current densities, radiances, and power efficiencies. When nanoscale titania and silica were added to MEH-PPV, the current density and light output increased by an order of magnitude at low driving voltages [358]. Lifetime was sacrificed slightly. The explanation for the improvement at low voltages was that the nanoparticles created a high surface area contact with the cathode and thus improved charge injection into the polymer. The improvement at high voltages may be due to the thin regions of MEH-PPV that exist between the nanoparticles. This creates regions of higher current density and thus higher light output. Nanoparticles can also lead to changes in the color of emitters. For example, CdSe/polymer nanocomposites emit blue light [359].

2.5.5.5 Other Optical Activity

The dispersion of metal particles into liquids or polymers leads to interesting absorption spectra that can be controlled by changing the metal used, the particle size, degree of dispersion, and the polydispersity of the particle size [360]. This has provided an impetus to look for new applications of nanocomposites in areas such as optoelectronic devices, nonlinear optical devices, and color filters. In these applications, the metal particle and/or the polymer matrix can be active optical components.

If the particle is the optically functioning component, then the optical properties of the composite material can be tuned by adjusting the particle size and the average separation between particles. The polymer in these situations plays the role of stabilizing the nanoparticles. It is often possible to create ordered arrays (2D or 3D) of nanoparticles in the polymer matrix. This involves size-selective deposition of nanoparticles, as well as surface modification of the particles to prevent agglomeration during processing. An example of a transparent, mechanically stable nanocomposite film is a continuous periodic array of self-assembled thiol-passivated Ag nanocrystals in a polystyrene matrix. The film is formed at an air–water interface by the Langmuir–Blodgett technique. The spatial distribution of particles in the film is altered by changing the thickness of the film. The films so produced have applications in optical filters [361].

The addition of relatively monodisperse silver particles to polyethylene, followed by drawing them into films has led to anisotropic absorption of light and a color that depends on the polarization of incident light [362]. In addition, annealing the composite led to further changes in color, giving and added dimension of control. The silver particles formed pearl-necklace type strings upon drawing, which gave rise to anisotropy. The films were easy to process, strong, and flexible.

Interesting nonlinear optical activity has also been obtained with nanoparticle-filled polymers. For example, by using Cr:forsterite (Cr:Mg_2SiO) in a polymer with a matched index of refraction, relative optical amplification was achieved [343]. Unfortunately, large particles, cracks in the films, and poor coupling to the electronics limited the gain obtained. The authors suggested that cubic materials with isotropic optical properties might be most appropriate for such applications.

The optical limiting behavior of polymer nanocomposites has been investigated, a good example being nanotube filled polymer composites [363]. Polymerization of

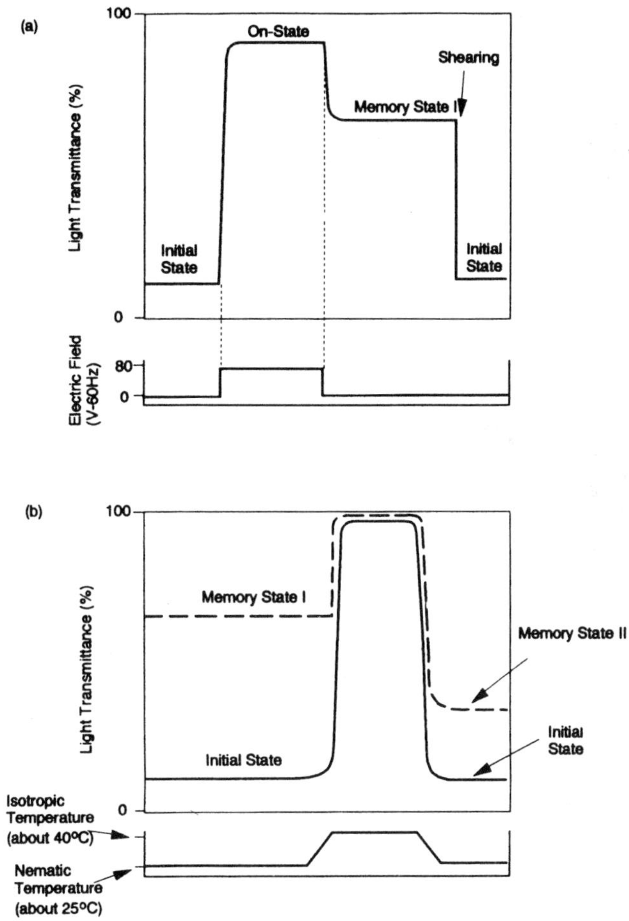

Fig. 2.46 Typical change in transmittance of clay-filled nematic liquid crystals (cell gap 12 μm) due to various fields: (a) electric field and shearing liquid at liquid crystalline temperature (about 25 °C), (b) thermal change. Reprinted with permission from [193]

phenylacetylene in the presence of short nanotubes yields a solvated product of tubules wrapped with poly(phenylacetylene). The optical limiting properties of this composite are excellent and far better than those of the pristine polymer and of polymer filled with other carbon structures. These composites also fare well in photovoltaic applications, suggesting that such nanocomposites could find an array of potential applications in optics-related and laser-based technologies [363].

Liquid crystalline materials have an interesting optical behavior that changes with electric field and with temperature. The addition of intercalated smectic clay to nematic liquid crystals added a memory effect to their behavior [193, 194]. Upon imposition of an electric field, the initially opaque composite became transparent. When the electric field was removed, the composite remained transparent (with a slight decrease in transparency). When heated so as to return to the initial state, the composite instead went into a new state (Memory State II) with light transmittance between that of the initial state and the first memory state (Figure 2.46).

2.6 Summary

The primary purpose of this chapter was to introduce common types of fillers used in nanocomposites, to emphasize the role of the interfacial region, to provide a basic level of understanding of how nanocomposites are currently processed, and to excite the reader about the multifunctional and as yet incompletely explored properties of nanofilled polymers.

Today, our ability to control the structures and properties of nanocomposites is limited only by our current understanding of how to manipulate these nanoscale structures. This area of research will clearly lead to fruitful commercial applications, with significant economic effect, driven by materials with new combinations of properties.

Acknowledgements

Sincere thanks go to the funding agencies that gave me the opportunity to enter this field, including the National Science Foundation, the Office of Naval Research, the Army Natick facility, Eastman Kodak Inc., General Electric, ABB, and Albany International. In addition, sincere thanks to several people for helpful discussions and reviews as well as for contribution of figures: Benjamin J. Ash, Elena Petrovicova, Sanford S. Sternstein, Brian C. Benicewicz, Richard W. Siegel, Keith Nelson, Jung Il Hong, Tao Li, and Quan Chen. In addition, I would like to thank several editors, including Kathy Westbrook, Thomas Feist, Margaret Schadler, and Harvey Schadler.

References

1 A. M. Bueche, J. Polym. Sci., 25, 139 (1957).
2 B. Kuriakose, S. K. De, S. S. Bhagawan, R. Sivaramkrishnan, S. K. Athithan, J. Appl. Polym. Sci., 32, 5509 (1986).
3 M. Sumita, T. Shizuma, K. Miyasaka, K. J. Shikawa, Macromol. Sci., Phys., B22, 601 (1983).
4 P. C. LeBaron, A. Wang, T. J. Pinnavaia, Appl. Clay Sci., 15, 11 (1999).
5 M. Sumita, Y. Tsukumo, K. Miyasaka, K. Ishikawa, J. Mater. Sci., 18, 1758 (1983).
6 P. B. Messersmith, E. P. Giannelis, Chem. Mater., 6, 1719 (1994).
7 P. B. Messersmith, E. P. Giannelis, J. Polym. Sci., Part A: Polym. Chem., 33, 1047 (1995).
8 K. Yano, A. Usuki, A. Okada, T. Kuraychi, O. Kamigaito, J. Polym. Sci., Part A: Polym Chem., 31, 2493 (1993).
9 S. Iijima, Nature, 354, 56 (1991).
10 D. T. Colbert, R. E. Smalley, *Past, Present, and Future of Fullerene Nanotubes: Buckytubes*, Proc. Int. Fullerenes Workshop, Tokyo Japan, February 20–21, 2001, Kluwer, Dordrecht, Netherlands.
11 L. T. Drzal, M. J. Rich, M. F. Koenig, J. Adhes., 16, 133 (1983).
12 R. Bacon, J. Appl. Phys., 31, 283 (1960).
13 T. W. Ebbesen, *Carbon Nanotubes: Preparation and Properties*, CRC Press: Boca Raton, FL, USA, 1997; M. S. Dresselhaus, G. Dresselhaus, P. C. Eklund, *Science of Fullerenes and Carbon Nanotubes*; Academic Press: New York, NY, USA, 1996.
14 P. M. Ajayan, T. W. Ebbesen, Rep. Prog. Phys., 60, 1025 (1997).
15 S. Iijima, T. Ichihashi, Nature, 363, 605 (1993).
16 D. S. Bethune, C. H. Kiang, M. S. de Vires, G. Gorman, R. Savoy, J. Vazquez, R. Beyers, Nature, 363, 605 (1993).
17 J. W. Mintmire, B. I. Dunlap, C. T. Carter, Phys. Rev. Lett., 68, 631 (1992).
18 N. Hamada, S. Sawada, A. Oshiyama, Phys. Rev. Lett., 68, 1579 (1992).
19 D. L. Carroll, P. Redlich, P. M. Ajayan, J.-C.Charlier, X. Blase, A. De Vita, R. Car, Phys. Rev. Lett., 78, 2811 (1997).
20 H. W. Kroto, J. R. Heath, S. C. O'Brien, S. C. Curl, R. E. Smalley, Nature, 318, 162 (1985).
21 M. S. Dresselhaus, G. Dresselhaus, K. Sugihara, I. L. Spain, H. A. Goldberg, *Graphite Fibers and Filaments*, Springer-Verlag, New York, NY, USA, 1988.
22 D. Qian, G. J. Wanger, W. K. Liu, Appl. Mech. Rev., 55, 495 (2002).
23 J.-P. Salvetat, G. A. D. Briggs, J.-M. Bonard, R. R. Bacsa, A. J. Kulik, T. Stockli, N. A. Burnham, L. Forro, Phys. Rev. Lett., 82, 944 (1999).
24 D. A. Walter, L. M. Ericson, M. J. Casavant, J. Liu, D. T. Colbert, K. A. Smith, R. E. Smalley, Appl. Phys. Lett., 74, 3803 (1999).
25 J. R. Wood, M. D. Frogley, E. R. Meurs, A. D. Prins, T. Peijs, D. J. Dunstant, H. D. Wagner, J. Phys. Chem., 103, 10388 (1999).
26 C. M. Leech, Int. J. Mechan. Sci., 44, 621 (2002).
27 M.-F. Yu, B. S. Files, A. Sivaram, R. S. Rodney, Phys. Rev. Lett., 84, 5552 (2000).
28 M.-F. Yu, O. Lourie, M. J. Dyer, K. Moloni, T. F. Kelly, R. S. Ruoff, Science, 287, 637 (2000).
29 H. D. Wagner, O. Lourie, Y. Feldman, R. Tenne, Appl. Phys. Lett., 72, 188 (1998).
30 C. Bower, R. Rosen, L. Jin, J. Han, O. Zhou, Appl. Phys. Lett., 74, 3317 (1999).
31 B. I. Yakabson, C. J. Brabec, J. Bernholc, J. Comput.-Aided Mater. Des., 3, 173 (1996).
32 B. I. Yakabson, Appl. Phys. Lett., 72, 918 (1998).
33 P. M. Ajayan, T. W. Ebbesen, Rep. Prog. Phys., 60, 1025 (1997).
34 J. W. G. Wildoer, L. C. Venema, A. G. Rinzler, R. E. Smalley, C. Dekker, Nature, 391, 59 (1998).
35 M. S. Dresselhaus, G. Dresselhaus, P. C. Eklund, *Science of Fullerenes and Carbon Nanotubes*; Academic Press, New York, NY, USA, 1996.
36 T. W. Ebbesen, *Carbon Nanotubes: Preparation and Properties*, CRC Press, Boca Raton, FL, USA, 1997.
37 R. Andrews, D. Jacques, A. M. Rao, F. Derbyshire, D. X. Qian, E. C. Dickey, J. Chen, Chem. Phys. Lett., 303, 467 (1999).
38 M. J. Bronikowski, P. A. Willis, D. T. Colbert, K. A. Smith, R. E. Smalley, J. Vac. Sci. Technol., 19, 1800 (2001).
39 P. Nikolaev, M. J. Bronikowski, R. K. Bradley, F. Rohmund, D. T. Colbert, A.

Smith, R. E. Smalley, Chem. Phys. Lett. 313, 91 (1999).
40 T. W. Ebbesen, P. M. Ajayan, Nature, 358, 220 (1992).
41 W. Krätschmer, L. D. Lamb, K. Fostiropoulos, D. R. Huffman, Nature, 347, 354 (1990).
42 E. G. Gamaly, T. W. Ebbesen, Phys. Rev. B, 52, 2083 (1995).
43 C. N. R. Rao, R. Sen, B. C. Satishkumar, A. Govindaraj, Chem. Commun., 15, 1525 (1998).
44 C. Niu, E. K. Sichel, R. Hoch, D. Moy, D. H. Tennet, Appl. Phys. Lett. 7, 1480 (1997).
45 Z. J. Zhang, B. Q. Wei, G. Ramanath, P. M. Ajayan, Appl. Phys. Lett., 77, 3764 (2000).
46 W. Z. Li, S. S. Xie, L. X. Qian, B. H. Chang, B. S. Zou, W. Y. Zhou, R. A. Zhao, G. Wang, Science, 274, 1701 (1996).
47 G. Che, B. B. Lakshmi, E. R. Fisher, C. R. Martin, Nature, 393, 346 (1998).
48 C. Journet, W. K. Maser, P. Bernier, A. Loiseau, M. Lamy de la Chappelle, S. Lefrant, P. Deniard, R. Lee, J. E. Fischer, Nature, 388, 756 (1997).
49 A. Thess, R. Lee, P. Nikolaev, H. Dai, P. Petit, J. Robert, C. Xu, Y. H. Lee, S. G. Kim, A. G. Rinzler, D. T. Colbert, G. E. Scuseria, D. Tomanek, J. E. Fischer, R. E. Smalley, Science, 273, 483 (1996).
50 H. Ago, T. Kugler, F. Cacialli, W. R. Salaneck, M. S. P. Shaffer, A. H. Windler, R. H. Friend, J. Phys. Chem. B, 103, 8116 (1999).
51 K. Tohji, H. Takahashi, Y. Shinoda, N. Shimizu, B. Jeyadevan, I. Matsuoka, Y. Saito, A. Kasuya, S. Ito, Y. Nishina, J. Phys. Chem. B, 101, 1974 (1997).
52 G. S. Duesberg, M. Burghard, J. Muster, G. Phillip, S. Roth, Chem. Commun. 3, 435 (1998).
53 S. Curran, A. P. Davey, J. Coleman, A. Dalton, B. McCarthy, S. Maier, A. Drury, D. Gray, M. Brennan, K. Ryder, M. Lamy de la Chapelle, C. Journet, P. Bernier, H. J. Byrne, D. Carrol, P. M. Ajayan, S. Lefrant, W. Blau, Synth. Met., 103, 2559 (1999).
54 A. Eitan, personal communication.
55 T. W. Ebbesen, P. M. Ajayan, H. Hiura, K. Tanigaki, K., Nature, 367, 519 (1993).
56 S. C. Tsang, P. J. F. Harris, M. L. H. Green, Nature, 362, 520 (1993).
57 M. Terrones, N. Grobert, J. Olivares, J. P. Zhang, H. Terrones, K. Kordatos, W. K. Hsu, J. P. Hare, P. D. Townshend, K. Prassides, A. K. Cheetham, H. W. Kroto, Nature, 388, 52 (1997).
58 C. Niu, E. K. Sichel, R. Hoch, D. Moy, H. Tennet, Appl. Phys. Lett. 7, 1480 (1997).
59 J. Liu, A. G. Rinzler, H. Dai, J. H. Hafner, R. K. Bradley, P. J. Boul, A. Lu, T. Iverson, K. Shelimov, C. B. Huffman, F. Rodriguez-Macias, Y. Shon, T. R. Lee, D. T. Colbert, R. E. Smalley, Science, 280, 1253 (1998).
60 R. Tenne, A. K. Zettl, pp. 81 – 109 in *Carbon Nanotubes: Synthesis, Structure, Properties and Applications*, ed M. S. Dresselhaus, G. Dresselhaus, P. Avouris, Springer, New York, NY, USA, 2000.
61 O. Stephan, P. M. Ajayan, C. Colliex, P. Redlich, J. M. Lambert, P. Bernier, P. Lefin, Science, 266, 1683 (1994).
62 X. Blase, J.-C. Charlier, A. de Vita, R. Car, P. Redlich, M. Terrones, W. K. Hsu, H. Terrones, D. L. Carroll, P. M. Ajayan, Phys. Rev.Lett., 83, 5078 (1999).
63 M. Terrones, H. Terrones, N. Grobert, W. K. Hsu, Y. Q. Zhu, H. W. Kroto, D. R. M. Walton, P. Kohler-Redlich, M. Rühle, J. P. Zhang, A. K. Cheetham, Appl. Phys. Lett., 75, 3932 (1999).
64 R. Tenne, Adv. Mater., 7, 965 (1995).
65 P. M. Ajayan, O. Stephan, P. Redlich, C. Colliex, Nature, 375, 564 (1995).
66 G. John, M. Masuda, T. Shimizu, Adv. Mater., 13 (10), 715 (2001).
67 R. A. Vaia, E. P. Giannelis, MRS Bull., 26, 394 (2001).
68 R. E. Grim, *Clay Mineralogy*, 2nd edit., p. 362, McGraw Hill, New York, NY, USA, 1968.
69 Y. Fukushima, S. Inagaki, J. Inclusion Phenom., 5, 473 (1987).
70 A.Okada, A. Usuki, T. Kurauchi, O. Kamigaito, Chapter 6 in *Hybrid Organic – Inorganic Composites*, ed. J. E. Marck, C. Y.-C. Lee, P. A. Bianconi, ACS Symp. series 585.
71 Z. Wang, T. J. Pinnavaia, Chem. Mater., 10, 1820 (1998).
72 W. Schwieger, D. Heidemann, K. H. Bergk, Rev. Chim. Miner., 12, 639 (1985).
73 W. T. Riechle, J. Catal., 94, 547 (1985).
74 P. B. Messersmith, S. I. Stupp, J. Mater. Res., 7, 2599 (1992).

75 C. O. Oriakhi, I. V. Farr, M. Lerner, J. Mater. Chem., 6, 103 (1996).
76 D. Yang, P. Westreich, R. F. Frindt, Nanostruct. Mater., 12, 467 (1999).
77 V. Sanchez, E. Benaventa, M. Angelica Santa Ana, G. Gonzalez, Chem Mater., 11, 2296 (1999).
78 F. R. Gamble, J. H. Osiecki, M. Cais, R. Pisharody, F. J. DiSalvo, T. H. Geballe, Science, 174, 4939 (1971).
79 L. Hernan, P. Lavela, J. Morales, L. Sanchez, J. L. Tirado, J. Mater. Chem. 6, 861 (1996).
80 P. Xiao, M. Xiao, K. Gong, Polymer, 42, 4813 (2001).
81 W. Stober, A. Fink, E. Bohn, J. Colloid Interface Sci., 26, 62 (1968).
82 Dupont Commercial Process, E.I. du Pont de Nemours, Wilmington, DE, USA.
83 P. A. Buining, L. M. Liz-Marzan, A. P. Philipse, J. Colloid and Interface Sci., 179, 318 (1996).
84 M. K. Wu, R. S. Winderler, C. K. R. Steiner, T. Bors, S. K. Friedlander, Aerosol Sci. Technol., 19, 527 (1993).
85 S. K. Friedlander, *Smoke, Dust and Haze: Fundamentals of Aerosol Behavior,* Wiley Interscience, New York, NY, USA. (1977).
86 J. C. M. Marinjnessen, S. Pratsinis, *Synthesis and Measurement of Ultrafine Particles: Proceedings of the International Workshop on the Synthesis and Measurement of Ultrafine Particles,* Delft University Press, Delft, Netherlands (1993).
87 Proc. Joint NSF–NIST Conf. Nanoparticles, May 12–13, 1997, Arlington, VA, USA.
88 S. K. Friedlander, "Synthesis of nanoparticles and their agglomerates," in *R&D Status and Trends,* ed. R. W. Siegel, E. Hu, M. C. Roco, Kluwer, Dordrecht, Netherlands (1998).
89 K. Kimoto, Y. Kamiya, M. Nonoyama, R. Uyeda, Jpn. J. Appl. Phys. 2, 702 (1963).
90 C. G. Granqvist, R. A. Buhrman, J. Appl. Phys., 47, 2200 (1976).
91 Nanophase Technologies Corp., http://www.nanophase.com/
92 H. Gleiter, Prog. Mater. Sci., 33, 223 (1989).
93 R. P. Andres, R. A. Averback, W. L. Brown, L. E. Brus, W. A. Goddard III, A. Kaldor, S. G. Louie, M. Moskovits, P. S. Peercy, S. J. Riley, R. W. Siegel, F. Spaepen, Y. Wang, J. Mater. Res. 4, 704 (1989).
94 R. W. Siegel, Mater. Sci. Technol., 15, 583 (1991).
95 R. W. Siegel, in *Encyclopedia of Applied Physics,* vol 11, ed. G. L. Trigg, VCH, Weinheim, Germany, pp. 173–200 (1994).
96 M. Kato, Jpn. J. Appl. Phys., part 1, 15, 757 (1976).
97 K. Ogawa, T. Vogt, M. Ullman, S. Johnson, S. K. Friedlander, J. Appl. Phys., 87, 63 (2000).
98 G. P. Johnston, R. Muenchausen, D. M. Smith, W. Fahrenholtz, J. Am. Ceram. Soc., 75, 3293 (1992).
99 T. Sasaki, S. Terauchi, N. Koshizaki, H. Umehara, Appl. Surf. Sci., 127, 398 (1998).
100 R. Whitlock, G. Frick, J. Mater. Res., 9, 2868 (1994).
101 E. H. Hu, D. T. Shaw, "Synthesis and Assembly," in *WTEC Panel Report on Nanostructure Science and Technology: R&D Status and Trends in Nanoparticles, Nanostructured Materials, and Nanodevices,* ed. R. W. Siegel, E. Hu, M. C. Roco (1998).
102 J. Hyeon-Lee, G. Beaucage, S. E. Pratsinis, Chem. Mater., 9, 2400 (1997).
103 M. Kishida, T. Fujita, K. Umakoshi, H. Ishiyama, H. Nagata, K. Wakabayashi, Chem. Commun., 763 (1995).
104 F. Grohn, B. J. Bauer, Y. A. Akpalu, C. L. Jackson, E. J. Amis, Macromolecules, 33, 6042 (2000).
105 M. Benaissa, K. E. Gonsalves, S. P. Rangarajan, Appl. Phys. Lett., 71, 3685 (1997).
106 M. P. Pileni, L. Motte, C. Petit, Chem. Mater., 4, 338 (1992).
107 M. Olshavsky, H. R. Allcock, Chem. Mater., 9, 1367 (1997).
108 V. Pillai, P. Kumar, J. J. Hou, P. Ayyub, D. O. Shah, Adv. Colloid Interface Sci., 55, 231 (1995).
109 L. Spahnhel, M. A. Anderson, J. Am. Chem. Soc., 113, 2826 (1991).
110 L. Guo, S. Yang, C. Yang, P. Yu, J. Wang, W. Ge, G. K. L. Wong, Appl. Phys. Lett., 76, 2901 (2000).
111 K. S. Suslick, T. Hyeon, M. Fang, Chem. Mater., 8, 2172 (1996).
112 J. E. Sunstrom IV, W. R. Moser, B. Marshik-Guerts, Chem. Mater., 8, 2061 (1996).
113 J. Mattsson, J. A. Forrest, L. Börjesson, Phys. Rev. E, 62, 5187 (2000).
114 J. L. Keddie, R. A. L. Jones, R. A. Covy, Europhys. Lett., 27, 59 (1994).

115 G. Reiter, Europhys. Lett., 23, 579 (1993).
116 K J. A. Forrest, K. Dalnoki-Veress., J. R. Stevens, J. R. Ducher, Phys. Rev. Lett., 77, 2002 (1996).
117 J. L. Keddie, R. A. L. Jones, R. A. Cory, Faraday Discuss. Chem. Soc. 98, 219 (1994).
118 J. H. van Zanten, W. E. Wallace, W.-L. Wu, Phys. Rev. E., 53 R2053 (1996).
119 K. F. Mansfield, D. N. Theodorou, Macromolecules, 24, 6283, (1991).
120 J. Baschnagel, K. Binder, MRS Symp. Proc., 543, 157 (1999).
121 K. Kendall, F. R. Sherliker, Br. Polym. J., 12, 85 (1980).
122 M. Y. Boluk, H. P. Schreiber, Polym. Compos., 7(5), 295 (1986).
123 W. Hergeth, U. Steinau, H. Bittrich, G. Simon, K. Schmutlzer, Polymer, 30, 254 (1989).
124 G. Tsagaropoulos, A. Eisenberg, Macromolecules, 28, 6067 (1995).
125 X. Zheng, M. H. Rafailovich, J. Sokolov, Y. Strzhemechny, S. A. Schwarz, B. B. Sauer, M. Rubinstein, Phys. Rev. Lett., 79, 241 (1997).
126 J. Zhu, S. S. Sternstein, MRS Symp. Proc., 661, KK4.3.1 (2001).
127 B. J. Ash, J. Stone, D. F. Rogers, L. S. Schadler, R. W. Siegel, B. C. Benicewicz, T. Apple, MRS Symp. Proc., 661, KK2.10.1 (2001).
128 C. Becker, H. Krug, H. Schmidt, MRS Symp. Proc., 435, 237 (1996).
129 C. Becker, P. Mueller, H. Schmidt, SPIE, 3469, 88 (1998).
130 G. Carotenuto, L. Nicolais, X. Kuang, Z. Zhu, Appl. Comp. Mater., 2, 385 (1995).
131 R. Krishnamoorti, R. A. Vaia, E. P. Giannelis, Chem. Mater., 8, 1728 (1996).
132 K. Gautam, A. Dhinojwala, Macro molecules, vol. 34, 1137 (2001).
133 K. Gautam, A. D. Schwab, A. Dhinojwala, D. Zhang, S. M. Dougal, M. S. Yeganeh, Phys. Rev. Lett., 85, 3854 (2000).
134 M. Vacatello, Macromolecules, 34, 1946 (2001).
135 P. Harder, M. Grunze, R. Dahint, G. M. Whitesides, P. E. Laibinis, J. Phys. Chem. B., 102, 426 (1998).
136 T. J. Webster, C. Ergun, R. H. Doremus, R. W. Siegel, R. Bizios, Biomaterials, 21, 1803 (2000).
137 T. J. Webster, C. Ergun, R. H. Doremus, R. W. Siegel, R. Bizios, J. Biomed. Mater. Res., 51, 475 (2000).
138 T. J. Webster, L. S. Schadler, R. W. Siegel, R. Bizios, Tissue Eng., vol. 7, 299–301 (2001).
139 E. Assouline, S. Pohl, R. Fulchiron, J.-F. Gerard, A. Lustiger, H. D. Wagner, G. Marom, Polymer, 41, 7843 (2000).
140 Y. Kojima, A. Usuki, M. Kawasumi, A. Okada, T. Kurauchi, O. Kamigaito, K. Kaji, J. Polym. Sci., Part B: Polym. Phys., 32, 625 (1994).
141 C. Saujanya, S. Radhakrishnan, Polymer, 42, 6723 (2001).
142 E. Petrovicova, R. Knight, L. S. Schadler, T. Twardowski, J. Appl. Polym. Sci., 77, 1684 (2000).
143 F.-D. Kuchta, P. J. Lemstra, A. Kellar, L. F. Batenburg, H. R. Fischer, MRS Symp. Proc., 576, 363 (1999).
144 Y. Kojima, A. Usuki, M. Kawasumi, A. Okada, Y. Fukushima, T. Kurauchi, O. Kamigaito, J. Mater. Res. 8, 1185 (1993).
145 Y. Kojima, T. Matsuoka, H. Takahashi, T. Kurauchi, J. Appl. Polym. Sci. 51, 683 (1994).
146 J. L. Kardos, Trans. NY Acad. Sci., 35, 136 (1973).
147 C. B. Ng, B. J. Ash, L. S. Schadler, R. W. Siegel, Polym. Composites, 10, 101 (2001).
148 D. Qian, E. C. Dicky, R. Andrews, T. Rantell, Appl. Phys. Lett., 76, 2868 (2000).
149 R. W. Siegel, S. K. Chang, B. J. Ash, J. Stone, P. M. Ajayan, R. W. Doremus, L. S. Schadler, Scripta Materialia, 44, 2061 (2001).
150 J. Liu, A. G. Rinzler, H. J. Dai, J. H. Hafner, R. K. Bradley, P. J. Boul, A. Lu, T. Iverson, K. Shelimov, C. B. Huffman, F. Rodriquez-Macias, Y. S. Shon, T. R. Lee, D. T. Colbert, R. E. Smalley, Science, 280, 1253 (1998).
151 J. Chen, M. A. Hamon, H. Hu, Y. Chen, A. M. Rao, P. C. Eklund, R. C. Haddon, Science, 282, 95 (1998).
152 E. T. Mickelson, I. W. Chiang, J. L. Zimmerman, P. J. Boul, J. Lozano, J. Liu, R. E. Smalley, R. H. Hauge, J. L. Margrave, J. Phys. Chem. B., 103, 4318 (1999).
153 P. J. Boul, J. Liu, E. T. Mickelson, C. B. Huffman, L. M. Ericson, I. W. Chiang, K. A. Smith, D. T. Colbert, R. H. Hauge, J. L. Margrave, R. E. Smalley, Chem. Phys. Lett., 310, 367 (1999).

154 K. D. Ausman, R. Piner, O. Lourie, R. S. Ruoff, M. Korobov, J. Phys. Chem. B., 104, 8911 (2000).
155 O. Lourie, D. M. Cox, H. D. Wanger, Phys. Rev. Lett., 81, 1638 (1998).
156 O. Lourie, H. D. Wagner, Comp. Sci. Technol., 59, 975 (1999).
157 O. Lourie, H. D. Wagner, J. Mater. Res., 13, 2418 (1998).
158 O. Lourie, H. D. Wagner, Appl. Phys. Lett., 72(24), 1 (1998).
159 X. Gong, J. Liu, S. Baskaran, R. D. Voise, J. S. Young, Chem. Mater., 12, 1049 (2000).
160 Q. Zhao, J. R. Wood, H. D. Wagner, J. Polym. Sci., Part B: Polym. Phys., 39, 1492 (2001).
161 Z. Jia, Z. Wang, C. Xu, J. Liang, B. Wei, D. Wu, S. Zhu, Mater. Sci. Eng., A271, 395 (1999).
162 L. S. Schadler, S. C. Giannaris, P. M. Ajayan, Appl. Phys. Lett., 73(26), 3842 (1998).
163 J. Scobbo, D. Grimes, T. Lemmen, V. Umamaheswaren, "Conductive thermoplastic resin for electrostatically painted applications," Proc. 1998 SAE Int. Cong. & Exposition (1998).
164 R. Andrews, D. Jacques, A. M. Rao, T. Rantell, F. Derbyshire, Appl. Phys. Lett., 75(9), 1329 (1999).
165 US Patent 2,951,087 (1960).
166 H. Z. Friedlander, Polym. Prepr., 4, 301 (1963).
167 US Patent 4,739007 (19 Apr 1988).
168 A. Okada, A. Usuki, T. Kurauchi, O. Kamigaito, Chapter 6 in *Hybrid Organic–Inorganic Composites*, ed. J. E. Marck, C. Y.-C. Lee, P. A. Bianconi, ACS Symp. series 585.
169 D. C. Lee, L. W. Jan, J. Appl. Polym. Sci., 61, 1117 (1996).
170 S. Bandyopadhyay, A. Hsieh, E. P. Giannelis, in *Polymer Nanocomposites*, ed. R. A. Vaia, R. Krishnamoorti, American Chemical Society, Washington, DC, USA (2001).
171 A. Okada, M. Kawasumi, A. Usuki, K. Yoshitsugu, T. Kurauchi, O. Kamigaito, MRS Symp. Proc., 171, 45 (1990).
172 M. G. Kanatzidis, L. M. Tonge, R. J. Marks, M. O. Marcy, C. R. Kannewurf, J. Am. Chem. Soc., 109, 3797 (1987).
173 M. W. Weiner, H. Chen, E. P. Giannelis, D. Y. Sogan, J. Am. Chem. Soc., 121, 1615 (1999).
174 E. Manias, H. Chen, R. Krishnamoorti, J. Genzer, E. J. Kramer, E. P. Giannelis, Macromolecules, 33, 7955 (2000).
175 J. Y. Lee, A. R. C. Baljon, R. F. Loring, A. Z. Panagiotopoulos, J. Chem. Phys., 102, 10321 (1998); J. Y. Lee, A. R. C. Baljon, R. F. Loring, J. Chem. Phys., 111, 9068 and 9754 (1999).
176 R. Bharadwaj, B. F. Farmer, R. A. Vaia, J. Chem. Phys., in press.
177 V. V. Ginzburg, O. V. Gendelman, L. I. Manevitch, Phys. Rev. Lett., 86, 5073 (2001).
178 R. A. Vaia, E. P. Giannelis, Macromolecules, 30, 8000 (1997).
179 R. A. Vaia, E. P. Giannelis, Macromolecules, 30, 7990 (1997).
180 N. Hasegawa, H. Okamoto, M. Kawasumi, A. Usuki, J. Appl. Poly. Sci., 74, 3359 (1999).
181 J. S. Shelly, P. T. Mather, K. L. DeVries, Polymer, 42, 5849 (2001).
182 A. Usuki, A. Koiwai, Y. Kojima, M. Kawasumi, A. Okada, T. Kurauchi, O. Kamigaito, J. Appl. Polym. Sci., 55, 119 (1995).
183 Y. Kojima, A. Usuki, M. Kawasumi, A. Okada, T. Kurauchi, O. Kamigaito, K. Kaji, J. Polym. Sci., Part B: Polym. Phys., 33, 1039 (1995).
184 M. Okamoto, P. H. Nam, P. Maiti, T. Kotaka, N. Hasegawa, A. Usuki, Nano Lett., 1, 295 (2001).
185 Y. Kojima, A. Usuki, M. Kawasumi, A. Okada, T. Kurauchi, O. Kamigaito, J. Polym. Sci., Part A: Polym. Chem., 31, 983 (1993).
186 L. Liu, A. Qi, X. Zhu, J. Appl. Poly. Sci., 71, 1133 (1999).
187 K. Yano, A. Usuki, A. Okada, T. Kuraychi, O. Kamigaito, Polym. Prepr. (Am. Chem. Soc., Div. Polym. Chem.), 32, 65 (1991).
188 K. Yano, A. Usuki, A. Okada, J. Polym. Sci., Part A: Polym. Chem., 35, 2289 (1997).
189 T. Lan, P. D. Kaviratna, T. J. Pinnavaia, Chem. Mater., 6, 573 (1994).
190 H.-L. Tyan, Y.-C. Liu, K.-H. Wei, Polymer, 40, 4877 (1999).
191 M. Kawasumi, N. Hasegawa, M. Kato, A. Usuki, A. Okada, Macromolecules, 30, 6333 (1997).

192 M. Kato, A. Usuki, A. Okada, J. Appl. Polym. Sci., 66, 1781 (1997).
193 M.Kawasumi, A. Usuki, A. Okada, T. Kurauchi, Mol. Cryst. Liq. Cryst., 281, 91 (1996).
194 M. Kawasumi, N. Hasegawa, A. Usuki, A. Okada, T. Kurauchi, MRS Symp., 425, 311 (1996).
195 D. C. Lee, J. W. Lee, J. Appl. Polym. Sci., 61, 1117 (1996).
196 M. Okamoto, S. Morita, H. Taguchi, Y.H. Kim, T. Kotaka, H. Tateyama, Polymer, 41, 3887 (2000).
197 G. Chen, X. Chen, S. Lin, W. Ye, J. Mater. Sci. Lett., 18, 1761 (1999).
198 X. Fu, S. Qutubuddin, J. Mater. Lett., 42, 12 (2000).
199 F. Dietsche, Y. Thomann, R. Thomann, R. Mulhaupt, J. Appl. Polym. Sci., 75, 396 (2000).
200 X. Fu, S. Qutubuddin, Mater. Lett., 42, 12 (2000).
201 R. Vaia, H. Ishii, E. Giannelis, Chem. Mater., 5, 1694 (1993).
202 T. Lan , T. J. Pinnavaia, Chem. Mater., 6, 2216 (1994).
203 Z. Wang, T. Lan, T. J. Pinnavaia, Chem. Mater., 8, 2200 (1996).
204 J. Massam, T. J. Pinnavaia, MRS Symp. Proc., 20, 223 (1998).
205 I. J. Chin, T. Thurn-Albrecht, H.-C. Kim, T. P. Russell, J. Wang, Polymer, 42, 5947 (2001).
206 T. Lan, P. D. Kaviratna, T. J. Pinnavaia, Chem. Mater., 7, 2144 (1995).
207 Z. Wang, T. J. Pinnavaia, Chem. Mater., 10, 1820 (1998).
208 Z. Wang, T. J. Pinnavaia, Chem. Mater., 10, 3769 (1998).
209 P. Aranda, E. Ruiz-Hitzky, App. Clay Sci., 15, 119 (1999).
210 K. G. Fournaris, M. A. Krakassides, D. Petridis, K. Yiannakopoulou, Chem. Mater., 11, 2372 (1999).
211 W. Chen, Q. Xu, .Z. Yuan, J. Mater. Sci. Lett., 18, 711 (1999).
212 R. A Vaia, S. Vasudevan, W. Krawwiec, L. G. Scanlon, E. P. Giannelis, Adv. Mater., 7, 154 (1995).
213 Y. Kojima, K. Fukumori, A. Usuki, T. Kurauchi, J. Mater. Sci. Lett., 12, 889 (1993).
214 E. N. Kresge, D. J. Lohse, US Patent 5,576,372 (1996).
215 M. Laus, O. Francescangeli, F. Sandrolini, J. Mater. Res., 12, 3134 (1997).
216 S. D. Burnside, E. P. Giannelis, Chem. Mater., 7, 1597 (1995).
217 S. Wang, C. Long, X. Wang, Q. Li, Z. Qi, J. Appl. Poly. Sci., 69, 1557 (1998).
218 Y. Wang, L. Zhang, C. Tan, D. Yu, J. Appl. Polym. Sci., 78, 1879 (2000).
219 C. O. Oriakhi, X. Zhang, M. M. Lerner, Appl. Clay Sci., 15, 109 (1999).
220 X. Huang, S. Lewis, W. J. Brittain, R. A. Vaia, Macromolecules, 33, 2000 (2000).
221 Y. Ke, C. Long, Z. Qi, J. Appl. Poly. Sci., 71, 1139 (1999).
222 Q. Wu, Z. Xue, Z. Qi, F. Wang, Polymer, 41, 2029 (2000).
223 M. Z. Rong, M. Q. Zhang, Y. X. Zheng, H. M. Zeng, K. Friedrich, Polymer, 42, 3301 (2001).
224 J. I. Hong, K. S. Cho, C. I. Chung, L. S. Schadler, R. W. Siegel, J. Mater. Res., 17(5), 940 (2002).
225 L. S. Schadler, personal communication.
226 G. Carotenuto, Y.-S. Her, E. Matijevic, Ind. Eng. Chem. Res., 35, 2929 (1996).
227 F. Yang, Y. Ou, Z. Yu, J. Appl. Poly. Sci., 69, 355 (1998).
228 Y. Ou, F. Yang, Z.-Z. Yu, J. Polym. Sci., Part B: Polym. Phys., 36, 789 (1998).
229 P. Hajji, L. David, J. F. Gerard, H. Kaddami, J. P. Pascault, G. Vigier, MRS Symp. Proc., 576, 357 (1999).
230 B. J. Ash, L. S. Schadler, R. W. Siegel, Polym. Prepr., 42, 52 (2001).
231 Y. Zhang, G. Zhou. L. Li, Y. Zhang, Chin. J. Mater. Res., 12 (1998).
232 M. Avella, M. E. Errico, E. Martuscelli, Nano Lett., 1, 213 (2001).
233 G.-H. Hsiue, W.-J. Kuo, Y.-P. Huang, R.-J. Jen, Polymer, 41, 2813 (2000).
234 S. Wang, M. Wang, Y. Lei, L. Zhang, J. Mater. Sci. Lett., 18, 2009 (1999).
235 P. Judeinstein, C. Sanchez, J. Mater. Chem, 6, 511 (1996).
236 K. A. Mauritz, P. R. Start, Polym. Prepr., 40, 2, 717 (1999).
237 C. J. T. Landry, B. K. Coltrain, M. R. Landry, J. J. Fitzgerald, V. K. Long, Macromolecules, 26, 3702 (1993).
238 S. P. Nunes, K. V. Peinemann, K. Ohlrogge, A. Alpers, M. Keller, A. T. N. Pires, J. Membr. Sci., 157, 219 (1999).

239 K. F. Silveira, I. V. Yoshida, S. P. Nunes, Polymer, 36, 1452 (1995).
240 D. W. McCarthy, J. E. Mark, D. W. Schaeffer, J. Polym. Sci., Part B: Polym. Phys., 36, 1167 (1998).
241 J. M. Breiner, J. E. Mark, G. Beaucage, J. Polym. Sci., Part B: Polym. Phys., 37, 1421 (1999).
242 Y. Wei, D. Yang, L. Tang, M. K. Hutchins, J. Mater. Res., 8, 1143 (1993).
243 B. M. Novak, Adv. Mater., 5, 422 (1993).
244 P. Gomez-Romero, Adv. Mater., 13, 163 (2001).
245 J. J. Senkevich, S. B. Desu, Chem. Mater., 11, 1814 (1999).
246 D. E. Collins, E. B. Slamovich, Chem. Mater., 11, 2319 (1999).
247 J. W. Burdon, P. Calvert, MRS Symp., 218, 203 (1991).
248 C. Saujanya, P. S. Ashamol, S. Radhakrishnan, Polymer, 42, 2255 (2000).
249 S. T. Selvan, T. Hayakawa, M. Nogami, M. Moller, J. Phys. Chem. B., 103, 7441 (1999).
250 A. B. R. Mayer, Mater. Sci. Eng. C,. 6, 155 (1998).
251 A. Roescher, M. Moller, PMSE Prep., 72, 283 (1995).
252 M. Antonietti, E. Wenz, L. M. Bronstein, M. Seregina, Adv. Mater., 7, 1000 (1995).
253 M. Seregina, L. M. Bronstein, O. A. Platonova, D. M. Chernyshov, P. M. Valetsky, E. Wenz, M. Antonietti, Chem. Mater., 9, 923 (1997).
254 M. Antonietti, C. Goltner, Angew. Chem. Int. Ed. Engl., 36, 910 (1997).
255 M. Antonietti, F. Grohn, J. Hartmann, L. Bronstein, Angew. Chem., 109, 2170 (1997).
256 A. B. R. Mayer, J. E. Mark, R. E. Morris, Polym. J., 30, 197 (1998).
257 E. Wenz, M. Antonietti, H. Weller, unpublished
258 U. Kriebig, M. Vollmer, *Optical Properties of Metal Clusters*, Springer-Verlag, Berlin, Germany, 1995.
259 V. M. Rudoy, I. V. Yaminskii, O. V. Dement'eva, V. A. Ogarev, Colloid J., 61, 800 (1999).
260 J. A. Becker, R. Schafer, R. Festag, W. Ruland, J. H. Wendorff, J. Pebler, S. A. Qaiser, W. Helbig, M. T. Reetz, J. Chem. Phys., 103, 2520 (1995).
261 D. L. Carroll, Phys. Rev. Lett., 81, 2332 (1998).
262 X. Blasé, Phys. Rev. Lett., 83, 5078 (1999)
263 R. Czerw, M. Terrones, J.-C. Charlier, X. Blasé, B. Foley, R. Kamalakran, N. Grobert, H. Terrones, D. Tekleab, P. M. Ajayan, W. Blau, W. Ruhle, D. L. Carroll, Nanolett. (2001).
264 A. Eitan, L. S. Schadler, J. Hansen, P. M. Ajayan, R. W. Siegel, M. Terrones, R. Andrews, Proc. Japan–US Conf. Composites Mater., San Jose, CA, USA, September, 2002
265 A. Eitan, K. Jiang, R. Andrews, L. S. Schadler, submitted Chem. Mater. (2002).
266 K. Jiang, A. Eitan, L. S. Schadler, P. M. Ajayan, R. W. Siegel, N. Grobert, M. Mayne, M. Reyes-Reyes, H. Terrones, M. Terrones, submitted Nano Lett. (2002)
267 J. Chen. M. A. Hamon, H. Hu, Y. Chen, A. M. Rao, P. R. Eklund, R. C. Haddon, Science, 282, 95 (1998).
268 C. N. R Rao, A. Govindaraj, B. C. Satishkumar, Chem. Commun., 1525 (1996).
269 S. S. Wong, E. Joselvich, A. T. Woolley, C. L. Cheung, C. M. Lieber, Nature, 394, 52 (1998).
270 J. Liu, A. G. Rinzler, H. Dai, J. Hafner, R. K. Bradley, P. Boul, A. Lu, T. Iverson, K. Shelimov, C. B. Huffman, F. Rodriquez-Macias, Y.-S. Shon, T. R. Lee, D. T. Colbert, R. E. Smalley, Science, 280, 1253 (1998).
271 J. L. Bahr, J. Yang, D. V. Kosynkin, M. J. Bronikowski, R. E. Smalley, J. M. Tour, J. Am. Chem. Soc., 123, 6536 (2001).
272 E. T. Mickelson, I. W. Chiang, J. L. Zimmerman, P. J. Boul, J. Lozano, J. Liu, R. E. Smalley, R. H. Hauge, J. L. Margrave, J. Phys. Chem. B, 103, 4318 (1999).
273 P. Allongue, M. Delamar, B. Desbat, O. Fagebaume, R. Hitmi, J. Pinson, J.-M. Salvetat, J. Am. Chem. Soc., 119, 201 (1997).
274 B. Ortiz, C. Saby., G. Y. Champagne, D. Belanger, J. Electroanal. Chem., 455, 75 (1998).
275 J. Chen, A. M. Rao, S. Lyuksyutov, M. E. Itkis, M. A. Hamon, H. Hu, R. W. Cohn, P. C. Eklund, D. T. Colbert, R. E. Smalley, R. C. Haddon, J. Phys. Chem. B, 105, 2525 (2001).
276 J.-M. Bonnard, T. Stora, J.-P. Salvetat, F. Maier, T. Stockli, C. Duschl, L. Forro, W. A.

De Herr, A. Chatelain, Adv. Mater., 9, 827 (1997).
277 A. Koshio, M. Yudaska, M. Zhang, S. Iijima, Nanolett. (2001).
278 Q. Chen, L. Dia, M. Gao, S. Huang, A. Mau, J. Phys. Chem. B., 105, 618 (2001).
279 J. Chen, M. A. Hamon, H. Hu, Y. Chen, A. M. Rao, P. C. Eklund, R. C. Haddon, Science, 282, 95 (1998).
280 R. J. Chen, Y. Zhang, D. Wang, H. Dai, J. Am. Chem. Soc., 123, 3838 (2001).
281 R. Czerw, Z. Guo, P. M. Ajayan, Y.-P. Sun, D. L. Carroll, Nano Lett. (2001).
282 F. Caruso, Adv. Mater., 13, 11 (2001).
283 O. Makarova, T. Rajh, M. C. Thurnauer, A. Martin, P. A. Kemme, D. Cropek, Environ. Sci. Technol., 34, 4797 (2000).
284 T. Konovalova, L. D. Kispert, V. V. Konovalov, J. Phys. Chem. B, 103 4672 (1999).
285 Paul Braun, personal communication
286 J. Wang, S. Kara, T. E. Long, T. C. Ward, J. Polym. Sci., Part A: Polym. Chem., 38, 3742 (2000).
287 C. H. M. Hofman-Caris, New. J. Chem., 18, 1087 (1994).
288 L. Quaroni, G. Chumanov, J. Am. Chem. Soc., 121, 10642 (1999).
289 N. Tsubokawa, A. Kogure, K. Maruyama, Y. Sone, M. Shimomura, Polym. J., 22, 827 (1990).
290 M. Z. Rong, M. Q. Zhang, Y. X. Zheng, H. M. Zeng, R. Walter, K. Friedrich, Polymer, 42, 167 (2001).
291 Y. Mao, B. M. Fung, Chem. Mater., 10, 509 (1998).
292 T. VonWerne, T. E. Patten, J. Am. Chem. Soc., 121, 7409 (1999).
293 F. Caruso, Chem. Eur. J., 6, 413 (2000).
294 O. Kalinina, E. Kumacheva, Macromolecules, 32, 4122 (1999).
295 M. Ohmori, E. Matijevic, J. Colloid Interface Sci., 150, 594 (1992).
296 H. Giesche, E. Matijevic, J. Mater. Res., 9, 436 (1994).
297 A. Hanprasopwattana, S. Srinivasan, A. G. Sault, A. K. Datye, Langmuir, 12, 3173 (1996).
298 R. Partch, Y. Xie, S. T. Oyama, E. Matijevic, J. Mater. Res., 8, 2014 (1993).
299 R. C. Stephenson, R. E. Partch, Proc. Mater. Res. Soc., 458, 435 (1996).
300 N.A. Dhas, A. Zaban, A. Gedanken, Chem. Mater., 11, 806 (1999).
301 C. Brinson, personal communication.
302 R. M. Christensen, *Mechanics of Composite Materials*, reprinted by Krieger, Malabar, FL, USA (1991), orig. pub. Wiley, New York, NY, USA (1979).
303 A.Oya, Y. Kurokawa, H. Yasuda, J. Mater. Sci., 35, 1045 (2000).
304 N. Hasegawa, H. Okamoto, M. Kato, A. Usuki, J. Appl. Polym. Sci., 78, 1918 (2000).
305 Y. T. Lim, O. O. Park, Macromol. Rapid Commun., 21, 231 (2000).
306 H. Shi, T. Lan, T. J. Pinnavaia, Chem. Mater., 8, 1584 (1996).
307 N. Hasegawa, M. Kawasumi, M. Kato, A. Usuki, A. Okada, J. Appl. Polym. Sci., 67, 87 (1998).
308 T. Agag, T. Koga, T. Takeichi, Polymer, 42, 3399 (2001).
309 C. B. Bucknall, "Deformation mechanisms in rubber-toughened polymers" in *Polymer Blends, Vol. 2: Performance*, ed D. R. Paul, C. B. Bucknall. Wiley, New York, NY, USA (2000).
310 S. Wu, Polymer, 26, 1855 (1985).
311 J. M. Gloagun, J. M. Lefebvre, Polymer, 42, 5841 (2001).
312 T.-K. Chen, Y.-I. Tien, K. Wei, Polymer, 41, 1345 (2000).
313 D. B. Zax, D.-K. Yang, R. A. Santos, H. Hegemann, E. P. Giannelis, E. Manias, J. Chem. Phys., 112, 2945 (2000).
314 E. Hackett, E. Manias, E. P. Giannelis, Chem. Mater., 12, 2161 (2000); M. R. Nyden, J. Gilman, Comp. Theor. Polym. Sci., 7, 191 (1997).
315 D. Cole, K. R. Shull, P. Baldo, L. Rehn, Macromolecules, 32, 771 (1999).
316 B. J. Ash, L. S. Schadler, R. W. Siegel, Mater. Lett., 53, 83–87 (2002).
317 R. Krishnamoorti, E. P. Giannelis, Macromolecules, 30, 4097 (1997).
318 S. S. Sternstein, A.-J. Zhu, Macromolecules, 35, 7262 (2002).
319 E. Petrovicova, R. Knight, L. S. Schadler, T. Twardowski, J. Appl. Polym. Sci., 78, 2272 (2000).
320 Q. Wang, J. Xu, S. Weichang, Q. Xue, Wear, 209, 316 (1997).
321 Q. Wang, J. Xu, S. Weichang, Tribology Int., 30, 193 (1997).
322 Q. Wang, Q. Xue, H. Liu, S. Weichang, J. Xu, Wear, 198, 216 (1996).

323 Q. Wang, J. Xu, S. Weichang, L. Weimin, Wear, 196, 82 (1996).
324 D. E. George, J. S. Bloom, T. P. Feist, US Patent 5,789,523 (1996).
325 L. Nielsen, J. Macromol. Sci. Chem., A1, 929 (1967).
326 E. L. Cussler, S. E. Hughes, W. J. Ward III, R. Aris, J. Membr. Sci., 38, 161 (1988).
327 Modern Plastics, February, 1998.
328 Y. Kojimoto, A. Usuki, M. Kawasumi, A. Okada, T. Kurauchi, O. Kamigaito, J. Appl. Poly. Sci., vol. 49, 1259-1264 (1993).
329 Modern Plastics, February, 1999.
330 Y. Kojima, A. Usuki, M. Kawasumi, A. Okada, T. Kurauchi, O. Kamigaito, J. Appl. Polym. Sci., 49, 1259 (1993).
331 G. Kraus, *Reinforcement of Elastomers*, Interscience, New York, NY, USA (1965).
332 S. S. Sternstein, J. Macromol. Sci., Phys., B6/1, 243 (1972).
333 N. Salahuddin, M. Shehata, Polymer, 42, 8379 (2001).
334 A. Blumstein, J. Polym. Sci., Part A: Polym. Chem., 3, 2665 (1965).
335 J. Lee, E. Giannelis, Polym. Prepr., 38, 688 (1997).
336 J. Gilman, T. Kashiwagi, J. Lichtenhan, SAMPE J., 33, 40 (1997).
337 J. Gilman, T. Kashiwagi, J. E. T. Brown, S. Lomakin, "Flammability studies of polymer layered silicate nanocomposites", Proc. 43rd Int. SAMPE Symp., 1053 (1998).
338 J. Gilman, Appl. Clay Sci., 15, 31 (1999).
339 S. Bourbigot, M. Le Bras, F. Dabrowski, "Intumescence and polymer blending: an approach for flame retardancy?" Proc. Annu. Conf. Fire Res., NISTIR 6242, 43 (1998).
340 H. Inoue, T. Hosokawa, Japanese Patent JP 10 81,510 (98 81,510) (1998).
341 R. A. Vaia, G. Price, P. N. Ruth, H. T. Nguyen, J. Lichtenhan, Appl. Clay Sci., 15, 67 (1999).
342 H. Fong, R. A. Vaia, J. H. Sanders, D. Lincoln, P. J. John, A. J. Vreugdenhil, J. Bultman, C. A. Cerbus, H. G. Jeon, Polym. Prepr., 42, 354 (2001).
343 D. Barber, C. R. Pollock, L. L. Beecroft, C. K. Ober, Opt. Lett., 22, 1247 (1997).
344 J. I. Hong, L. S. Schadler, E. Martensson, personal communication.
345 H. Sabuni, J.K. Nelson, J. Mater. Sci., 11, 1574 (1976).
346 S. M. H. Sabuni, J. K. Nelson, J. Mater. Sci., 14, 2791 (1979).
347 J. Coleman, A. Dalton, S. Curran, A. Rubio, A. Davey, A. Drury, B. McCarthy, B. Lahr, P. Ajayan, S. Roth, R. Barklie, W. Blau. Adv. Mater., 12, 213 (2000).
348 C. Brinker, E. Giannelis, R. Laine, C, Sanchez (ed), *Hybrid Materials*, Proc. MRS Symp. Proc. Series, 519 (1999).
349 M. Kerker, *The Scattering of Light*, Academic, New York, NY, USA (1969), p. 414.
350 T. Li, Q. Chen, L. S. Schadler, R. W. Siegel, J. Mendel, G. C. Irvin, Polym. Composites, in press.
351 L. Zimmerman, M. Weibel, W. Caseri, U. Suter, J. Mater. Res., 8, 1742 (1993).
352 T. Kyprianidou-Leodidou, W. Caseri, U. W. Suter, J. Phys. Chem., 98, 8992 (1994).
353 T. Kyprianidou-Leodidou, H.-J. Althaus, Y. Wyser, D. Vetter, M. Buchler, W. Caseri, U. W. Suter, J. Mater. Res., 12, 2198 (1997).
354 H. Ago, M. S. P. Shaffer, D. S. Ginger, A. H. Windle, R. H. Friend, Phys. Rev. B, 61, 2286 (2000).
355 H. S. Woo, R. Czerw, S. Webster, D. L. Carroll, J. Ballato, A. E. Strevens, D. O'Brien, W. J. Blau, Appl. Phys. Lett., 77, 1393 (2000).
356 C. O. Oriahkhi, X. Zhang, M. M. Lerner, Appl. Clay Sci., 15, 109 (1999).
357 T.-W. Lee, O. O. Park, J. Yoon, J.-J. Kim, Adv. Mater., 13, 211 (2001).
358 S. A. Carter, J. C. Scott, P. J. Brock, Appl. Phys. Lett., 71, 1145 (1997).
359 B. A. Dabbousi, M. G. Bawendi, O. Onitduka, M. F. Rubner, Appl. Phys. Lett., 66, 1316 (1996).
360 F. Zimmerman, A. Wokaun, Mol. Phys., 73, 959 (1991).
361 M. Lin, D. G. Ast, Adv. Mater., 13, 719, 2001.
362 Y. Dirix, C. Bastiaansen, W. Caseri, P. Smith, J. Mater. Sci., 34 3859 (1999).
363 B. Z. Tang, H. Xu, Macromolecules, 32, 2569 (1999).

3
Natural Nanobiocomposites, Biomimetic Nanocomposites, and Biologically Inspired Nanocomposites

Paul V. Braun

3.1
Introduction

Self-organization and directed assembly of biological macromolecules and inorganic materials plays an important role in the creation of the nanostructured and nanocomposite materials so commonly found in biology. Over the past 10 years, materials scientists and chemists have made considerable efforts to create synthetic analogs of biological materials by attempting to mimic biology and, importantly, to learn the design rules of biological systems. Considerable efforts have also been made to use what has been learned from biology to create new materials with properties not found in biological systems. All this effort has been made because natural materials unquestionably have exquisite properties not found in synthetic materials. And additionally, biological systems can produce these exquisite materials at or near room temperature in aqueous environments, whereas most synthetic schemes that produce materials often inferior to natural biomaterials require elevated temperature and pressure and harsh chemicals.

Biological nanocomposite materials can be entirely inorganic, entirely organic, or a mixture of inorganic and organic materials. Even where the final material may be entirely one class of material, multiple classes of materials may have been involved in the synthetic process, which may or may not remain in the final structure. A good example of a biological nanocomposite in which the organic material does not remain in the final product is the enamel of the mature human tooth, which is 95 % by weight hydroxyapatite. During tooth formation, enamel consists of a composite of proteins (primarily amelogenin and enamelin) and hydroxyapatite; however, the proteins are removed as the tooth develops. The presence of the proteins, and the self-assembled structures they form with other biological macromolecules, do however help generate the mineral cross-ply structure of the enamel, which plays a large part in its observed toughness. The best known example of an inorganic/organic structural composite for which both phases remain in the final product is the aragonitic nacreous layer of the abalone shell, which is exceptionally strong because of its organic/inorganic layered nanocomposite structure, in which crystalline ceramic layers are separated by highly elastic organic layers. However, there is much more to the exceptional properties of the abalone shell than simply its layered structure, as many researchers are discovering. Synthetic efforts have been made for about 10 years to create such a structure in the laboratory, and although some synthetic structures resemble the abalone shell, to date their properties have been inferior. Thus the practical value of the synthetic structures is limited. With the many exciting scientific discoveries that have been made, there is great potential for new materials in the not too distant future.

Many more examples of attempts to copy biology have not been as successful in generating engineering materials as one would have hoped. In hindsight, this is not surprising. Biological materials generally form over a period of days to years, use a limited set of elements, and are designed to be used within a limited temperature range. Practical engineering materials must be made rapidly (hours or minutes) and generally must operate over a wide range of temperature and other environmental conditions. The disconnect between the needs of engineering materials and of biological materials has led many scientists to conclude that, rather than attempting to directly copy biology, a much better philosophy is to learn from biology and use this knowledge to create synthetic materials. This may or may not involve the use of some biological molecules, but no attempt is made to 'copy' specific biological processes.

This chapter covers natural nanobiocomposites, biomimetic nanocomposites, and biologically inspired synthetic nanocomposites. Natural nanobiocomposites are exactly what the name implies, that is, natural composite materials with structure on the nanoscale; biomimetic nanocomposites are synthetic nanocomposite materials formed through processes that mimic biology as closely as possible; and biologically inspired nanocomposites are composite materials with nanoscale order created through processes that are inspired by a biological process or a biological material, but without attempting to mimic, or directly copy, the mechanism of formation of the biological material. There are many areas I do not attempt to cover, including mixtures of two or more materials in which the materials form a homogeneous mixture (and thus are not composites) or in which the organic molecule is simply dispersed in an organic host, such as organic molecules in a sol–gel matrix. I also only minimally cover the widely publicized work on mesoporous silica and related materials, because several very good reviews on this subject already exist [1–5].

The formation of solids containing defined nanometer-scale structures and features is a challenging goal of great interest in the synthesis of novel materials. Significant efforts have been made to generate nanostructured materials by techniques ranging from scanning microscopy to molecular self assembly [6–14]. Through careful control of molecular architecture, self-assembly can yield many different nanostructured organics. The potential for creating new materials by mineralizing, polymerizing, or otherwise replicating these organized structures is very interesting. For example, mineral growth within and on lipid aggregates has resulted in unusual morphologies, including mineralized tubules and disks [15, 16]. Another approach to controlling mineral growth through self-assembly involves the coassembly of molecular species and mineral-phase precursors into mesoporous structures. This methodology has been successful in the now widely studied formation of mesoporous oxides [17–23].

3.2
Natural Nanocomposite Materials

In biology, examples abound of structures with nanoscale dimensions, and virtually all the functionality provided by these materials is a direct consequence of the nanoscale dimensions of the structure. A few examples of nanoscale materials in biology are lipid cellular membranes, ion channels, proteins, DNA, actin, spider silk, and so forth. In all these structures, the characteristic dimension, at least in 1D, and often in 3D, is on the order of a few nanometers. Although a materials scientist does not typically consider such materials to be composites, in truth, the properties of many biological materials are driven by structure on the nanoscale and, in the sense that the larger material is composed of discrete nanoscale building blocks, most of these biological materials can be considered nanocomposite materials. In their active form, that is, when folded, proteins are composed of domains with varying hydrophilic and hydrophilicity, as well as domains with such structural features as alpha helixes, beta sheets, and turns. The assembly of proteins into these complex structures containing nanometer-sized domains of varying chemical properties gives proteins, as well as many other biomolecules, their intriguing properties. Because these chemically diverse regions can exhibit acidic, basic, hydrogen-bonding, hydrophilic, or hydrophobic behavior, they can interact in exceedingly diverse ways with precursors for mineral compounds and the final mineral product. The inorganic/organic composite materials formed by these exceedingly complex biomolecular structures are the primary focus of the rest of this section.

However, before discussing inorganic/organic nanocomposites, a short discussion of at least one completely organic nanocomposite with outstanding properties, namely spider silk, is in order. Dragline spider silk, which makes up the spokes of a spider web, is five times tougher than steel by weight and can stretch 30%–40% without breaking. However, to be fair to steel, I should point out that, although the work-to-failure is greater for silk than steel, the elastic modulus of silk is significantly less than that of steel. For applications in which flexibility and toughness are the primary need, a synthetic route to creating a material with properties equivalent to spider silk would be exceedingly valuable. The strong core of spider silk is composed primarily of two protein components that self-assemble into crystalline and amorphous regions. The crystalline regions consist of alternating alanine-rich crystalline forming blocks, which impart hardness [24–26], and glycine-rich amorphous blocks, which are believed to provide elasticity [27]. This nanocomposite structure of organic nanocrystals in an amorphous organic matrix generates the exceptional properties of spider silk. Spiders cannot be kept in close quarters and harvested, because they eat one another; thus the only route to creating quantities of spider silk sufficient for application will be synthetic.

Knowledge of the molecular structure of spider silk is not sufficient for creating a synthetic material with its properties. Inside the spider, the silk precursor exists as a lyotropic liquid crystal that is approximately 50% silk. As the silk is excreted, the protein molecules that make up silk fold and are aligned as they approach and then pass through the spinneret, forming a complex insoluble nanostructured fiber. In fact, not only is the spider silk made up of crystalline and amorphous regions, but the crystalline regions are in turn composed of both highly oriented crystals and less oriented, but still crystalline, regions [28, 29]. These poorly oriented crystalline regions may serve to mechanically couple the highly oriented crystalline structure with the amorphous domains. This three-phase morphology presumably has properties exceeding those of a similar two-phase system because of better coupling between the phases, due to the smaller mismatch in mechanical properties at the interface between the various domains.

Recently, a synthetic spider silk-like material with properties approaching those of natural spider silk has been created. Because of their high complexity, the silk precursors could not be created by conventional organic synthesis and thus were created by expressing two of the dragline silk genes in mammalian cells [30]. The resulting soluble recombinant dragline silk proteins were wet-spun into fibers of diameters ranging from 10 to 40 micrometers. After a postspinning draw, fibers with mechanical properties approaching those of natural spider silk were obtained. The fibers were highly birefringent, indicating a high degree of chain orientation, but it is not yet known if the internal nanostructure of the fibers was similar to that of natural spider silk. Work is now ongoing to build up a herd of transgenic goats that express spider silk proteins and excrete them in their milk, which should result in the production of sufficient silk protein for manufacturing.

Much like the nanostructure of natural silk, which self-organizes in the biological organism, natural organic/inorganic nanocomposites are also formed through self-assembly. The two extremes of the mechanism for formation of natural nanocomposites are when the organic matrix forms first, followed by mineralization, and when the organic and inorganic materials coassemble into the nanostructured composite. There do not appear to be any examples in which the inorganic structure forms first, followed by organic structure formation. Most biological composites appear to form through the first route; a process whereby an organic structure first forms, followed by the biologically directed nucleation and growth of a mineral phase. However, it is also clear that the organic matrix restructures and reorganizes continuously as the mineral deposits, which resembles the second mechanism. With both mechanisms, the organic material is generally composed of an amazing diversity of macromolecules, including proteins and other biopolymers. Considerable effort has been made to understand the mechanism of formation of these composite materials, a process that requires an understanding of both the structure and function of the organic matrix and the interactions between the organic matrix and the inorganic precursors and product.

The simplest examples of biological nanocomposites are those in which the mineral phase is simply deposited onto or within an organic structure. The next level of complexity is shown by examples in which the structure of the mineral phase is clearly determined by the organic matrix. The greatest complexity is where the mineral is intimately associated with the organic phase to create a structure with properties that are superior to those of either the mineral or organic phase. In the first and second examples, the organic phase forms first, but in the third, the organic and inorganic phases almost certainly coassemble into the final structure.

3.2.1
Biologically Synthesized Nanoparticles

For the simplest example of a biological nanocomposite, one needs to look no farther than the grasses. Many species of grass precipitate SiO_2 nanoparticles within their cellular structures. Perhaps due to some internal structure in their cells, the silica nanoparticles generated within the cells can be found in sheet-like, globular, or rod-like morphologies, with characteristic dimensions ranging from a few nanometers to tens of nanometers (Figure 3.1) [31]. The exact reason for the silica deposition is not known, although it may simply be a way for the plant to sequester silica, which it continuously takes in as dissolved silicic acid and other silica-containing soluble species. If the plant could not sequester silica, it would build up and potentially limit growth. At high levels, silica may make the plant undesirable as food [32], for example, the dry weight of horsetails consists of 20%–25% silica.

Another example of a biologically simple organism that can generate internal nanocomposite structures is bacteria, the most famous of which are magnetic bacteria, which contain an internal chain of magnetite (Fe_3O_4) nanocrystals running down their long axis (Figure 3.2). These nanocrystals are about 40 nm in diameter and give the bacteria a net magnetic moment, which aligns them with respect to the earth's magnetic field. The presence of this magnetic chain of nanocrystals plays a role in defining

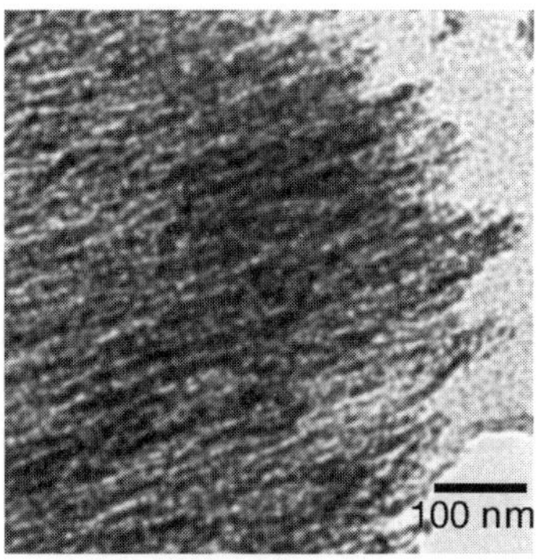

Fig. 3.1 Transmission electron micrograph of rod-like amorphous silica extracted from a grass. The rod-like nature is due at least in part to the local environment inside the plant cells. Adapted from [31]

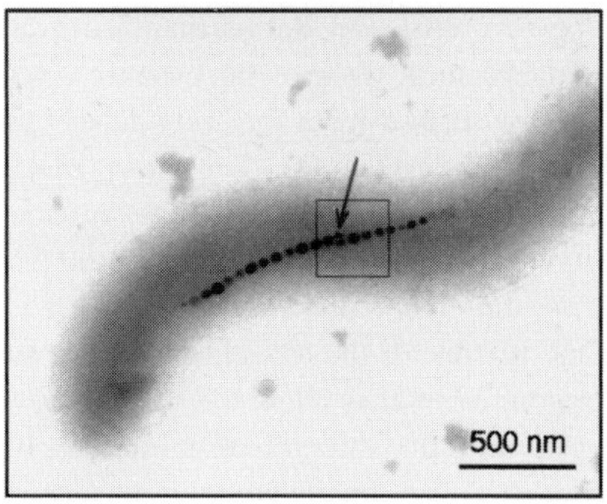

Fig. 3.2 Transmission electron microscope bright-field image of a single cell of *Magnetospirillum magnetotacticum* strain MS-1. Off-axis electron holography of the boxed region in the transmission electron microscope showed that the magnetite crystals were all single magnetic domains. Adapted from [161]

the direction of locomotion. Because an individual bacterium, on average, points in a specific direction, as it moves it continuously explores fresh territory. In contrast, the movement of a bacterium lacking a directing force follows a random walk, resulting in a substantially lower probability for movement into fresh territory. Besides the Fe_3O_4-containing bacteria, other bacteria deposit chains of nanoparticles of the magnetic mineral greigite (Fe_3S_4) within themselves. This mineral chain appears to serve the same function as the Fe_3O_4 chain in the previous bacteria.

3.2.2
Biologically Synthesized Nanostructures

Biological organisms commonly synthesize nanostructured and nanocomposite materials of much greater complexity than nanoparticles. These structures can come in many forms, ranging from needles and plates to complex 3D nanostructures, and can exhibit many unique and interesting mechanical, chemical, and optical properties. Although simplifying all biological nanostructure formation routes into a single mechanism is not possible, several general synthetic tenets tend to hold true. The first is that the biological structure forms first, and then the inorganic phase begins to form. Although truly cooperative assembly is possible, it is much more common that the organic matrix forms first, regulating the growth of the inorganic material. We should keep in mind, however, that the organic matrix is plastic and can deform or reshape in response to the growing inorganic material.

Besides forming nanoparticles, bacteria can form more complex nanostructures. Bacteria in general are responsible for a vast amount of mineral deposition and in fact can contribute greatly to mineral deposits on the bottoms of lakes and other aquatic environments. Mineral formation in the vicinity of bacteria is generally driven by a biologically induced change in the pH or ionic strength around individual bacteria. This change in the local environment reduces the solubility of certain mineral compounds, leading to precipitation of minerals. Generally, this results in an unfeatured mineral, so is not of interest here. However, in specific cases, the structure of the precipitated mineral is highly regulated by proteins on the surface of the bacteria, resulting in the formation of complex nanostructures. Specifically, some bacteria have on their exterior a layer of proteins called an S-layer. As these bacteria respire, the local pH and ionic strength change, driving, for example, the mineralization of gypsum ($CaSO_4$). Rather than being deposited as an unfeatured solid on the exterior of the bacteria, the growing gypsum is templated by the organic S-layer into a complex nanostructured form that very closely replicates the structure of the protein S-layer (Figure 3.3) [33, 34]. The exact mechanism of the mineral templating is not known, but clearly preferred sites must exist on the S-layer for heterogeneous nucleation of mineral, or a regular structure would not form. As these mineral nuclei grow, they impinge on one another, forming a thin, periodically structured film of gypsum. One reason for the complexity of this structure is that the resulting pores in the inorganic layer may allow the bacterium to exchange nutrients with its surroundings even after the mineral phase forms. If the gypsum had formed as a dense solid film, it would rapidly stop the bacterium's respiration, thus killing it.

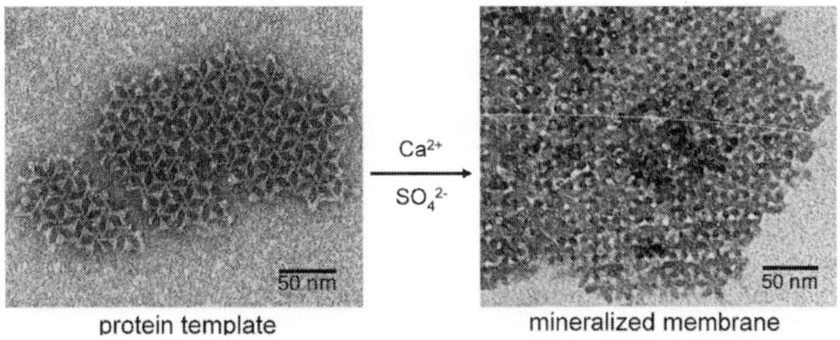

Fig. 3.3 Transmission electron micrograph of (left) bacterial S-layer, which serves as a protein template for the formation of (right) a thin film of mesostructured gypsum. Adapted from [33]

Although this specific S-layer templated structure as yet has no known applications, inorganic nanostructures of these length scales are of great interest for many high-technology applications. Patterning on these length scales is exceedingly difficult, so if bacteria could be genetically engineered to express proteins that result in S-layers having technologically significant nanostructures, there would be the potential to biomineralize materials for advanced applications. In fact, initial strides have been made in this regard and are covered later in this chapter.

Higher organisms also generate inorganic/organic composite structures. An excellent example is the sea urchin. The sea urchin spine is a single crystal essentially composed of calcite, containing only about 0.02% glycoproteins trapped within the crystal lattice of the spine [35]. However, this very small amount of organic matter leads to the extraordinarily high toughness and unique fracture properties of the spine. The spine is much tougher than a synthetic equivalent composed of polycrystalline or single-crystalline calcite. Most surprising is the fracture mode of the spine. Rather than fracturing along a specific crystallographic direction, as occurs for most single crystals, it fractures with conchoidal cleavage, that is, with a smooth rounded fracture surface, similar to that observed for glassy materials (Figure 3.4). The fact that it is a single crystal but does not fracture along any specific crystallographic direction is quite unlike almost all other crystalline ceramic materials. The mechanism for the toughening is still unknown but is clearly due at least in part to the nanoscale dispersion of protein molecules in the crystal lattice.

Fig. 3.4 Scanning electron micrograph of the fracture surface of a single-crystal spine from a sea urchin, demonstrating conchoidal cleavage. Adapted from [35]

Fig. 3.5 Scanning electron micrograph of a fracture surface of the nacre of the red abalone, showing a side view of the aragonite tablets. Adapted from [162]

Another sea creature that creates an organic/inorganic composite that is vital to the creature's survival is the abalone, as already mentioned. The nacreous (mother-of-pearl) layer of the abalone shell consists of alternating layers of 500-nm-thick aragonite platelets and ~30-nm-thick sheets of an organic matrix (Figure 3.5). The matrix forms first, then the mineral plates develop between the layers. The matrix defines the thickness of the crystals, and certainly also adds to the strength of the composite structure by forcing fractures to dissipate their energy as they travel a tortuous path through this elastic structure. The matrix also controls the crystallographic orientation of the mineral plates, which must also be important for strength. The resulting composite structure has a fracture toughness that is about 3000 times greater than that of inorganic aragonite [32]. Along with defining the size and shape of the aragonite plates, the organic matrix also likely serves to define their crystallographic orientation. The mechanism for formation of nacreous aragonite is not entirely clear, although it is known that the organic matrix forms first into a hollow layered structure, and then the voids in this structure are mineralized.

Another excellent example of a biological nanocomposite with complex structure and function is bone. Bone must perform multiple functions. It must have high strength, yet low weight; it must support remodeling in response to applied stresses, yet not deform under applied stress; it must contain pores to allow oxygen and nutrients to reach the cells within the bone, yet these same pores cannot lead to fractures; finally, it must act as a reservoir for minerals, but not allow itself to demineralize and thus weaken. Bone exists in several forms; lamellar bone provides the structural properties of long bones in mammals and other species and is the subject of the following discussion, adapted from [36, 37] (Figure 3.6). The basic structure of bone is a mineralized collagen fibril consisting of on average 65% mineral, the remainder being organic material and water. The collagen phase of bone is composed of individual fibrils about 80–100 nm in diameter. These fibrils are in turn composed of assemblies of a triple helix of polypeptide chains with an average diameter of 1.5 nm and length around 300 nm. The internal arrangement of this triple helix structure in the collagen fibril is not random: the helix aligns with its long axis parallel to the long axis of the bone and has additional fine structure. The mineral phase of the bone,

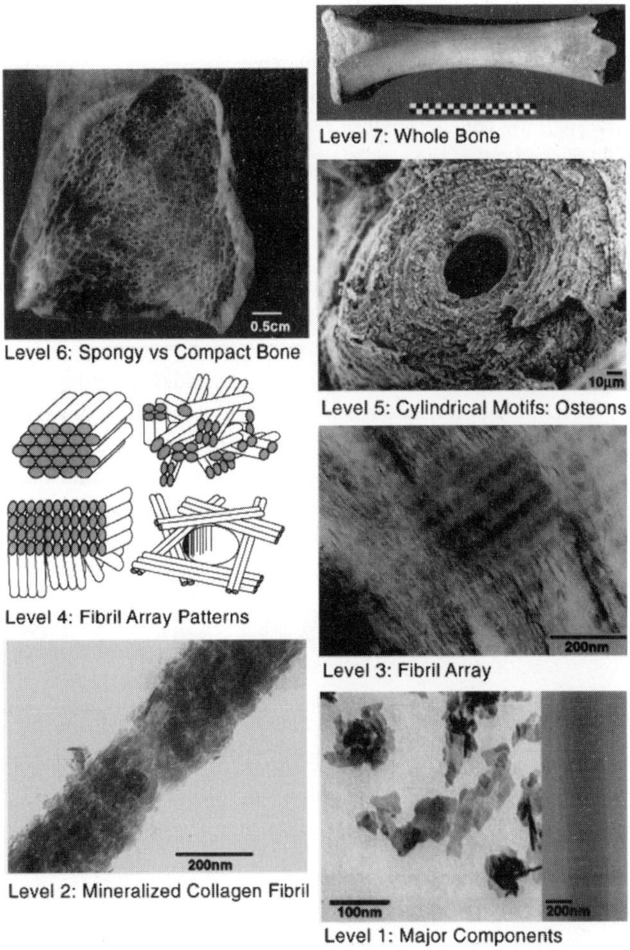

Fig. 3.6 The 7 hierarchical levels of organization of the bone family of materials, according to Weiner [37]. Level 1: Transmission electron micrographs of individual mineral crystals from human bone (left side) and an unmineralized, unstained collagen fibril from turkey tendon embedded in vitreous ice (right side). Level 2: Transmission electron micrograph of a mineralized collagen fibril from turkey tendon. Level 3: Transmission electron micrograph of a thin section of mineralized turkey tendon composed of multiple fibrils. Level 4: Four fibril packing motifs found in the bone. Level 5: Scanning electron micrograph of a single osteon from human bone. Level 6: Light micrograph of a fractured section through a 5500 year old fossilized human femur. Level 7: Whole bovine bone (scale: 10 cm). Adapted from [37]

carbonated apatite ($Ca_5(PO_4,CO_3)_3(OH)$), forms as 50-nm-long, 25-nm-wide platelets that are on average 1.5–4 nm thick (Figure 3.7). These platelets lie within the collagen fibrils, with only a small number of polypeptide chains between each platelet. The long axis of the platelets is generally parallel to the outer surface of the bone, most likely to maximize the compressive strength of the bone.

The mineral structure forms in bone after the collagen fibrils form. However, as the mineral forms, the collagen structure may reorganize to some extent. Thermodynamically, the carbonated apatite platelets should coarsen into larger single crystals over time (Ostwald ripening), but clearly this is prevented by the collagen matrix. Thus, although the collagen matrix reorganizes to some extent, it clearly cannot be completely pushed out of the way by the growing mineral. This level of control by biology over structural development during mineralization currently exceeds that found in synthetic materials. Additionally, after a bone is broken, new structures form which generate material that is as strong as the original bone; yet, the mineralization process stops when the healing process is complete. Bone thus is an excellent example of a self-healing biological nanocomposite in which an organic host phase is formed, followed by highly regulated mineralization processes. The mechanism for this process is not entirely clear, but substantial strides in understanding it are being made.

Clearly, biological systems can create a tremendous diversity of inorganic materials; however, biological systems are also very limited in terms of material selection. Metals, as well as many polymers, semiconductors, and ceramics cannot be synthesized by biological processes, because the precursors are not available in the natural environment, the materials or precursors are toxic, and/or the product is not stable in the

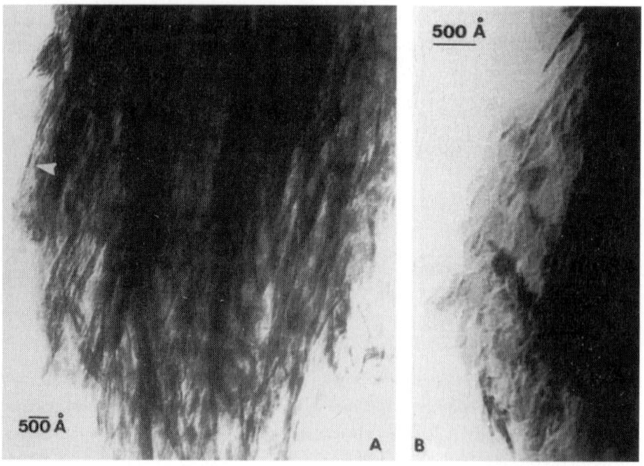

Fig. 3.7 Transmission electron micrograph of a fractured piece of mineralized bone from a 50-year-old human male femur. (a) Many mineral crystals are oriented edge-on and are parallel to the long direction of the collagen fibrils. (b) Higher magnification view of the bone fragment at the location in (a) marked with an arrowhead. Here, crystals are oriented face-on. The flat, plate-like nature of the mineral crystals in bone can be observed in these figures. Adapted from [163]

presence of water and/or oxygen. Because of this, there is serious interest in using elements of biological systems to create synthetic nanocomposites from materials not found in biology, thus achieving the best of both the biological and synthetic worlds.

3.3 Biologically Derived Synthetic Nanocomposites

Because of the limited variety of materials available by purely biological routes, there is great interest in utilizing the sophistication offered by biological systems in concert with synthetic procedures to create materials with otherwise unobtainable nanostructures and thus, potentially, unique properties. Biology offers a unique selection of building blocks that would be difficult or impossible to synthesize in the laboratory. These include proteins, DNA, RNA, and small but highly functional molecules.

3.3.1 Protein-Based Nanostructure Formation

As already mentioned, protein S-layers on the surface of bacteria create complex nanostructures. The protein S-layer present on the surface of some bacteria has now also been used to create entirely synthetic nanostructures. The diversity of S-layer structures, coupled with the potential for chemical functionalization, make them ideal starting points for nanostructure synthesis. To review, S-layers are 2D protein crystals that have oblique, square, or hexagonal lattice symmetry with lattice constants between 3 and 30 nm. Interesting for potential nanostructure formation, and to enable bacterial respiration, S-layers almost universally contain pores of identical size and identical surface chemistry – properties that make S-layers useful for nanostructure and nanocomposite fabrication. In an early example [38, 39], an S-layer was used to template a periodic structure into a thin metal film. Specifically, a 1-nm-thick metal (Ta/W) film containing 15-nm holes periodically arranged in a triangular lattice with a lattice constant of 22 nm was created by S-layer templating. This was accomplished by first depositing a suspension of S-layer fragments onto an amorphous carbon support film. After S-layer deposition, a 1.2-nm thick film of Ta/W was evaporated onto the S-layer at an angle of 40° from the normal of the substrate surface. When examined in the transmission electron microscope (TEM), the resulting film shows contrast indicating some thickness variation in the metal, but the contrast is not significant. To improve the contrast and to open up holes in the metal film, the film was argon-ion-milled for a short time. The result was a metal film that contained a periodic array of holes with the same symmetry as in the S-layer template [38, 39]. This work was done more than 15 years ago, yet creating such a nanostructure today would still be a challenge, even with a modern e-beam lithography system.

Much more recently [13], an S-layer was used to create nanostructured semiconductor films. The S-layer in this study had an oblique 2D lattice (space group p_1; $a = 9.8$

nm, $b = 7.5$ nm, $\theta = 80^\circ$). Depending on the mechanism of S-layer deposition, either its negatively charged inner face or its charge-neutral outer face was exposed to a 10 mM $CdCl_2$ solution. After drying, the cadmium-ion-doped S-layer was exposed to H_2S gas, resulting in formation of a nanocrystalline film of the semiconductor CdS on the S-layer. In the TEM, the CdS film exhibited a superlattice structure that was a direct copy of the structure of the S-layer [13]. The S-layer may have survived the entire process, and thus the actual mineral structure formed may be a nanocomposite of CdS and protein. It is still unclear what applications might be found for such materials, although robust metal or semiconductor nanoporous structures may find application as filter membranes, sensors, and optoelectronic devices.

Another protein assembly that has been used to form nanocomposite materials is ferritin (Figure 3.8). In its native form, ferritin consists of a supramolecular arrangement of proteins around an iron oxide core. The iron oxide core can be selectively dissolved without damaging the structure of the protein shell, yielding a hollow ball of protein about 10 nm in diameter. The demineralized protein (apoferritin) can be refilled with iron oxide, demonstrating that its structure was not greatly affected by the demineralization process. Interestingly, apoferritin can also be filled with other mineral nanoparticles. In one example [32], the iron oxide core of ferritin was converted to FeS by treating the ferritin with H_2S. Nanoparticles of MnOOH, UO_3, Fe_3O_4, and CdS can also be formed inside apoferritin through the appropriate chemi-

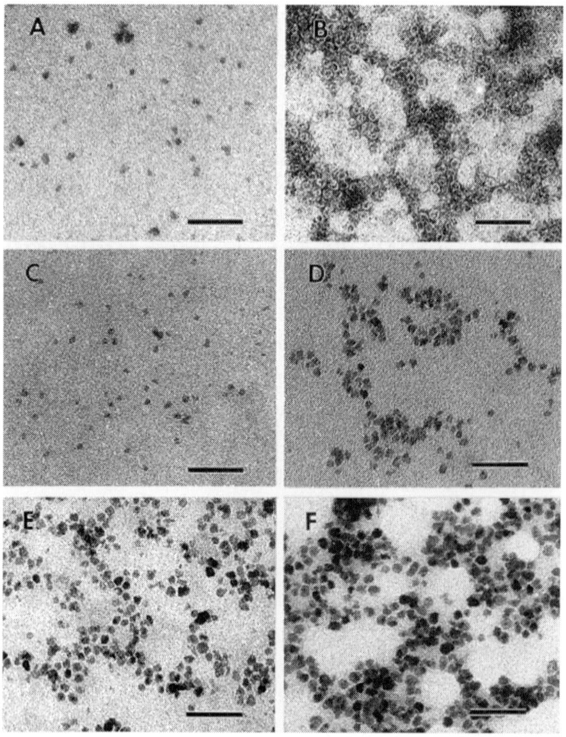

Fig. 3.8 Transmission electron micrographs of magnetite (Fe_3O_4)- and maghemite (γ-Fe_2O_3)-filled ferritins. (a) 260 Fe atoms/molecule, unstained; only the discrete electron-dense inorganic cores can be seen. (b) 260 Fe atoms/molecule, after staining with uranyl acetate, showing encapsulation of inorganic cores by intact protein shell (white halo around each particle). (c) 530 Fe atoms/molecule, unstained. (d) 1000 Fe atoms/molecule, unstained. (e) 2040 Fe atoms/molecule, unstained. (f) 3150 Fe atoms/molecule, unstained. Scale bars in all figures = 50 nm. Adapted from [164]

cal treatment. Because the protein shell is not disrupted by the mineralization process, the final product is truly a nanocomposite of protein and inorganic material. See [32] for an extended discussion of the use of apoferritin as a template for synthesizing nanocomposite materials.

3.3.2
DNA-Templated Nanostructure Formation

DNA offers great potential as a building block for nanocomposite materials. It can be tethered to a wide range of substrates, can direct assembly with specificities that greatly exceed that of any synthetic molecule, is relatively robust, can be synthesized in relatively large quantities, and can be functionalized with tags such as fluorescent molecules to enable rapid detection of binding events.

The use of DNA to assemble and create nanostructures and nanocomposite materials is only in its infancy, but even the preliminary work done so far indicates the great potential of DNA-based assembly techniques. A few of the approaches that have been studied to date include the mineralization of DNA, the use of DNA to assemble nanoparticles, and the use of DNA to assemble much larger colloidal particles. These three approaches are outlined in this chapter; however, the number of possibilities is vast, and significant work on DNA-mediated assembly of nanostructures is currently being done.

The possibility of using plasmid DNA as a template for mineralization was first explored in 1996 [40]. In this work, single strands of a 3455-basepair circular plasmid DNA were mineralized with CdS nanoparticles, by mixing the plasmid and cadmium perchlorate in solution, followed by spin coating this solution onto a polylysine-coated glass slide. The DNA/cadmium perchlorate-coated glass slide was then exposed to H_2S, converting the cadmium perchlorate to CdS, which preferentially mineralized the DNA. This procedure results in ring-like structures consisting of DNA embedded within a 5–10-nm-thick CdS strand that can be directly observed in the TEM (Figure 3.9). The diameter of the ring formed by the CdS-encrusted DNA is directly related to the diameter of the DNA plasmid, and the thickness of the CdS/DNA composite strand is about 10 nm. This represents the first example of DNA templating of semi-

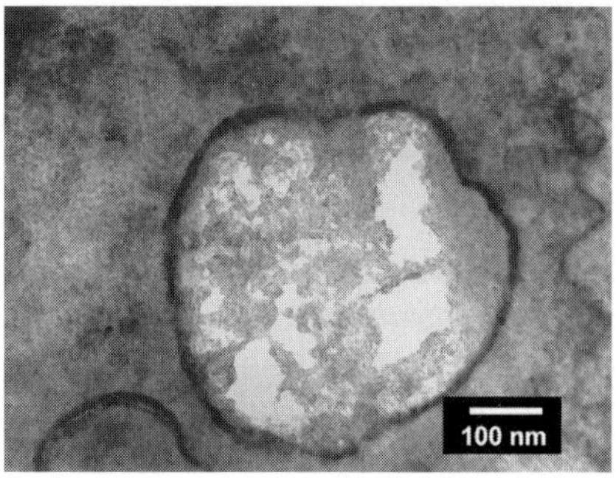

Fig. 3.9 Transmission electron micrograph of a nanostructure obtained by mineralization of a circular plasmid DNA with CdS nanoparticles. The ring diameter closely matches the predicted diameter of the plasmid DNA. Adapted from [40]

conductors but, given the vast array of structures that can be formed from DNA, should not be the last.

Another example of DNA-based nanostructure development can be seen in the work on DNA-based nanoparticle assembly. Several approaches have been explored, with the majority based on the Watson–Crick base-pairing of DNA strands attached to various nanoparticles. The first approach involved the functionalization of gold nanoparticles with multiple strands of thiol-terminated type-A or -C DNA in separate reactions [41] (where 'type' is used as a label for different nucleotide sequences). The thiol terminal group is important, because it enables covalent attachment of the DNA strand to the gold nanoparticles. Importantly, the A and C strands were designed so that they would not hybridize. The result, after mixing the solutions of each kind of DNA-functionalized nanoparticles, is a simple mixture of DNA-functionalized gold of type A and DNA-functionalized gold of type C. Then a single strand of DNA with one end complimentary to the type-A DNA and the other end complimentary to the type-C DNA was added. We label this strand B-D. Upon hybridization of the B-D strand with both nanoparticles of type A and nanoparticles of type C, a DNA-linked aggregate of nanoparticles was formed. Aggregate formation is thermally reversible by increasing the temperature above the melting point of the DNA, at which point the strands dehybridize. The assembly could be monitored by UV/visible spectroscopy, because the UV/visible absorption of gold nanoparticles changes greatly when they are close together, due to changes in the surface plasmon.

The formation of controlled aggregates, although interesting, is really a starting point for the creation of much more complex structure through a marriage of biological macromolecules and synthetic materials. In addition to placing multiple DNA molecules on a nanoparticle, it is possible to place a single or a finite number of DNA molecules on a nanoparticle. However, this requires a significant degree of synthetic effort, because one usually starts with a population of nanoparticles with various numbers of attached DNA molecules. Through careful separation, nanoparticles with exactly one or exactly two single-stranded DNA molecules on their surface can be recovered. These DNA-functionalized particles can then be assembled in highly controlled fashions through DNA hybridization reactions, by adding single-stranded DNA that is complimentary to the single-stranded DNA on the nanoparticles (Figure 3.10). This is similar to the approach used to make DNA-linked aggregates; however, because each nanoparticle contains only one DNA strand, not multiple strands, the result after hybridization is not the formation of an aggregated structure, but rather, individual dimers of gold nanoparticles. Trimers of gold nanoparticles were also created by a similar approach, except that nanoparticles with one strand of DNA of three different sequences was mixed with a strand of DNA complimentary to all three of the DNA strands on the surface of the nanoparticles [42, 43].

Recently, it was demonstrated that DNA can assemble much larger colloidal particles into 3D assemblies of controlled shape. The basic approach is very similar to the assembly of nanoparticles; however, here the results after hybridization are assemblies of colloidal particles with defined connectivity [44].

The overall power of DNA-based assembly of nanostructures is based on several powerful aspects of DNA hybridization. First, hybridization is reversible with tempera-

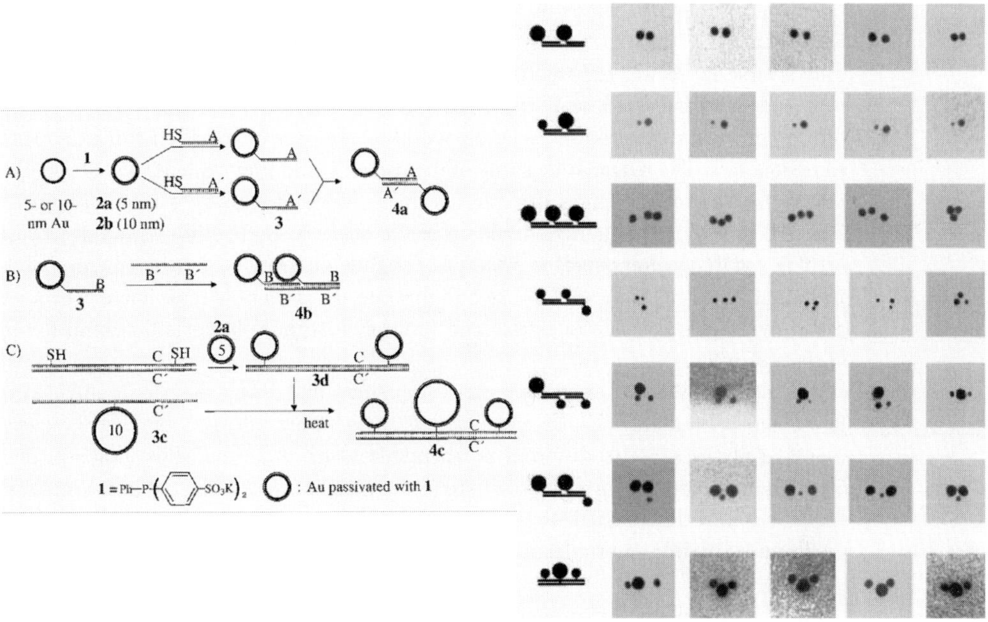

Fig. 3.10 DNA allows for precise manipulation of the order and spatial arrangement of nanoparticles. The large and small nanoparticles can be designed to hybridize or react with a single site on the DNA backbone. See examples A, B, and C on the left and the resulting nanostructures on the right. Adapted from [43]

ture; thus it is possible to anneal structures that at least partially eliminate kinetic traps that may be present in the assembly process. For example, if during nanostructure assembly, a particle does not hybridize in the correct location, the structure can be gently heated, releasing the particle, and then the particle can be retrapped by cooling below the melting point of the DNA. Second, DNA hybridization is very specific. Unlike most organic linking chemistries, which cannot distinguish small changes in molecular structure, DNA can be designed to hybridize only to its exact complimentary strand. Thus it may be possible to combine many DNA-functionalized nanoparticles into very specific arrays by hybridizing to a long strand of DNA that is encoded so as to react with each DNA functionalized nanoparticle at only a very specific location.

3.3.3 Protein Assembly

To achieve a level of complexity similar to what can be obtained by DNA-based assembly, one can engineer or select biological organisms that recognize and bind with high specificity to specific minerals, semiconductors, and metals. In the first example [45], *Escherichia coli* containing genetic sequences specific for recognizing and binding iron oxide but not other metal surfaces was identified and multiplied. Bacteria that would specifically adhere to iron oxide were identified via serial enrichment from a population of bacteria. The experimental procedure was as follows: A population of genetically diverse *E. coli* was exposed to iron oxide particles. The bacteria that bound to the iron oxide were collected, and the remainder were discarded. After repeating this procedure for several generations, bacteria with high degrees of specificity for iron oxide were obtained. The reason only some of the bacteria bind iron oxide is that *E. coli*, like many other bacteria, express proteins on their outer surface whose sequences are a function of the unique genetic makeup of the bacteria. By starting with a library of $\sim 10^7$ *E. coli* that express surface proteins of slightly different sequences, the probability that at least one of the *E. coli* has a highly specific protein is reasonable. Through enrichment, it is possible to extract and multiply just those bacteria that bind to the surface of interest.

It is also possible to select for protein sequences that attach to specific metal surfaces, and presumably also to metal nanoparticles. Through such biological experiments, a library of $\sim 10^7$ different polypeptides 14 or 28 amino acids long was created. Then, the polypeptides that adhere to the metal surface were isolated, and their sequences were determined [46]. This approach is exceedingly powerful because of the vast number of different molecular sequences that can be studied in parallel in a single experiment.

From a technological standpoint, it is interesting to consider the possibility of recognizing semiconductor surfaces and semiconductor nanoparticles with this procedure. If an appropriate genetic sequence can be identified, it would be possible to assemble nanoparticles via biological organisms that are designed to present the appropriate biological macromolecules on their surface. Phage display enables the simultaneous testing of many peptide sequences for specificity to a given surface. A phage library containing 10^9 polypeptide sequences was exposed to a surface, and all the phages that did not bind to the surface were washed away. The phages that stuck were then removed from the surface by lowering the pH and were amplified by infecting *E. coli* bacteria with them. This process was repeated until only phages that stuck strongly to the surface were present. The DNA of these phages was then sequenced to determine the specific peptide sequence(s) that bind with such high affinities to the specific surface of interest [47].

A phage that identifies and binds to a surface of a specific material also binds strongly to nanoparticles of this material. The result of mixing a nanoparticle-binding phage with nanoparticles is 'decoration' of the phage with nanoparticles, which occurs only at the end of the phage that contains the specific binding polypeptide, and nowhere else on the phage [48]. Depending on the experimental conditions and phage design, it may even be possible to bind single nanoparticles to an individual phage or to assemble the nanoparticles into defined structures, although in the present state of the art, multiple nanoparticle are bound in a fairly random fashion to each phage.

3.4
Biologically Inspired Nanocomposites

The properties of biocomposites and synthetic pathways for their formation have inspired wide ranging research. However, early on it was recognized that it is not always necessary or even desirable to use biologically derived materials for many applications and that it may be possible simply to use biology as an inspiration for totally synthetic nanocomposite systems. It is interesting to consider biological systems as an inspiration for nanocomposite materials, because biological systems exhibit many characteristics that would be attractive in synthetic materials, but it is also clear that direct mimicking of biology will be limited to a specific small subset of materials and to specific nanostructures. However, many lessons can be learned from biology on how to form complicated nanostructures and on the potential properties of these synthetic nanostructured materials, should one be successful in synthesizing them. Of course, just because a synthetic material resembles a natural process and the process of forming that material resembles a natural synthesis does not always mean that the scientists and engineers who performed the work were inspired by biology. Often it is not stated whether biology was an inspiration for the work, and thus some care must be taken before assuming that, just because something appears biologically inspired, it is biologically inspired. In this chapter, I do not attempt to make this distinction and by necessity assume that, if the work has a biological analog, then it can be considered 'biologically inspired', although certainly in some cases this assumption may be wrong. In this chapter I take the liberty to lump such work under the heading of 'Biologically Inspired Nanocomposites'.

Much can be learned from biological systems to further the development of synthetic approaches to the formation of complex inorganic structures. Each of the routes to nanostructure formation that is discussed in this section – liquid crystal templating, colloidal particle templating, block copolymer templating, and surfactant-inorganic self-assembly (mesoporous silica being the most famous of this approach to nanostructure formation) – invokes many of the tenets of biologically directed mineral growth. As already discussed, biological systems rely on self assembly and mineralization in the synthesis of hard inorganic structures such as shells, teeth, and bone [11], and their approaches to materials fabrication can provide guidance and direction to synthetic systems. Often, the term 'biomimetic' has been applied to any approach using self assembly in the synthesis of nanocomposite materials. As mentioned above, biological systems do indeed use self-assembling molecules, and high levels of molecular organization are a very important part of an organism's inorganic structure development. However, it would be naive to expect to simulate this process in the laboratory except on the most basic level. Biological processes are extremely dynamic, involving huge numbers of very specific proteins and other molecules being generated and transported to very specific locations, with temporal control. The best synthetic systems are indeed very simple approximations of living systems and generally are much too simple to be considered to be truly mimicking biology. Although the synthetic systems are only simple approximations of life, still much can be learned by attempting to mimic living systems [8], even if biomimetic is perhaps not the best term. For example, many attempts centered around the synthesis of mineral phases in a self-organized matrix have indeed mimicked the mineralization processes of many biological systems. Examples of matrix-mediated biological mineralization processes include the reliance of bacteria, plants, shells, and even mammals on organically mediated growth of mineral phases to eliminate byproducts (bacteria and plants) [31, 33, 34], create exoskeletons (shells) [49], and grow teeth (mammals) [35]. If during the synthetic process, organic molecules are incorporated into the mineral phase, the resulting material may resemble the spicule of a sea urchin, which contains much less than 1% protein intercalated into the crystal lattice [35].

Despite the low degree of sophistication of synthetic systems, they do have very distinct advantages over biologically based schemes. Biological systems operate with only a limited subset of elements and compounds, but synthetic systems can be designed to use a wider range of elements and compounds, many of which would be toxic to most living organisms. Biological systems form and operate near room temperature in the presence of water and oxygen (for the most part), but synthetic systems can be formed and operate under a wide range of temperature and conditions. Finally, biological materials generally take days to years to form, but synthetic systems may be formed rapidly. For biology, systems that form slowly in response to external stimuli have significant advantages, for example, bone remodels to meet the demands of applied loads. However, in general, the long time scales required for formation of biological nanocomposites limit the application of direct biological synthesis of engineering materials to a few very specific cases.

Through self-assembly of organic molecules, biological systems have succeeded in synthesizing a wide range of composite and inorganic nanostructures. Biological systems contain large quantities of lipids, or soap-like molecules, which self-assemble (along with many other biological macromolecules, including proteins) to form the external membranes of cells, as well as smaller vesicles within the cell. These membranes serve to protect the interior contents, as well as to provide synthetic microreactors for biological processes. The concept of performing syntheses in self-assembled microreactors is in this sense biologically inspired, has been exploited by many researchers to create nanoparticles and composite nanostructures, and is the topic of this section.

The simplest examples of nanomaterials are zero-dimensional materials, known as 'nanoparticles'. The study of small semiconductor particles in general is of significant interest, because the electronic and optical properties of semiconductor nanoparticles change drastically as the characteristic dimension of a particle is reduced to the nanoscale regime, largely due to the quantum confinement of electrons within the particle, although surface-area effects may also play a role (Figures 3.11, 3.12) [50–52]. The most popular nanoparticles for basic studies have been II–VI semiconductor particles, because of their scientifically interesting and technologically important properties. In this context, synthetic methodologies for the formation of metal sulfide and selenide quantum dots and their assembly into higher-order structures have been widely studied [53–57]. For nanostructuring to dominate the properties of metal sulfides and other semiconductors, it is generally necessary for the characteristic dimension of the nanoparticle to be <10 nm [58]. Semiconducting nanoparticles are generally synthesized through one of three routes. The first, grinding of large chunks is rarely done for nanoparticle preparation, because the grinding process is poorly regulated, generally generating a very polydisperse population of particles, and grinding introduces too many contaminates for most applications. The other two methods, gas-phase and solution-phase synthesis, are much more common. Gas-phase synthesis is essentially a vaporization and condensation process – a crucible containing the desired semiconductor (or other material) is heated until it starts to sublime, and then an inert carrier gas is flowed over the material. The carrier gas then heads into a cool region where the gaseous semiconductor atoms or molecules condense into nanoparticles and are collected [59]. Although this method is fairly versatile, it operates only under conditions of high temperature and vacuum and generally produces solid spherical particles.

Solution-based synthetic routes for nanoparticle formation have ranged from simple precipitation reactions to much more complex self-assembly-based routes. In general, simple precipitation results in agglomerates of nanoparticles, and the size distribution generally varies widely. These problems led to research into synthetic procedures that would result in nanoparticles that are stable against aggregation and have narrow size distributions. A primary route to preventing both these problems is to use self-assembly-based techniques, which in many respects resemble nanostructure development in biological systems, including biomineralization, cell membrane development, and other biological structure formation. Solution-phase synthesis of semiconductors is often preferred over other techniques, because it is generally mild (even being carried out at room temperature and pressures) and can be used to create reasonable volumes of materials. Solution-phase synthesis has been widely used to grow semiconductor

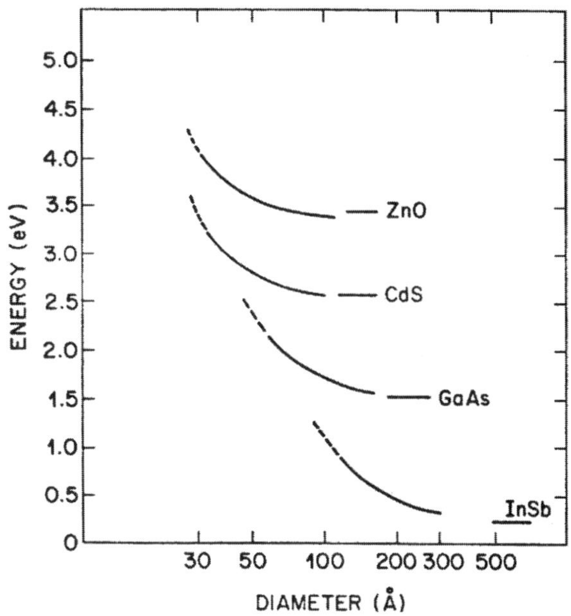

Fig. 3.11 Theoretical calculation of band-gap energy as a function of particle diameter for several different semiconductors. Adapted from [58]

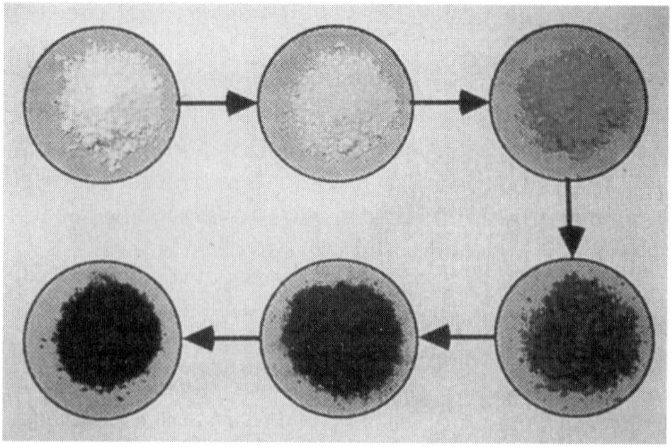

Fig. 3.12 Cd_3P_2 quantum dots. Particle size increases in the direction of the arrows. The white particles are about 1.5 nm, and the black particles are larger than 5 nm. Adapted from [50]

quantum dots, yielding particles with low polydispersities and novel optical properties [50, 51, 56, 57, 60]. Solution-based chemical syntheses are very attractive because they allow for direct control over the actual concentrations of the chemical precursors. Depending on actual conditions, it is even possible to cap the surface of the particles with organic molecules, which allows for further solution-based processing [53, 55].

The more conventional route to creating nanostructured materials is of course through top-down lithographic methods. Some examples of top-down techniques for generating very small features are extreme UV ($\lambda << 200$ nm) lithography [61, 62], electron beam writing [63], focused ion-beam lithography [64], x-ray-lithography [65], scanning probe lithography [66], and microcontact printing [67]. In general, all these techniques can form structures on the scale of tens to hundreds of nanometers, although generally only on very flat substrates, and they can be quite slow and often very expensive. The self-assembly-based route to nanostructure formation has the significant advantage that it is not limited to feature generation on flat surfaces, and it can be massively parallel. Of course, the general problem with self-assembly is that it is not possible to highly regulate the exact spatial position of the nanostructure, and thus we are still many years away from creating highly functional self-assembled electronic circuits.

In the micellar routes to nanoparticle formation, micelles are self-assembled from surfactant molecules and a solvent that contains at least one of the precursors for the inorganic nanoparticles in solution. The result is a solution that contains vast numbers of discrete nanoreactors which individually contain only a finite number of precursor species for the inorganic phase. When these ions are converted to mineral, generally by reduction or oxidation, the result can be one nanoparticle per micelle. The polydispersity in nanoparticle diameter is thus directly related to the polydispersity in the initial micelle size. If it were possible to create a suspension of monodisperse micelles, it would be a straightforward process to create nanoparticles with a very narrow size distribution. Taking a lesson from biology, this might be possible through the use of complex macromolecules that organize into particles of only a specific size. A very good biological example of a potential nanoreactor with a tight size distribution is a virus particle. It may be possible to use virus particles to synthesize nanoparticles with tight size distributions if a way to load the interior of the virus particles with precursors for nanoparticles can be developed.

Most of the early nanoparticle synthesis routes produced solid semiconductor particles with morphologies never far from spherical. The creation of nanoparticles with complex morphologies is most interesting, because even such sophisticated patterning techniques as e-beam lithography are limited to ~10-nm features, which are often too large to result in the desired quantum confinement and other properties [63]. Nanoparticles having certain complex morphologies cannot be created through e-beam patterning or other top-down processing routes. The possibility of generating complicated

Fig. 3.13 Transmission electron micrograph of nanocrystalline CdS structure grown under an arachidic acid monolayer at room temperature. Scale bar = 200 nm. Adapted from [165]

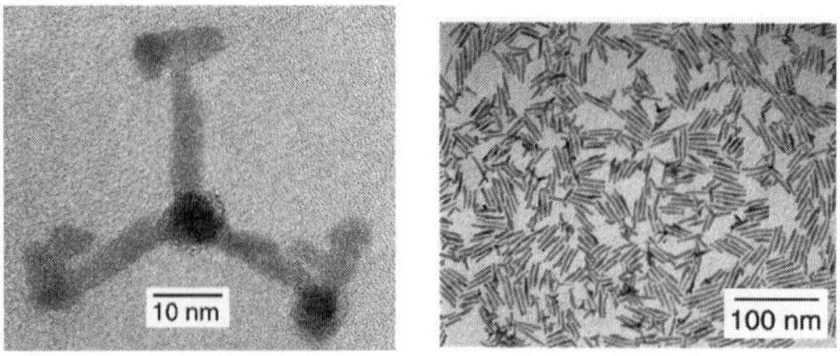

Fig. 3.14 Examples of CdSe nanoparticles with complex shape and form that can be created by solution synthesis from a mixture of surface passivating agents. Adapted from [70]

morphologies was examined in studies of CdS synthesized under Langmuir monolayers, in which dendritic structures were generated (Figure 3.13) [68]. Subsequently, a range of strategies have resulted in rod-like and even complex nanoparticles with complex morphologies that are based on self-assembly regulating the growth of semiconductor particles.

As examples of self-assembly processes which create nanoparticles with complex shapes, Alivisatos and coworkers [69, 70] demonstrated the formation of CdSe nanorods with aspect ratios of 30:1, as well as arrow-, teardrop-, tetrapod-, and branched tetrapod-shaped nanocrystals of CdSe (Figure 3.14). These highly shaped nanoparticles result from using a mixture of hexylphosphonic acid and trioctylphosphine oxide as passivating agents in the synthesis. Apparently, it is possible to block the growth of specific crystallographic faces while encouraging the growth of other faces. Thus, particles that have one crystallographic direction as their long axis, and their short axis as another crystallographic direction can be formed. The advantage of high aspect ratio particles over normal, spherical particles has not yet been proven, but the electronic or physical properties of the particles might be impacted by the change in shape. In addition, the particles might be able to self-assemble into higher-order structures because of their high aspect ratios; for example, they may form liquid crystalline phases, much like molecules with high aspect ratios [69, 70].

Semiconductor nanostructures may also have potential nonquantum confinement-based properties not found in the bulk equivalent. Besides quantum confinement, one way to greatly modify the properties of nanostructured semiconductors may be to design synthetic methodologies that result in dispersion of organic molecules within the inorganic phase at the molecular level. These composite materials could exhibit novel properties significantly enhanced over those of either the inorganic or organic phase alone, as has actually been observed in a wide range of materials. Composite materials that are tougher [71–73], have increased thermal stability [74], are electronically more sophisticated [75,76], or have enhanced chemical selectivity [77] than either of the constituent parts have been created. Even without the incorporation of organic material, periodically nanostructured semiconductors have great potential in solid-state science and technology, for example because of their potential for both electronic and catalytic activity. For example, a periodically nanostructured semiconductor might behave as an array of antidots (a material with a regular array of scattering centers spaced closer than the mean free path of electrons traveling through them) [78, 79]. A nanostructural material is necessary because, if the mean free path of the electrons is shorter than the spacing between the scattering centers, the antidot lattice does not operate. At high magnetic fields (>2 Tesla), quantum steps in the conductivity as a function of electric field may be observed if the lattice spacing of the antidots is on the order of the cyclotron diameter of the electrons. In contrast to forming a solid of quantum dots, the formation of an antidot lattice requires the semiconducting structures to be continuous, with a periodic array of nanocavities.

Early on, the greatest emphasis was on creating nanoparticles with narrow size distributions, not on creating superlattice structures. However, through careful control of size distribution and chemical functionality, CdSe nanocrystals, and now many other nanocrystals, have been observed to order into superlattice structures [55]. These structures may present properties beyond simple quantum effects. The individual crystallites in this system do not form a continuous mineral structure, but are actually separated by thin layers of organic molecules, which are composed of the self-assembling molecules used in the synthesis to regulate the diameter and polydispersity of the particles. During synthesis, the organic self-assembles into a shell around the nanoparticles, imparting organic solubility to the nanoparticles, which enables them to be processed similar to organic compounds. Because of the high degree of regularity in size and shape, these organically coated nanoparticles assemble into a crystal of nanoparticles with long range periodicity, much in the way that organic molecules and atoms can crystallize.

The next level of complexity in nanostructure formation is the creation of nanostructures with complex, predefined morphologies. Here, the biological concepts of self-assembly and nanostructure formation become most applicable. For example, in biology, it is common to have complex predefined structures on the nanometer scale, yet this length scale is exceedingly difficult to regulate in synthetic materials. However, if the power of self-assembly is coupled with the materials synthesis strategies known today, there is great potential for the formation of complex composite nanostructures.

Liquid-crystal templating of inorganic nanostructures, an approach in which the periodic structure of liquid crystals is imparted to a mineral phase, is one such route. Liquid crystals present an ideal matrix for the creation of nanoscale composite materials, because the characteristic 1–10 nm length scales most often expressed in liquid crystals are similar to the size scale of interest for semiconductor nanostructures. Furthermore, the periodic structure in liquid crystals can be quite long-range, and thus the periodic nanostructure also has the potential for long-range order, which is of exceptional interest for many applications. A very important goal of several research groups is in fact the creation of long-range nanoperiodic order in materials such as semiconductors through liquid-crystal templating.

Semiconductors with long-range periodic composite structures could come in several forms, for example as a particle containing a periodic array of embedded second-phase material, as a thin film with a periodic topography, or as a periodically porous material. Highly porous, periodically nanostructured semiconductor particles could be quite interesting for solution-based chemistry. For example, the photochemical nature of their semiconducting phase and zeolite-like pore structure could make them highly applicable for photochemical degradation of toxic compounds or for performing shape-selective chemistry, that is, chemistry that operates only on molecules of specific shape and size. Thin films may be even more technologically important, given the wide range of potential uses for both supported and freestanding thin films. The ability to predefine a nanoperiodic array of features in semiconductor thin films may open up many applications, including electronic devices, sensors, and filter membranes. Three-dimensional semiconductor structures may have unique optical or electronic properties, depending on the characteristic length scale of the structure. As the length approaches hundreds of nanometers, the materials may even exhibit photonic bandgap effects. Because direct top-down patterning of <10-nm-long nanostructures is difficult or impossible, and the patterning of 3D nanostructures of almost any length scale is difficult, templating of nanostructures through self-assembly processes has significant promise. Essentially, self-assembly-based templating can take place in either 2 or 3D, and the templating agent may or may not be removed, as desired. Here, we discuss several templating methodologies with potential for nanocomposite formation, the properties of such templated materials, and their potential applications.

3.4.1
Lyotropic Liquid-Crystal Templating

Lyotropic liquid-crystal templating is an approach for nanostructure and nanocomposite formation that utilizes the self-assembled structure of a liquid crystal to regulate the structure of a growing inorganic material. When processed correctly, the structure of the inorganic phase directly replicates the structure of the liquid crystal; thus the liquid crystal is a 'template' for the inorganic. The most important aspect of liquid-crystal templating of inorganic material is the lyotropic liquid crystal; thus, we should to review some of the basics of lyotropic liquid crystals. Lyotropic liquid crystals are composed of at least two covalently linked components, one of which is usually an amphiphile, which is a molecule that has two or more physically distinct components, and the other a solvent. Typically, one of the components making up the amphiphile is hydrophobic, and the other component is hydrophilic, as in common soaps, although this is not the necessary distinction. The dual solvent properties of an amphiphile lead to the interesting self-assembly of these molecules in solution by means that include surface segregation, formation of micelles [80,81] and vesicles [82,83], as well as the formation of a wide range of LC structures [84,85].

Most amphiphilic molecules contain two or occasionally three segments, of which at least one is water soluble. Clearly, if half a molecule dissolves in a solvent and the other half does not, interesting self-assembled structures may result. Within the general class of amphiphilic molecules are four distinct subclasses: cationic, anionic, zwitterionic, and nonionic. Cationic, anionic, and zwitterionic amphiphiles all contain a formally charged polar moiety, typically called the head group, and a nonpolar moiety, typically termed the tail. As the names imply, cationic amphiphiles contain a cationic head group, such as a quaternary ammonium salt, and anionic amphiphiles, an anionic head group such as a sulfonate salt. Zwitterionic amphiphiles contain a head group having both positive and negative charge, for example trimethylammonium phosphonate. Nonionic amphiphiles do not contain charged functional groups, but instead contain polar segments such as oligo(ethylene oxide) or oligo(vinyl alcohol). Examples of anionic and nonionic amphiphiles, sodium octanoate and oligo(ethylene oxide) oleyl ether, respectively, are shown below.

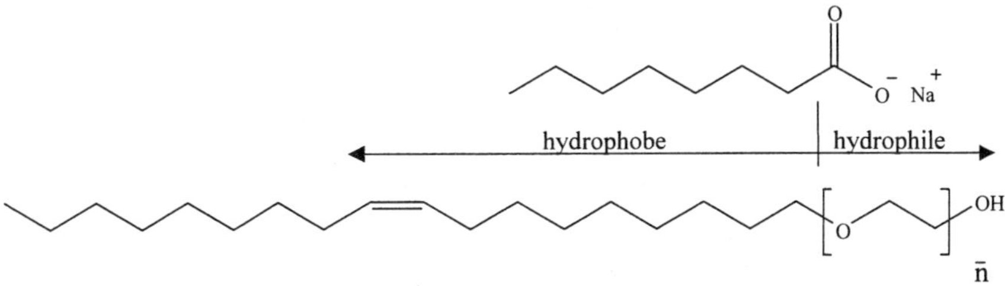

Amphiphiles exhibit very rich, complex phase behavior as a function of solvent concentration. In the dilute amphiphile limit, structures such as micelles form, and in the concentrated amphiphile limit (even solvent free), some amphiphiles show liquid crystalline or crystalline phases. Between these endpoints, a wide range of phases and structures can form. A typical nonionic amphiphile, oligoethylene oxide ($\overline{10}$) oleyl ether [(EO)$_{\overline{10}}$ oleyl], forms micelles, micellar rods, and hexagonal, cubic, and lamellar liquid crystals as the amphiphile:water ratio increases. This rich self assembling behavior occurs because the polar segment is readily solubilized by water, and the non-polar tail is not. For a single molecule in solution, other than curling up on its self (which is entropically very unfavorable), there is no physical way to reduce the unfavorable interactions, but if multiple molecules are allowed to associate, or self-assemble, the unfavorable interactions can be reduced. Determination of the structure of this minimum-energy aggregate is beyond the scope of this book, but essentially one can calculate the free energies of each possible phase for a particular concentration and temperature and then select the lowest-energy phase [86].

The three most common phases observed in mixtures of water and amphiphile are hexagonal, lamellar and cubic. Correctly stated, the cubic phase encompasses a wide range of potential phases, all exhibiting cubic symmetry but with differing degrees of continuity in the hydrophilic or hydrophobic phases. Similar to the structures observed for block copolymers [87], the structures observed in lyotropic liquid crystals can be rationalized as a function of the volume fraction of the various components. Figure 3.15 shows representative phase diagrams for two nonionic amphiphiles: the first liquid crystalline phase observed as the amphiphile concentration is increased is the hexagonal phase, which is in essence the hexagonal close packing of rod-like micelles (Figure 3.16). These rod-like micelles develop in solution as the concentration of the amphiphile increases, but below a critical concentration do not pack closely. As the concentration of amphiphile is increased further, the spacing between the rod centers decreases, and eventually a critical point is reached where a bicontinuous cubic phase is formed (Figure 3.17). Finally, at high enough concentration of amphiphile, a lamellar phase forms (Figure 3.18). In some systems, especially those composed of nonionic triblock amphiphiles, a close-packed cubic phase is observed, usually at lower concentrations of amphiphile than the hexagonal phase (Figure 3.19) [88, 89].

Rather then relying on coassembly, which is the process by which nanostructures develop in many mesoporous oxide systems, many biological processes utilize the order present in a preformed structure to form nanostructured inorganics. Substantial efforts have been made to utilize the order present in an organic mesophase to

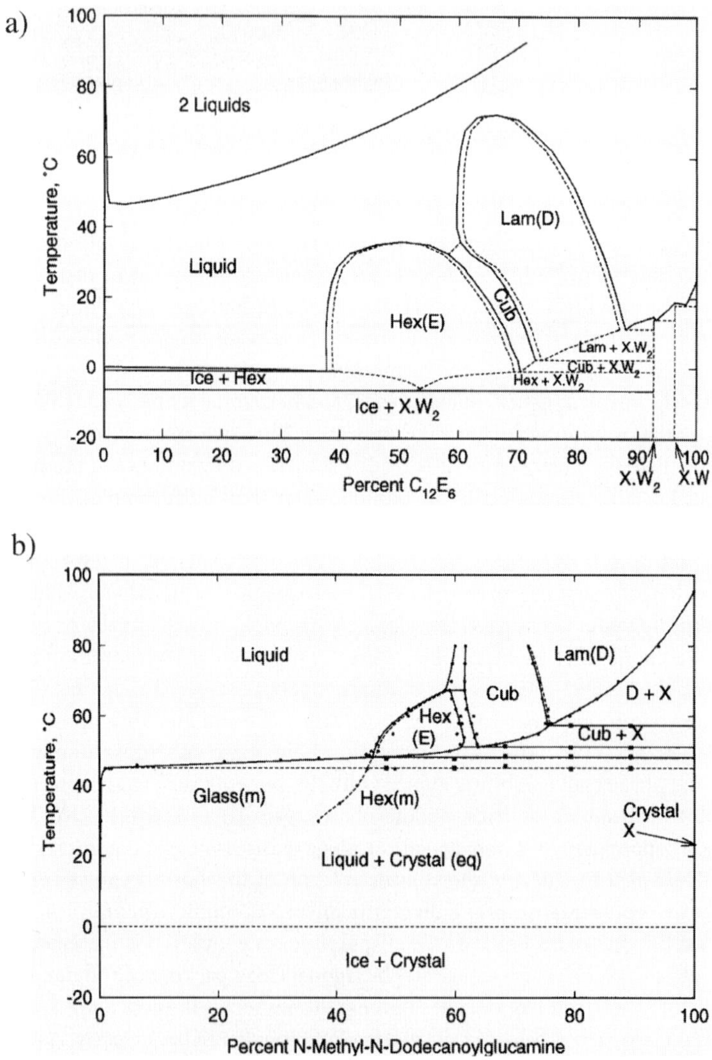

Fig. 3.15 Phase diagrams for nonionic amphiphile/water systems. (a) $C_{12}E_6$ (C_{12} represents the number of carbons in the tail, and E_6 represents the number of ethylene oxide groups in the head), (b) N-Methyl-N-dodecanoylglucamine. Adapted from [84]

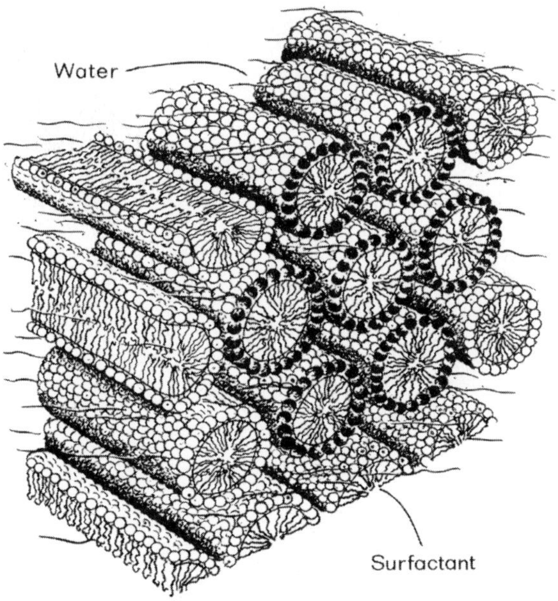

Fig. 3.16 Schematic illustration of the hexagonal phase. The polar head group of the surfactant points out into the water phase, and the hydrophobic tail inserts into the center of the rod-like micelles. Adapted from [166]

directly template the growth of an inorganic phase; liquid crystal templating has been one of the most successful of these approaches. Early efforts in this area resulted only in oblong or cubic crystallites or microporous reticulated structures [90–92]. More recently, through liquid-crystal templating, the successful synthesis of periodically nanostructured semiconductors that copied directly the symmetry and dimensionality of the precursor liquid crystal was demonstrated [11, 93–96]. Liquid-crystal templating appears to be a general route to the synthesis of semiconductor nanostructures. The general concept of liquid-crystal templating is to first form a liquid crystal that contains at least one of the precursors of the mineral phase, and then to induce a mineral phase to precipitate in only one chemical region of the liquid crystal by applying an outside

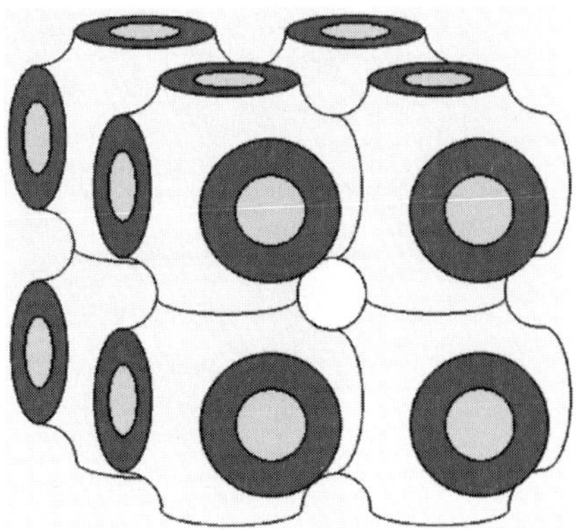

Fig. 3.17 Schematic illustration of a bicontinuous phase. The light gray region in the center of each circle indicates the hydrophobic segments, and the dark gray, the hydrophilic segments

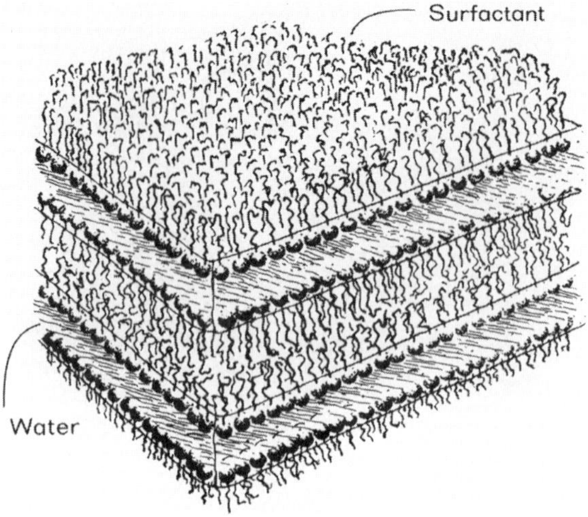

Fig. 3.18 Schematic illustration of the lamellar phase. Adapted from [166]

perturbation. Obviously, it is very important to select the correct synthetic conditions, mesophase, and mineral phase to be successful in this process. The versatility of this process has been demonstrated by the fact that materials have been formed in liquid-crystal phases, including the hexagonal, lamellar, and cubic phases, and the materials have included the already mentioned II–VI semiconductors; periodically nanostructured metals, both as thin films and in bulk [97–102]; as well as films of the chacogenides selenium and tellurium [103, 104]. Not only are the characteristic dimensions of the materials synthesized by liquid-crystal templating smaller than those obtainable

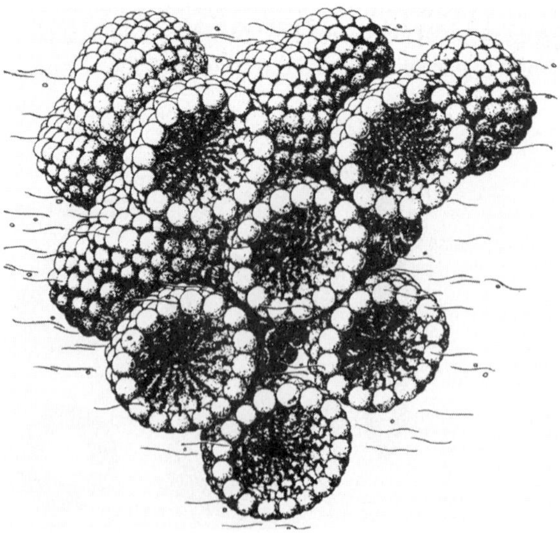

Fig. 3.19 Schematic illustration of the simple cubic phase. Adapted from [166].

by lithographic techniques, but they are often attainable through bulk synthesis, which is obviously not possible via lithography.

The direct templating of materials in the preordered environment of a nonionic amphiphilic mesophase generates semiconductor/organic superlattices containing both the symmetry and the long-range order of the precursor liquid crystal. In the templated growth of II–VI semiconductors, the semiconductor is grown in a water-containing liquid crystal by reaction of H_2S or H_2Se with a dissolved salt such as $Cd(NO_3)_2$. Both the chemical nature and the structure of the amphiphile are important for direct templating. For example, the order obtained in the nanostructured systems was even observed to be dependent on the counterion for the metal [95].

One advantage of direct templating with liquid crystals as a route to nanostructure formation, as was already discussed, is that there are a large number of amphiphilic liquid crystals, with lattice constants ranging from a few nanometers to tens of nanometers, and which include lamellar, hexagonal, cubic, and bicontinuous phases [84, 105, 106]. Potentially, many of these systems can be mineralized, generating materials with an array of novel structures and properties.

Specifically, II–VI semiconductors have been directly templated by hexagonal liquid crystals based on $(EO)_{\overline{10}}$ oleyl [11, 93, 95], and a lamellar liquid crystal formed from oligo(vinyl alcohol) $(\overline{23})$ oleyl ester [93]. As expected, the hexagonal liquid crystal

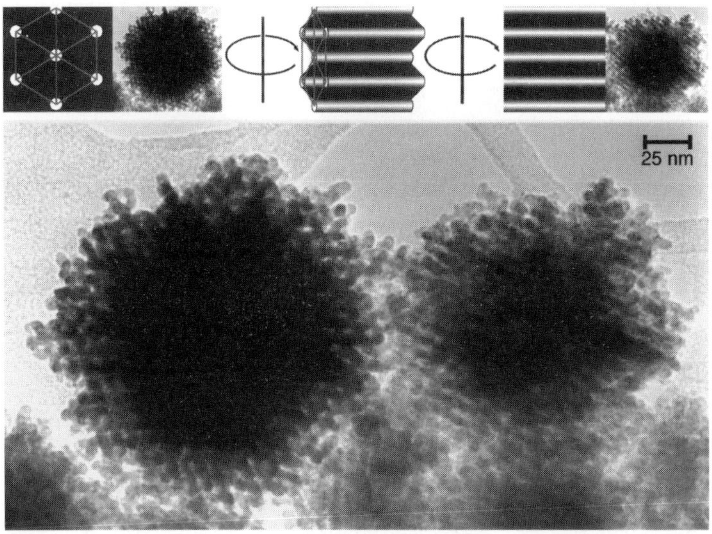

Fig. 3.20 The structure on the left in the transmission electron micrograph (bottom) is nanostructured CdS viewed with the hexagonally packed cylindrical templated pores parallel to the electron beam; the one on the right has the cylindrical structures perpendicular to the electron beam. Top: model of the nanostructure as a hexagonal arrangement of cylindrical pores of low electron density, corresponding to organic material in a solid matrix of semiconductor. The left and right schematic representations correspond to the adjacent micrographs; the central view shows the cylindrical assemblies in an intermediate state of rotation

yielded product with a hexagonal nanostructure, and the lamellar liquid crystal yielded a lamellar structure. $(EO)_{\overline{10}}$ oleyl can also form a lamellar liquid crystal, which also was successful in templating a lamellar CdS product. Other lyotropic phases, such as a variety of bicontinuous and cubic liquid crystals [84, 105, 107], may also yield interesting mineral nanostructures. In this regard, CdS has been grown in a body-centered cubic phase by using a triblock copolymer of poly(oxyethylene)-poly(oxypropylene)-poly(oxyethylene) $[(EO)_{\overline{106}}(PO)_{\overline{70}}(EO)_{\overline{106}}]$ as the amphiphile. When mixed with water, the PO segment is only weakly solvated, but the EO is highly solvated. Thus, when this molecule is hydrated it forms micelles which pack closely, forming a cubic phase [106]. The result of precipitation in this cubic phase was the formation of hollow nanospheres of CdS, which are covered later in this chapter.

It is remarkable that a soft organic liquid crystal can directly template a hard covalent mineral phase. As a specific example, a hexagonal mesophase consisting of 50 vol. % aqueous 0.1 M $Cd(OAc)_2$ and 50 vol. % $(EO)_{\overline{10}}$ oleyl templated an inorganic/organic nanocomposite of CdS and amphiphile when exposed to H_2S gas (Figure 3.20) [93]. The composite material contained an internal nanostructure that replicated the symmetry and dimensions of the liquid crystal in which it was grown. Interestingly, CdS and ZnS exhibited a superlattice morphology when formed in a liquid crystal from their respective nitrate salts and H_2S, but Ag_2S, CuS, HgS, and PbS did not (Figure 3.21). CdSe is also nanostructured by this method (Figure 3.22). An additional virtue of the $(EO)_{\overline{10}}$ oleyl system is that, when combined with H_2O, it forms a hexagonal mesophase at 25 °C over the range of ~35 to ~65 vol. % amphiphile and a lamellar me-

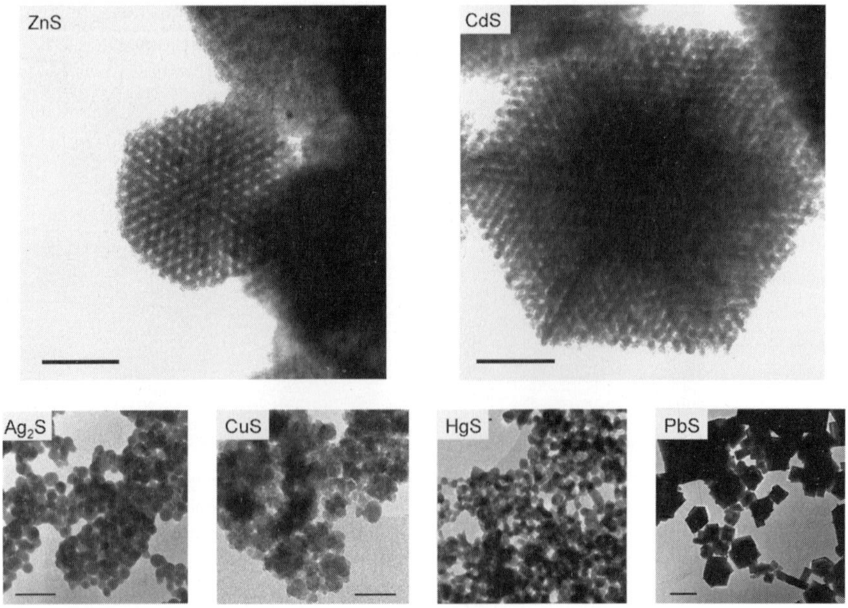

Fig. 3.21 Transmission electron micrographs of mineralized structures grown in hexagonal mesophases from their respective nitrate salt and H_2S (bar = 50 nm, except for PbS where the bar is 250 nm)

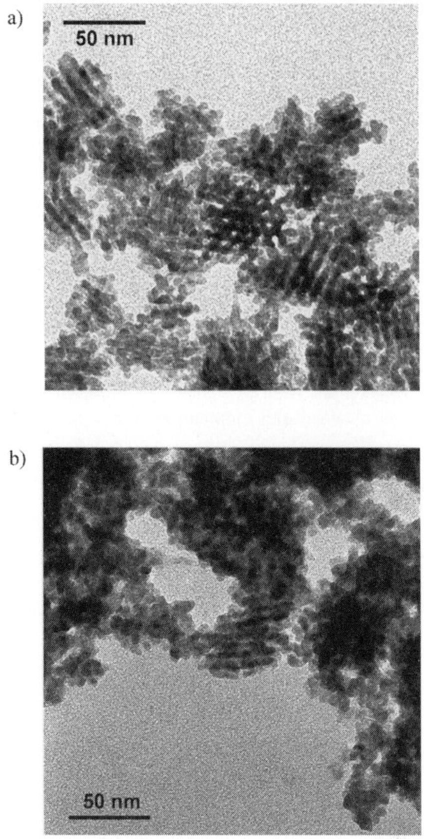

Fig. 3.22 Transmission electron micrographs of CdSe mineralized in hexagonal mesophases doped with (a) cadmium nitrate and (b) cadmium acetate

sophase from ~70 to ~85 vol. % amphiphile [108, 109], and it can directly template the growth of mineral over almost this entire range.

As shown in Figure 3.21, the nanostructures of the semiconductors CdS and ZnS synthesized by precipitation in hexagonal mesophases doped with their respective nitrate salts have hexagonal symmetry with a periodicity and dimensionality commensurate with that of the template. The hexagonal nanostructure is not always evident in TEM micrographs, due to random orientation of particles in the field of view. Presumably, if properly oriented, all the particles would show hexagonal symmetry. This was partially observed by tilting the samples on the TEM stage, revealing many more particles with hexagonal nanostructures.

CdS is also templated when grown from its acetate salt, although the order of its nanostructures is not nearly as well defined as in the nitrate systems (Figure 3.23) [95]. When ZnS is generated from its acetate salt, only spherical polycrystalline particles with a porous appearance are formed. Another difference between the product obtained from the acetate and nitrate salts is the average particle diameters of the semiconductor product: both CdS and ZnS grown from their respective nitrate salts are approximately five times larger than when grown from their acetate salts. This size

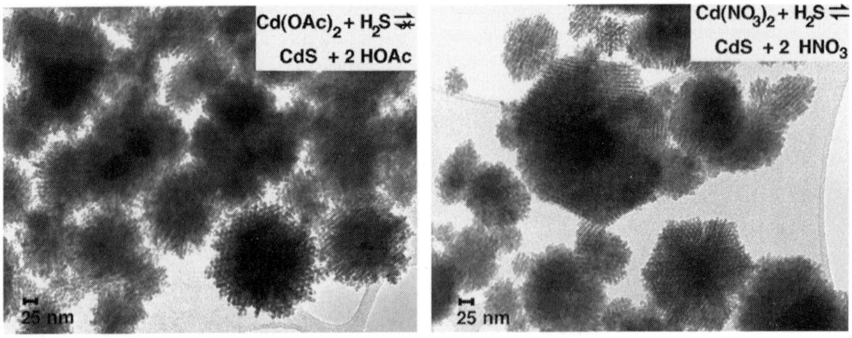

Fig. 3.23 Transmission electron micrographs of CdS grown in identical hexagonal mesophases except for precursor salt. Note the significantly improved order when cadmium nitrate is used as the precursor over the order obtained with cadmium acetate

difference can be clearly observed in low magnification electron micrographs. As controls, CdS was also grown in aqueous environments from both the nitrate and acetate precursors, and as expected, no nanostructure was generated. The counterion of the metal did not affect the templating of the other mineral systems studied: Ag_2S, CuS, HgS, and PbS were not generated with a superlattice morphology when grown in a hexagonal mesophase, irrespective of whether the acetate or nitrate salts were used. The reason may be that the byproduct of the synthesis from the nitrate salt is nitric acid, whereas the byproduct of the synthesis from the acetate salt is acetic acid. Nitric acid is a much stronger acid, and apparently enables the mineral phase to reform around the template during growth so as to remove any structural defects.

As already mentioned, in addition to the hexagonal mesophase, a lamellar mesophase of $(EO)_{\overline{10}}$ oleyl also can template a precipitated mineral (Figure 3.24). The la-

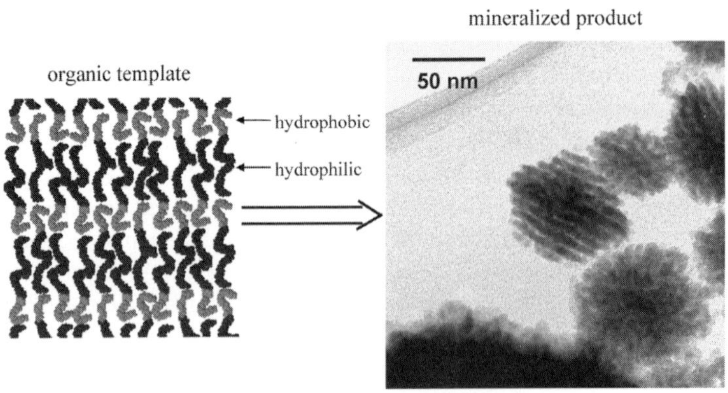

Fig. 3.24 Schematic representation of the lamellar organic template and transmission electron micrograph of the resulting product after mineralization. The dark bands are CdS that mineralized in the hydrophilic region of the template liquid crystal; mineral growth is prohibited in the hydrophobic regions

mellar periodicity in the resulting CdS is ~7 nm, which agrees very well with the periodicity of the lamellar template. The lamellar morphology can be confirmed by careful tilting of particles within the TEM. If a particle is tilted about an axis perpendicular to the stripes, no change is observed in the pattern. However, if it is tilted on an axis parallel to the stripes (and perpendicular to the electron beam), the pattern quickly disappears. These observations constitute strong evidence for a lamellar morphology within the particles and agree closely with results obtained from lamellar nanocomposites formed in a poly(vinyl alcohol)-based liquid crystal [94]. As shown in Figure 3.24, the mineralized product consists of disk-like particles with a long axis in the plane of the layers that is ~1.5 times the maximum width perpendicular to the layers. Very interestingly, particles having lamellar morphology do not disperse, even with repeated ultrasonication, perhaps due to mineral or organic bridging between the CdS layers. Presumably, CdS nucleates within a hydrophilic layer of the mesophase and grows rapidly in the plane, but concurrently an occasional finger forms perpendicular to this layer, piercing the hydrophobic region. This finger then nucleates another layer of CdS, resulting in a mineral bridge between layers. Another possibility is tethering of organic molecules in the mineral phase, tying the layers together. Of course, growth in the plane of the layers is faster than growth perpendicular to the layers, generating the disk-like habit observed. Not surprisingly, since it seems to enhance the templating effect, lamellar-nanostructured CdS was seen only when the nitrate salt was used, while the acetate salt afforded only small particles.

The layered structure formed by templating with a lamellar liquid crystal is in fact reminiscent of the layered abalone shell structure; however, in the templated material the characteristic dimension is just a few nanometers, not hundreds of nanometers to micrometers. It still remains to be seen if the properties of the layered material are significantly improved over those of the solid equivalent, but if the abalone shell is any guide, advanced mechanical properties are a possibility.

As described, mineral growth in the hexagonal and lamellar phases yields interesting, controlled nanostructures. Mineral growth in the cubic phase formed from $(EO)_{\overline{106}}(PO)_{\overline{70}}(EO)_{\overline{106}}$ was also done, using cadmium acetate and H_2S as precursors [110]. Although the mechanism is not entirely understood, the result was hollow spheres 20–200 nm in diameter (Figure 3.25), which can be observed by both TEM and SEM. When the sample is tilted in the stage, the shape and observed structure do not change, as expected for a hollow sphere (Figure 3.26). If the particles had been corpuscular in shape, their appearance would change as a function of sample tilt. The strongest evidence for their hollow nature is the dark edges of the spheres observed in TEM micrographs. If each sphere were solid, TEM would show greater scattering from the center than the edges, making the center appear darker. In the SEM, the spheres also scatter the most electrons from their edges and appear somewhat transparent in their centers, providing further proof of their hollow nature (Figure 3.25). Unlike materials produced by direct templating in hexagonal and lamellar liquid crystals, the hollow spheres are not of a size commensurate with the structure of the liquid crystal, but are rather 1–10 times the size of the characteristic dimension of the liquid crystal in which they were formed.

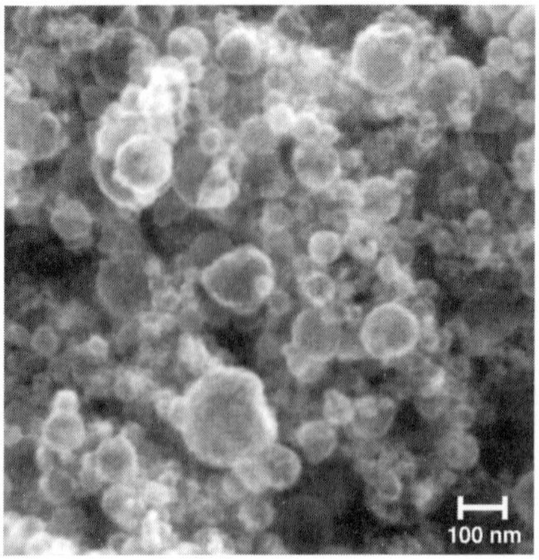

Fig. 3.25 Scanning electron micrograph of hollow spheres of CdS mineralized within a cubic mesophase

Because the characteristic dimension (diameter) for hollow CdS spheres obtained from the $Cd(CH_3CO_2)_2$-doped cubic phase is 20 – 200 nm, (1 – 10 times the diameter of the micelles making up the cubic phase), it does not appear that the mineral nanostructure is directly templated by the liquid crystal. In addition, this nanostructure is entirely absent from the CdS when the cubic phase is doped with $Cd(NO_3)_2$ as the precursor salt. This result is important when taken in the context of previous results in which the use of the nitrate salt led to a nanostructure with enhanced order [95]. In essence, the nitrate salt allowed the growing mineral to access a thermodynamically lower-energy morphology, which was a nanostructure commensurate with the structure of the liquid crystalline matrix. The inability of the nitrate salt to 'sharpen' the order in the cubic system is not surprising, because, as observed in previous studies, the nitrate salt only improves the registry between the nanostructure and the liquid crystal and does not result in a new nanostructure. In the cubic system, the spheres are not a copy of the liquid crystal, so there is no registry to improve. The mineralization of the cubic phase must lead to local rearrangements of the liquid crystal, leading to the hollow-sphere morphology observed; however, the detailed mechanism is still not understood.

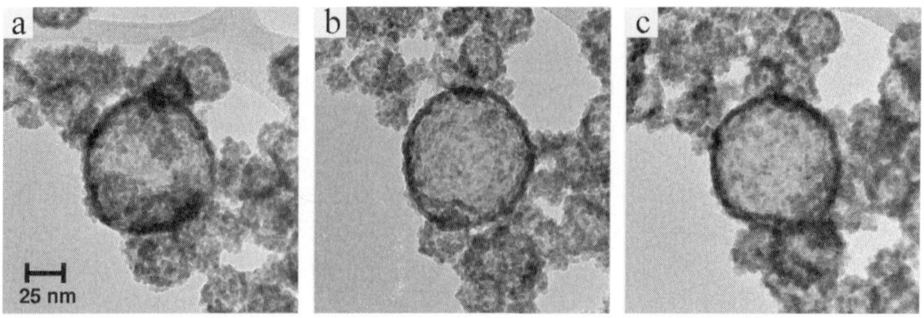

Fig. 3.26 Transmission electron microscope tilt series of hollow CdS spheres grown in an acetate-doped cubic phase. The tilt axis was diagonal running from upper left to lower right of each micrograph. (a) −45°, (b) 0°, (c) 45°

A potentially significant difference between the cubic phase and the other phases studied (hexagonal and lamellar) is the connectivity of the hydrophilic and hydrophobic portions of the liquid crystal. In the lamellar and hexagonal phases, both these regions are continuous in at least one direction. In the cubic phase, however, the hydrophobic regions are confined to discrete micelles. This confinement may cause a difference in templating ability relative to the other phases. For all templating phases, as the mineral nucleates and grows, it is necessary to expel some molecules from the volume occupied by the mineral. For both the lamellar and hexagonal phases, a molecule can diffuse away from the growing mineral without ever exposing its hydrophobic (oleyl) or hydrophilic (EO) segments to domains of the opposite nature. In contrast, in the cubic phase, when a molecule is forced away from the growing mineral it must leave its micelle and expose its hydrophobic (PO) segment to the polar surroundings (EO + H_2O), a high-energy situation. As already stated, the nitrate salt allows the mineral phase to access a lower-energy configuration, which in the lamellar and hexagonal systems results in a high degree of fidelity between the template and the semiconductor nanostructure. That a hollow-sphere morphology is not observed in the cubic system when the nitrate salt is used is not surprising, given that the hollow morphology is not directly templated by the liquid crystal and that formation of the spheres requires the mesophase to go through a high-energy intermediate state. The full reason for the spherical morphology when $Cd(CH_3CO_2)_2$ is used as the semiconductor precursor must be due to a subtle kinetic balance that is not yet understood. At the very least, the energy difference between the lamellar, hexagonal, and cubic systems is rather small, and thus the fact that templating is successful in the first two systems but unsuccessful in the last indicates that some fairly specific interactions are necessary for direct templating. This is similar to the action of many biological systems, in which very specific interactions between proteins and other macromolecules and growing inorganic phases are exceedingly important for structural development.

In addition to simple one-component systems, it is interesting to consider the result of templating binary mixtures of the precursor salts within a liquid crystal. In one system, a hexagonal mesophase containing 0.05 M $Cd(NO_3)_2$ and 0.05 M $Zn(NO_3)_2$ was mineralized. The resulting semiconductor ($Cd_xZn_{1-x}S$; $x \sim 0.5$) product's nanos-

Fig. 3.27 Transmission electron micrograph of the product obtained from precipitation in a hexagonal mesophase doped with 0.05 M $Cd(NO_3)_2$ and 0.05 M $Zn(NO_3)_2$. The templated product is a solid solution of CdS and ZnS

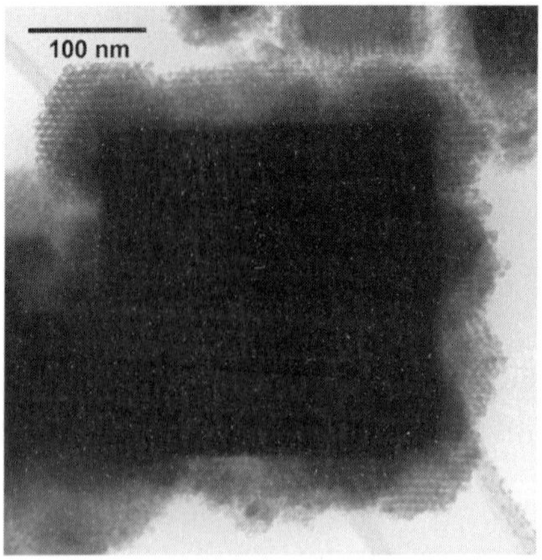

Fig. 3.28 Transmission electron micrograph of a composite product of PbS and ZnS grown in a hexagonal mesophase doped with 0.05 M $Pb(NO_3)_2$ and 0.05 M $Zn(NO_3)_2$. The single crystal cube at the core of the particle is PbS, and the shell is a periodically nanostructured solid consisting of ZnS

tructure exactly matched that of the template (Figure 3.27). In a system of 0.05 M $Pb(NO_3)_2$ and 0.05 M $Zn(NO_3)_2$, the result was a very different nanostructure, which consisted of a single-crystal core of PbS surrounded by a shell of nanostructured ZnS (Figure 3.28). From this it appears that, when mixed metal products are synthesized, there is a direct correspondence between the behavior of the constituent solids and the mixed solid. $Cd_xZn_{1-x}S$ was nanostructured (as were CdS and ZnS, Figure 3.21), but precipitation from Pb^{2+} and Zn^{2+} ions resulted in a single crystal of PbS surrounded by nanostructured polycrystalline ZnS, corresponding with the result seen for PbS and ZnS when grown discretely (Figure 3.21). The formation of mixed metal precipitates gives an insight into the growth processes and also opens the possibility of engineering a property. The system of $Cd_xZn_{1-x}S$ offers the possibility of band-gap engineering, although, due to the somewhat lower solubility of CdS than ZnS in water, the particles may be cadmium-rich in their center and zinc-rich on their exterior. The differing solubilities of PbS and ZnS in water play a very important role in the structure of the particles formed from this mixed system. PbS has a much lower solubility than ZnS and, as a result, upon exposure of the doped mesophase to H_2S, single-crystal PbS cubes nucleate and grow. Then, after most of the Pb^{2+} ions are locally exhausted, ZnS heterogeneously nucleates on the PbS particles. As expected, because ZnS is templated by liquid crystals, the shell of ZnS around the PbS single crystal contains a periodic superlattice structure.

Direct templating of an inorganic by an organic liquid crystal may depend on many factors, the most important of which is probably the thermodynamic stability of the mesophase throughout the mineral growth process. The mesophase must be stable to the addition of mineral precursors, and the mineral precipitation process must not disrupt the order of the liquid crystal. In studying direct templating, researchers have observed that the textures observed by polarized optical microscopy are the same for the pure mesophase and for a mesophase that contains the precursor salt, indicating that the doping did not lead to radical disruption of the order in the mesophase. Nuclear magnetic resonance (NMR) can also be utilized to verify the structure of liquid crystalline mesophases. To verify that the characteristic molecular order of the mesophase was not disrupted by ion doping, broadline ^2H NMR spectra were obtained from both cadmium ion-doped and undoped mesophases [96]. For both mesophases, the same quadrupole splitting was observed (Figure 3.29). If ionic doping had perturbed the structure of the mesophase, the splitting would have decreased [111, 112]. As additional proof of molecular order in the mesophase, x-ray diffractograms were collected to characterize both the mesophase's long period and symmetry. For systems containing 35%, 40%, 50%, and 60% amphiphile, the 100, 110, and 200 reflections are clearly observed, indicating that the liquid crystalline structure is hexagonal (Figure 3.30). A mesophase containing 78% amphiphile forms a lamellar liquid crystal, as indicated by the presence of 100 and 200 reflections and the absence of a 110 reflection (Figure 3.31). As expected, a strong correlation was found between the phase diagrams as determined by optical analyses and the x-ray data. Similar experiments have been performed

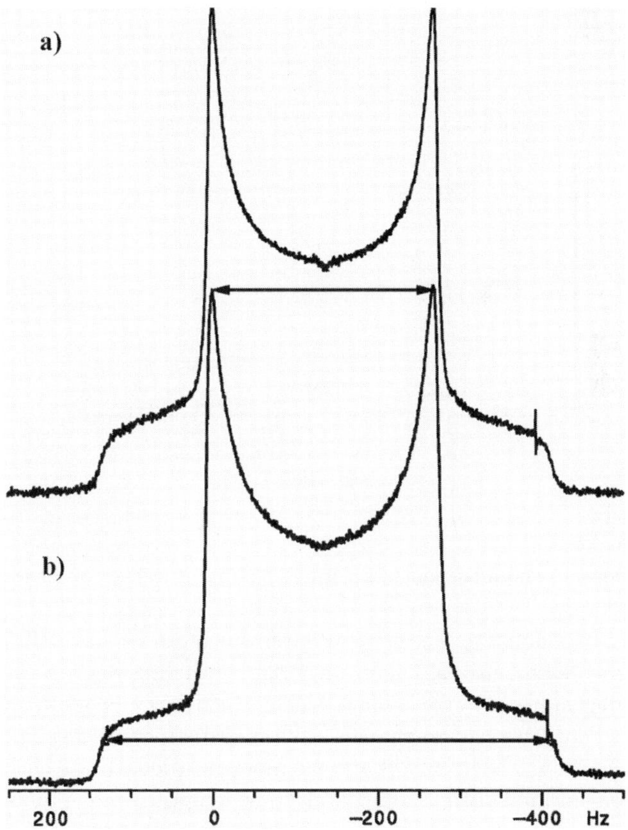

Fig. 3.29 Quadrupole splitting for a hexagonal mesophase (a) doped with 0.1 M cadmium acetate and (b) undoped. The splitting indicated by an arrow is the same for both samples

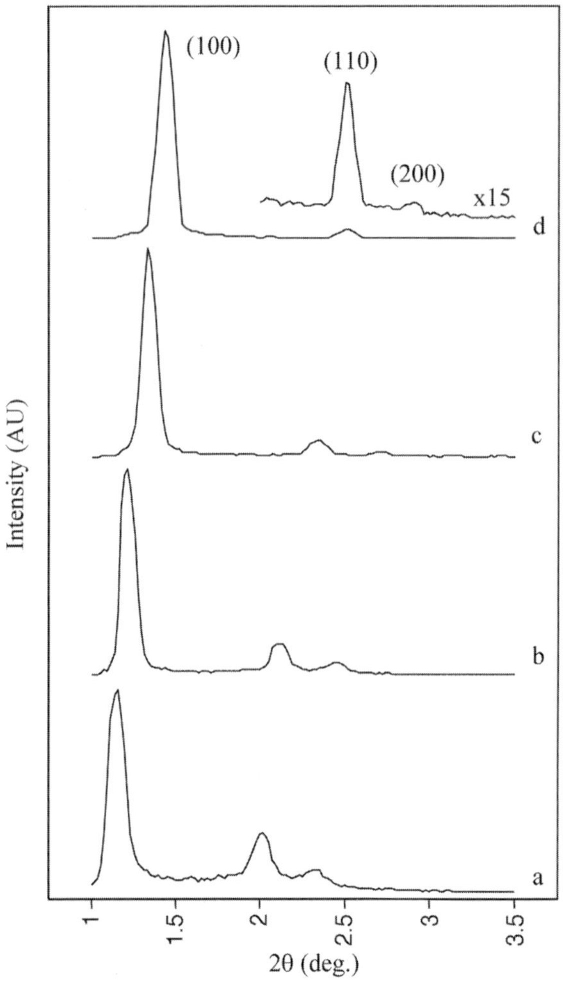

Fig. 3.30 Small-angle x-ray scattering of hexagonal aqueous mesophases containing (a) 35 %, (b) 40 %, (c) 50 %, and (d) 60 % amphiphile by volume

for other liquid-crystal templating systems, as well as for the formation of mesoporous silica.

In addition to the metal sulfides, which can be successfully templated as already mentioned, several sulfide materials, including Ag_2S, CuS, HgS, and PbS, were not templated by the liquid crystal in which they were grown, irrespective of the counterion. Design of amphiphiles with proper structures and binding constants for both the inorganic precursors and the inorganic product may enable a wide variety of inorganic and organic compounds to be templated in the future.

To better understand the scientific underpinnings of liquid-crystal templating, it is instructive to consider a few additional experiments. For example, liquid-crystal templating at elevated temperature sheds some light on the mesophase–ion–product interactions. As already mentioned, it appears the periodically nanostructured materi-

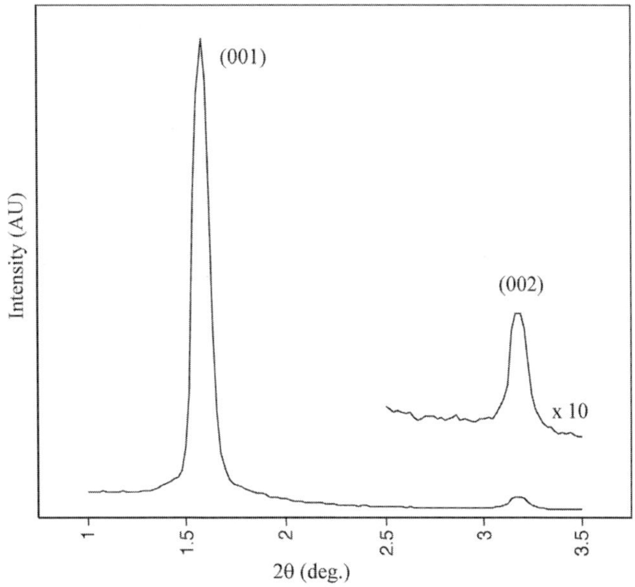

Fig. 3.31 Small-angle x-ray scattering of a lamellar mesophase containing 78% amphiphile and 22% water by volume

als are thermodynamically stable with respect to their solid equivalent. This implies that there is a critical energy balance between the energy gained by reducing the surface area of the mineral phase and the energy lost due to disruption of the mesophase structure. To study this further, CdS was precipitated in mesophases at both 35 °C and 50 °C. These temperatures are both below the isotropization temperature of the doped mesophases, so the reactions were carried out in a self-assembled medium. The CdS produced from the reaction at 35 °C did express the order of the mesophase, albeit poorly when compared with the order obtained at room temperature (22 °C). Precipitation at 50 °C resulted in mineral with no periodic order, clearly indicating that the energetic difference between periodically nanostructured and disordered product is small and that it only takes a small perturbation to result in nonnanostructured product.

As additional evidence for the direct templating mechanism, samples composed of 35%, 40%, 50%, and 60% by weight $(EO)_{\overline{10}}$ oleyl were mineralized with CdS. By varying the amphiphile content of the mesophase, the spacing between the cylindrical aggregates of amphiphilic molecules making up the hexagonal mesophase was varied. Assuming that CdS is directly templated by the liquid crystal in which it is grown, the hexagonal symmetry and associated length scale should be nearly identical to that found in the precursor hexagonal mesophase, which is what indeed happened. The result (Figure 3.32), demonstrating that the superlattice dimension in the precipitate can be varied by changing the lattice constant of the mesophase, is very strong evidence for direct templating. If the nanostructure present in the templated inorganic phase had been the result of a cooperative self-assembly process, much as occurs for

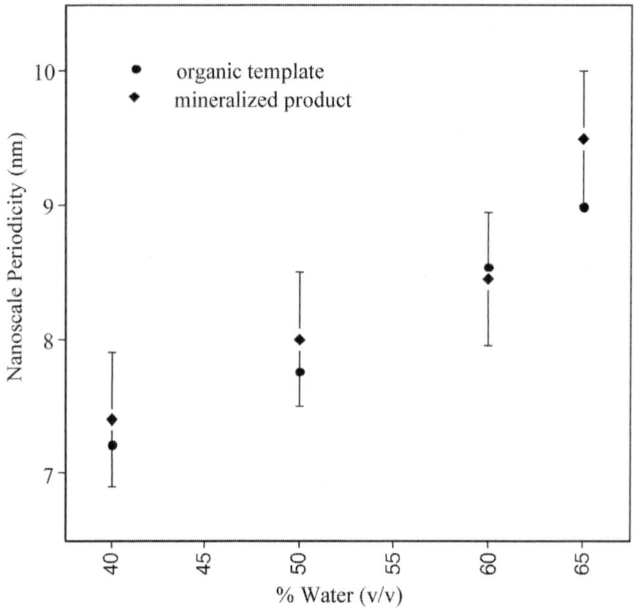

Fig. 3.32 Center-to-center spacing in a cylindrical assembly of amphiphilic molecules of hexagonal mesophases as determined by x-ray diffraction (•), and center-to-center pore spacing in templated CdS as measured by transmission electron microscopy (♦). Both are plotted as a function of the water content in the mesophase

most mesoporous silica systems, the result would have been that the periodicity of the nanostructure would not have changed with varying water content.

3.4.2
Liquid-Crystal Templating of Thin Films

The templating of thin films by self-organized organic structures should find significant application in both technology and scientific study. Thin-film templating is structurally related to the bulk templating of inorganic materials, which generally results in periodically structured particles; however, because the result is a periodically structured thin film, the potential for application is clearer. Templating with organic structures is especially intriguing because of the potential to create features much smaller than those feasible by almost any top-down technique, because it utilizes the nanoscale molecular order inherent in self-assembled organic structures to define the structure of the resulting thin film. A few key points must hold true for successful thin-film templating by liquid crystals. First, clearly, for templating to be successful, the self-assembled matrix must be compatible with the substrate. Then, via some process, the inorganic material must be deposited or grown on the substrate. Realistically, only chemical and electrochemical routes for materials deposition can meet these requirements. Other conventional methods of thin-film deposition require high vacuum, which is incompatible with lyotropic liquid crystals and furthermore cannot operate through a thick overlying layer of liquid crystal.

The synthetic routes for liquid-crystal templating of thin films are relatively straightforward. To date, most studies have used electrochemical techniques to drive the material deposition. First, a precursor containing lyotropic liquid crystal is interfaced with the substrate. Then, under an applied potential, material is electrochemically deposited at the liquid-crystal/substrate interface. Nanostructured materials that have been created through this process include a variety of metals, selenium, and tellurium [97, 99–104, 113]. It may also be possible to electrodeposit other interesting materials including semiconductors; however, no publications have yet appeared demonstrating success.

Throughout, it is quite interesting that all the successful templating experiments have relied on liquid crystals formed of nonionic amphiphiles. However, in biology, most preformed matrices are formed via ionic macromolecules. In part, this is likely because biology makes use of very specific interactions to create mineral structures, but synthetic systems do not have this degree of sophistication. Thus, the fact that nonionic amphiphiles are much more stable to varying concentrations of soluble salts is actually an advantage. Ionic amphiphiles (as well as biomolecules) are affected much more strongly by salts, because a single salt ion can bind to one or more of the polar amphiphile headgroups, greatly reducing their polarity. This was indeed observed for several of the anionic amphiphiles studied, and most likely was the reason that ionic systems were not successful in templating the growth of a mineral phase. In biology, where the molecular structures are designed to interact specifically with one salt under very specific conditions, the strong interaction of ionic groups with dissolved species is an advantage. However, for generalized synthetic systems, this may not be an advantage.

3.4.3 Block-Copolymer Templating

Block copolymers are a widely studied class of materials that organize into both 2D and 3D structures at slightly longer length scales than observed for liquid crystals. Similar phase behavior is observed, with systems transitioning between lamellar, hexagonal and cubic phases as the relative volume fraction of the two blocks changes [87]. Much more complex morphologies can form in triblock systems. The characteristic periodicity in block copolymer systems ranges from $\sim$5 nm to hundreds of nanometers and primarily depends on the molecular weight of the block copolymer. As expected, as the molecular weight of the block copolymer increases, the characteristic length scale increases. It was realized that if there was a way to impart this nanoscale order into a substrate, one might have a powerful technique for patterning materials with a periodic array of nanometer-sized structures. Unlike lyotropic liquid crystals, block copolymers are generally solvent-free and can be taken to elevated temperatures and under vacuum without destabilizing the self-assembled structure; thus high-vacuum material deposition and processing approaches can be used. Usually the chemistry of the block copolymer is designed so that one of the two bocks can be removed via a dry etch with ozone or other reactive compound to generate the porous structure, which will subsequently serve as a template for nanostructure formation.

The procedure for block copolymer templating of nanostructures usually is as follows. A thin film of some block copolymer is spun-coated from solvent onto a substrate and allowed to self-assemble. After this, one of the blocks of the polymer is removed by ozone etching. The result of this etching procedure is a substrate coated with a thin polymer film containing a periodic nanoscale void structure. After removal of any solid polymer film that overlies the void structure, the polymer film is used as a mask. Material can be evaporated through the polymer film onto the substrate, material can be electrochemically grown from the substrate up through the polymer film, or the polymer film can be used as an etch mask. In all cases, the result is material structured to be a replica of either the polymer film (when the polymer is used as an etch mask) or the pore structure of the polymer film (when material is deposited in the pores). Usually, at the end, the polymer film is removed with solvent or reactive ion etch, leaving behind nanostructured templated features on a substrate. With this approach it is possible to create features as small as ~20 nm holes or dots in a periodic array on a substrate of a wide range of materials, including oxides, semiconductors, magnetic materials, and of course the polymer itself (Figure 3.33) [114–121]. The power of block copolymer templating is further enhanced due to the facts that both the size and spacing of the feature can be modulated simply by varying the molecular weight and composition of the polymer and that the lattice structure can be modulated by varying the relative length of the two blocks. For example, both the lamellar and hexagonal phases can template nanometer-scale lines if they are oriented

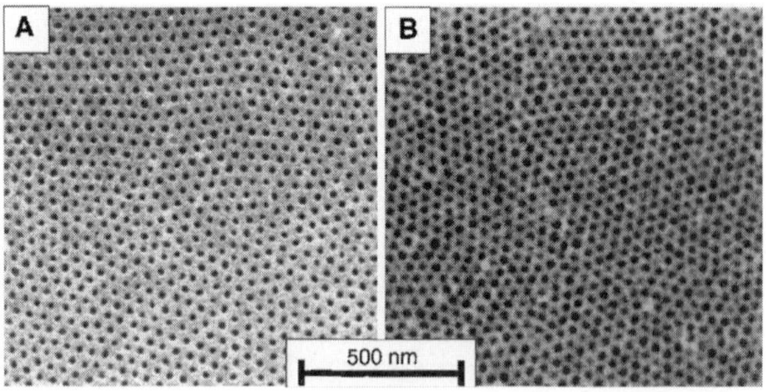

Fig. 3.33 (a) Scanning electron micrograph of a polystyrene/polyisoprene block copolymer template after it has been partially etched with ozone and the continuous polystyrene layer on the top has been removed. The now empty PI domains are now holes and are darker in the micrograph. (b) Scanning electron micrograph of hexagonally ordered arrays of holes in silicon nitride on a thick silicon wafer. This pattern is formed by using as a template a copolymer film such shown in (a). The darker regions are ~20-nm-deep holes in the silicon nitride layer, which are formed by etching through the overlying template. Adapted from [114]

properly on the substrate. I should point out that the periodic arrays of dots have attracted the greatest attention for applications such as magnetic storage media.

3.4.4
Colloidal Templating

In any discussion of biologically inspired nanocomposite materials, one must include recent developments on colloidal crystal templating of photonic materials. The basic premise behind this approach is to use the 3D periodic structure of synthetic opals to direct the structure of a second phase material. This approach is not biologically inspired, but should more accurately be described as 'naturally inspired', because opals, although natural, are geological, not biological, in origin. Furthermore, although the lower limit to the characteristic length scale of the material generated is 10 – 20 nm, the characteristic dimension is often relatively large, on the order of 500 nm. This is in fact intentional, since most of the applications for these materials are optical, and thus the characteristic length should not be much smaller than the wavelength of the light that one desires to modulate. Commonly, however, much smaller features are embedded within the templated structure, and these features may be as small as a few nanometers. Because of the clear natural inspiration for the colloidal templating of materials and the relationship of the templating process to many biological processes, I cover this approach in this chapter.

The interest in microperiodic 3D structures has grown tremendously due to the exciting potential of such materials, particularly in the area of photonics [122]. Such 3D structures, often termed photonic crystals, are the extension of the well-known dielectric stack into three dimensions. Although the colors that occur in opals, which stem from diffraction of white light by planes of highly ordered submicrometer silica spheres, are our inspiration; for practical application, synthetic approaches are needed to create materials and structures with the necessary refractive index and periodicity to meet the requirements for most optical applications, which opals simply do not have.

A particularly interesting class of optical structures are the so-called photonic band-gap materials. For example, a microperiodic material consisting of low-refractive-index spheres arranged in a face-centered-cubic array in a matrix with a high index of refraction, and having a lattice constant on the order of the wavelength of light (visible or infrared), could be such a photonic band-gap material [123]. Similar to how a dielectric stack has a stop-band for light in a given frequency range, this material would not allow light in a given frequency range to travel through it *in any direction*. In essence, it would be an omnidirectional, perfectly lossless, mirror.

The synthesis of these structures however is exceedingly difficult. Layer-by-layer fabrication of photonic crystals using state-of-the-art VLSI tools, e.g., deep UV photolithography, chemical vapor deposition (CVD), chemical-mechanical polishing, has been demonstrated [3], but formidable processing difficulties limit the formation of large area and truly 3D structures.

When appropriately formed, self-assembled colloidal crystals are natural candidates for the construction of photonic crystals. Good crystal quality is achieved only with colloids that have very low size polydispersity (<5 %), which currently limits the choice of materials to SiO_2 or polymers, both of which have a fairly low index of refraction around 1.5, which is much smaller than that required for most optical applications. This has led researchers to take a two-stage templating approach. In a first step the desired microperiodic structure is assembled by using colloids. In a second step this structure is used as a template to build a complementary structure with a material having a higher index of refraction [124].

Colloidal crystal templating is a very promising approach for production of high-resolution, micrometer-scale, 3D periodic photonic crystals, but as conventionally applied to the fabrication of photonic band gap structures it has serious optical limitations, unless materials with the necessary optical properties can be used. Typically, the most important point is to infill with a material with a sufficiently high refractive index to generate an optically interesting material. A range of approaches have been suggested to maximize the index contrast, including sol–gel [125–127], chemical vapor deposition [128–131], imbibing of nanoparticles [132–134], reduction of GeO_2 to Ge [135], electroless [136] and electrochemical deposition [137], and melt imbibing [138]. In addition, polymers have been used to infill colloidal crystals, and in one report, the colloidal particles were less than 100 nm in diameter, which, although perhaps not interesting from an optical standpoint, may have potential for separation membranes and confined chemical reactor spaces [139]. Although they are not the focus of the following discussion, the colloidal templates used in these attempts are commonly polycrystalline and can contain unacceptably high numbers of defects; thus, substantial effort has also gone into creating colloidal crystals with low defect densities. Each of these infilling techniques has various advantages and disadvantages, which are discussed below. In general, they all consist of approaches to infill the interstitial space of the colloidal template, after which the colloidal template is generally removed.

The sol–gel infilling of colloidal crystal templates is intriguing to consider as a route to 3D porous materials [125–127], although it is somewhat limited in application for photonic materials for several reasons. First, the refractive index of most materials that can be formed via sol–gel is < 2 (with the exception of TiO_2, which can have a refractive index of $\sim$2.5), second, there is considerable reduction in volume during the conversion of the sol to solid material, and third, the refractive index of most sol–gel-derived material is substantially less than that of a single crystal of the same material. The net effect is that most sol–gel-derived macroporous materials have relatively low refractive index contrast, and their long-range order is somewhat disrupted due to the uneven contraction of the matrix. However, if one is not interested in photonic materials, but rather is attempting to make a ceramic macroporous material, sol–gel infilling of colloidal crystals may be a very good route. The contraction of the matrix may in fact be an advantage, in that it may be possible to make structures with pore diameters on the order of 50 % of the diameter of the template. Because it is difficult to make colloidal crystal templates from spheres smaller than a few hundred nanometers, this may be valuable for creating nanoperiodic structures.

Another pathway to macroporous materials is to fill the interstitial space of a colloidal crystal with nanoparticles, followed by removal of the colloidal template. This has some advantages over sol–gel infilling, in that the contraction of the structure

Fig. 3.34 Left, transmission electron micrographs of 3% erbium-doped hydrothermally synthesized titania nanoparticles. Right, scanning electron micrographs of macroporous titania structure formed by imbibing these nanoparticles into a colloidal template formed from 466 nm polystyrene colloidal particles. The colloidal template was removed by calcination at 300 °C for one hour in air

upon removal of the template from a nanoparticle filled colloidal crystal is significantly less than that seen upon removal of the template from a sol-gel filled system, and a much larger subset of materials can be prepared as nanoparticles, including semiconductors, metals, and ceramics. The first example of semiconductor nanoparticle infilling of colloidal template used II–VI semiconductor nanoparticles [132]; since then, Er-doped TiO_2 nanoparticles, for example, have been filled into a colloidal template, followed by removal of the template to generate a macroporous solid (Figure 3.34) [140].

The use of CVD as a pathway to filling colloidal crystals at first may seem counterintuitive. After all, CVD generally is most efficient at coating planar surfaces, and it would seem almost impossible to fill structures with deep pores, such as the interstitial space of a 3D colloidal crystal. However, significant strides have been made in the past few years, and now virtually complete infilling of colloidal structures with both Si and Ge via CVD has been demonstrated [128–131]. After dissolution of the colloidal template, the result is an inverse structure with the necessary refractive index and structural conditions to exhibit a complete photonic band gap.

Electrodeposition-based infilling is intriguing for several reasons and has the potential to be general with respect to both characteristic lattice constant and material (Figure 3.35). The potential for high-refractive-index materials, large area structures, and the complete infilling of thick 3D colloidal templates, as well as the low cost of electrodeposition have led to interest in this area. To date, three different classes of materials that have been electrodeposited into self-assembled colloidal crystals, namely semiconductors, polymers, and metals.

Semiconductors are interesting candidates for photonic crystals, primarily because of their high refractive indices and generally robust nature. For example, CdS has a refractive index of 2.5, and materials such as GaP, Si, and Ge have indices of 3.4, 3.5, and 4.0, respectively. However, routes to creating periodic macroporous structures

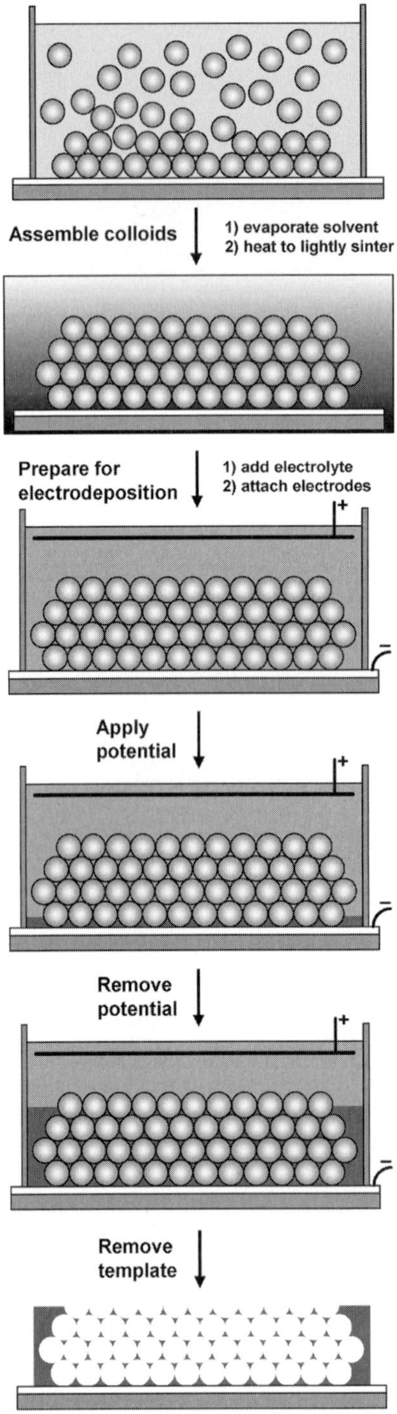

Fig. 3.35 Generalized procedure for creating 3D periodic macroporous materials by colloidal templating and electrodeposition. Monodisperse colloids sediment onto a conducting substrate, self-assembling into a crystal. The sample may be dried and sintered before electrolyte is added. A counterelectrode allows electrodeposition of the desired material (semiconductor, polymer, metal) into the interstitial spaces. In a final step the electrolyte and the templating colloid are removed. For polymeric colloids this can be done by treatment at elevated temperature or by dissolution with a solvent. For silica colloids, aqueous HF is effective for dissolving the template

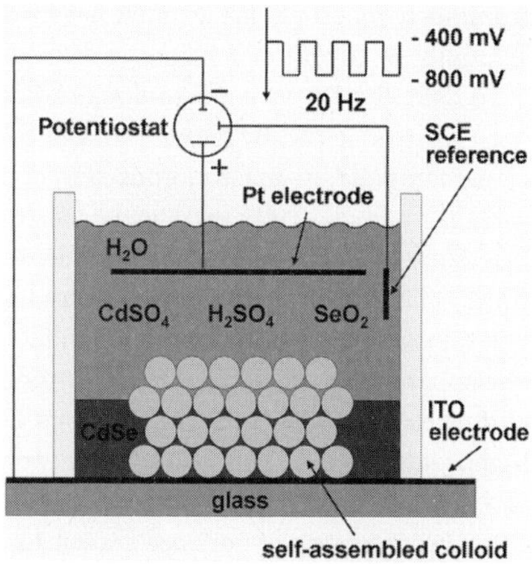

Fig. 3.36 Schematic representation of the experimental setup for potentiostatic deposition of CdSe through the interstitial space of a colloidal crystal

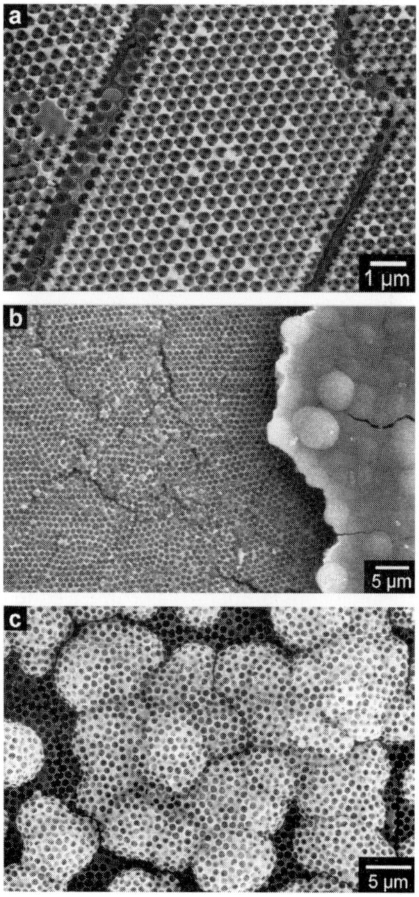

Fig. 3.37 Scanning electron micrographs of potentiostatically deposited CdSe (a) and galvanostatically deposited CdS (b, c) after removal of the polystyrene colloidal template. In the overdeposited system (b), the overlying solid CdS film can be clearly seen on the right. The apparent lack of periodic pore structure in the underdeposited system is not due to disorder in the colloid, but occurs because the nodular surface of the semiconductor cuts through multiple lattice planes of the template

from such materials are limited because of their very high melting points and low solubility in common solvents.

To date, the II–VI semiconductors CdS and CdSe [137, 141], and ZnO [142], have been electrochemically grown through colloidal templates, resulting, after dissolution of the template, in macroporous semiconductor films. For all systems, a conducting oxide film on glass was used as the substrate. Macroporous CdS films were generated by galvanostatic deposition through the interstitial spaces of a colloidal crystal formed from 1 μm SiO_2 spheres, and CdS and CdSe macroporous films were generated by potentiostatic deposition through a colloidal template generated from 466 nm polystyrene spheres (Figure 3.36). After electrodeposition, the SiO_2 and polystyrene colloidal templates were removed with aqueous HF and toluene, respectively. Because of the high rigidity of the semiconductor network, contraction upon removal of the template was limited to a few percent at most. The fine and gross morphologies of the electrodeposited semiconductors are shown in Figure 3.37.

Macroporous ZnO films were formed by potentiostatic deposition through a colloidal crystal formed from 368 nm polystyrene spheres, and the spheres were removed with toluene. Careful control of the electrodeposition conditions was necessary: if electrodeposition was done at a potential less negative than -1.0 V vs. Ag/AgCl, large crystalline grains of ZnO formed, which disrupted the structure of the colloidal template. Using a deposition potential more negative than -1.0 V suppressed the formation of large-grain ZnO, and the colloidal template was not disrupted.

All the electrodeposited semiconductor films are reported to be opalescent; however, detailed optical spectroscopy has yet to be performed. Real progress in optically interesting materials may await the electrochemical deposition of materials such as GaP, Ge, and Si, which, because they have refractive indices >3, may result in materials with 3D photonic band gaps. Routes to the electrodeposition of such materials have been demonstrated [143], but problems, such as hydrogen gas evolution and generally harsh conditions, need to be solved before success in these areas is likely.

Electrodeposition of conducting polymers (electropolymerization) through self-assembled colloidal crystals, followed by removal of the colloidal template, is a promising route to achieving active macroporous materials. Several significant advancements over the past few years have begun to demonstrate the potential of conducting polymer-based microperiodic photonic structures. Inherently, because of the low refractive index of polymeric materials, it is quite unlikely that a 3D photonic band-gap material will result from a polymer-based photonic crystal; however, conducting polymers have advantageous properties as compared to conventional polymers or inorganic materials: their optical properties can be electrochemically modulated, fine control over properties can be obtained through organic chemistry, and they are often mechanically flexible.

Electrochemical growth of conducting polymers is a fairly well developed field, and many procedures for growing solid films have been published [144]. There are, however, only a few reports on the growth of porous conducting polymer films. Fibers of polypyrrole, poly(3-methylthiophene), and polyaniline were formed in the early 1990s by electrodeposition from the appropriate monomer solution through a porous membrane [145]. The first example of electrochemical deposition of a conducting polymer around a colloidal template was in 1992, when polypyrrole was grown around latex particles [146]. However, no attempt was made to remove the colloidal particles, and the optical properties of the resulting films were not measured. Only in the past few years have researchers been exploring the possibility of the templated growth of conducting polymers for photonic applications.

To date, three reports on colloidal templating of conducting polymers have appeared, all of which followed the general procedure of (1) colloidal crystal formation on a conducting substrate, (2) electrochemical deposition from solution, and (3) dissolution of the colloidal template with an appropriate solvent. In the first example, polypyrrole was grown potentiostatically from a solution of pyrrole in acetonitrile through a colloidal crystal composed of SiO_2 spheres with a mean diameter of 238 nm assembled on F-doped SnO_2-coated glass, followed by removal of the colloidal template with aqueous HF [147]. Macroporous polypyrrole, polyaniline, and polybithiophene films have been potentiostatically polymerized through a colloidal crystal assembled from 500 nm and 750 nm polystyrene spheres, on a substrate of gold-coated glass. The polystyrene template was then removed with toluene [148]. In the most recent example, polypyrrole and polythiophene macroporous films were potentiostatically grown through colloidal crystals assembled from 150 nm and 925 nm polystyrene spheres, respectively, on glass coated with indium tin oxide; the polystyrene was removed with tetrahydrofuran [149]. A preliminary optical characterization [149] showed a weak dip in transmittance that appeared to be correlated with the periodic structure.

One significant issue is the contraction of the period structure upon removal of solvent for electrodeposited macroporous polymers. This was most clearly observed in polystyrene-templated systems, in which significant contraction, ranging from 13% to 40%, was observed for the macroporous polypyrrole and polyaniline. However, very little contraction was observed in polystyrene-templated macroporous polybithiophene or when SiO_2 spheres were used as the template. For example, poly(ethylenedioxythiophene) was templated by silica colloidal particles (Figure 3.38). The primary difference is that organic solvents are used to remove the polystyrene spheres and an aqueous HF solution is used to remove the SiO_2 spheres. This suggests that the organic solvent softens the electrodeposited polymer, allowing it to contract; however, there may be other systems similar to polybithiophene, in which contraction of the macroporous matrix does not occur. This is less of a problem for macroporous metals and semiconductors.

Metallic macroporous ordered replicas of colloidal assemblies are of potential interest for a wide range of applications including filtration, separation, and catalysis. In addition, they might have interesting electrical, magnetic, or optical properties. The tools and techniques for electrochemically plating metals have been well established for thin films and even bulk materials. It is thus fairly straightforward to develop recipes to backfill the interstitial space of a colloidal self-assembled crystal with almost any metal.

From a photonic standpoint, the properties of bulk metals are very poor, although of course templated structures may have many other applications. The imaginary components of the dielectric constants of bulk metals are large, hence they readily absorb

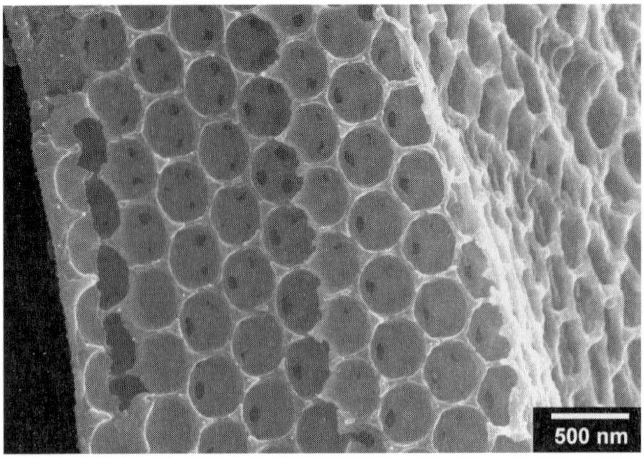

Fig. 3.38 Scanning electron micrograph of poly(ethylenedioxythiophene) electrodeposited around a colloidal crystal template, after dissolution of the template

light. When the metallic structures become small enough, however, strong optical resonances associated with plasmon frequencies of the conduction electron in the metals can lead to qualitatively new phenomena. A well-known example of this is the red color of a nanosized dispersion of gold colloid. A more recent manifestation of unexpected behavior is the anomalously high light transmission through small holes (<200 nm) in thin metallic films [150].

Theoretical calculations [151–153] on ordered 3D arrays of metallo-dielectric spheres show that these are promising for the construction of materials with full photonic band-gap in the visible part of the optical spectrum. The advantage of metallo-dielectric structures over purely dielectric structures is that it should be easier to achieve a full band-gap in the visible. A full band-gap in the visible is exceedingly difficult, if not impossible, to create with purely dielectric structures, because very few dielectric materials have an index of refraction >3 and very low absorption in the visible. This has led to the development of synthesis routes to produce metallo-dielectric colloidal core-shell particles with sizes in the submicrometer range [154, 155] and metallic shell thicknesses or cores that are small enough to show resonance effects.

Vos et al. [156] made gold replicas of colloidal crystals made of silica (radius 113 nm) and polystyrene (radius 322 nm). Prior to electrodeposition of the gold, the silica spheres were sintered by heat treatment at 600 °C. After electrodeposition of the gold, the silica template was removed by etching with aqueous HF, and the polystyrene spheres were removed by combustion at 450 °C (Figure 3.39). There are no dimensional changes between the dried, sintered colloid and the final replica, although some cracking is observed during the original drying and sintering process, indicating that the electrochemically formed gold is dense and structurally robust. This is a definite improvement over other methods of infilling macroporous structures with high-dielectric materials, e.g., liquid-phase or sol–gel chemistry [126, 127], and infiltration with nanosized particles [132], in which considerable contraction of the matrix is observed, which leads to serious crack formation and warping of the colloidal structure. Other electrodeposited materials include Ni, Pt, and a SnCo alloy [157, 158], Pd, Pt, and Co [159]; electroless deposition has also been attempted [136].

As just described, recent work has demonstrated that templating of the interstitial space of highly ordered colloidal crystals has promise for creating macroporous photonic crystals from a diverse set of materials including oxides, semiconductors, metals, polymers, and glasses. The resulting 3D macroperiodic materials have been formed with close-packed macropores ranging in diameter from 100 nm to a few micrometers, giving the potential to modulate light ranging from deep UV to the infrared. However, problems with the infilling process still need to be overcome before this approach to photonic structures comes to fruition. As outlined, filling the 3D interstitial space of a colloidal crystal with a high-index material has been problematic, because many techniques either only deposit material in the top few layers of a colloidal crystal or do not fully fill the colloidal crystal with a material of high enough refractive index.

Another route to infilling of colloidal crystals to generate a high-refractive-index structure is melt-imbibing of a chalcogenide glass such as selenium, followed by dissolution of the silica template. Selenium was selected because it has a high refractive index of 2.5, and thus can provide a nearly complete 3D photonic gap (Figure 3.40), a very low optical loss coefficient between 1 and 10 μm [160], low melting point (217 °C), and relatively low surface tension ($\sim$100 dynes cm^{-1}), which reduces the force necessary to infill the structure. Importantly, selenium vitrifies easily, forming an optically isotropic glass [160]. Other chalcogenide glasses certainly could be used to infill colloidal crystals; however, they have higher softening points and thus are not suitable for initial investigation. Through melt imbibing, essentially complete infilling of a colloidal crystal was demonstrated (Figure 3.41).

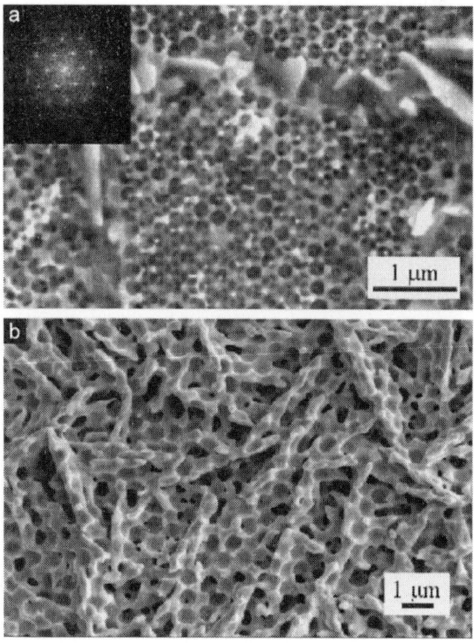

Fig. 3.39 (a) Scanning electron micrograph of a crystal of air spheres (radius 111 nm) in gold, made with a silica template. The inset is a Fourier transform of the image. (b) Scanning electron micrograph of macropores (radius 322 nm) in gold, made with a latex template. The structure has short-range order, but no long-range order. Adapted from [156]

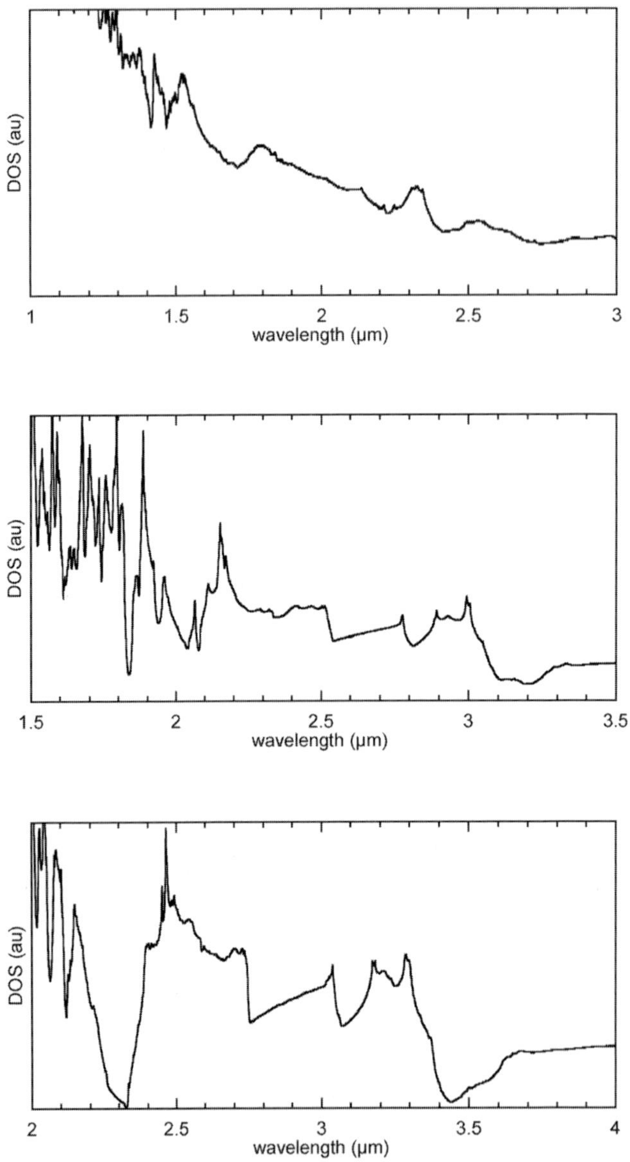

Fig. 3.40 Photonic density of states for inverse fcc structure for refractive index contrasts of (a) 1.45, (b) 2.5, and (c) 2.95; 1.45 corresponds to filling with a simple oxide such as silica, 2.5 to filling with selenium, and 2.95 to filling with a high-index chalcogenide glass such as Ge_{25},$As_{20}Se_{25}Te_{30}$. Note the deep photonic band gap for an index contrast of 2.5 at 2 μm and a complete gap at 2.3 μm for the system with an index contrast of 2.95. Adapted from [138].

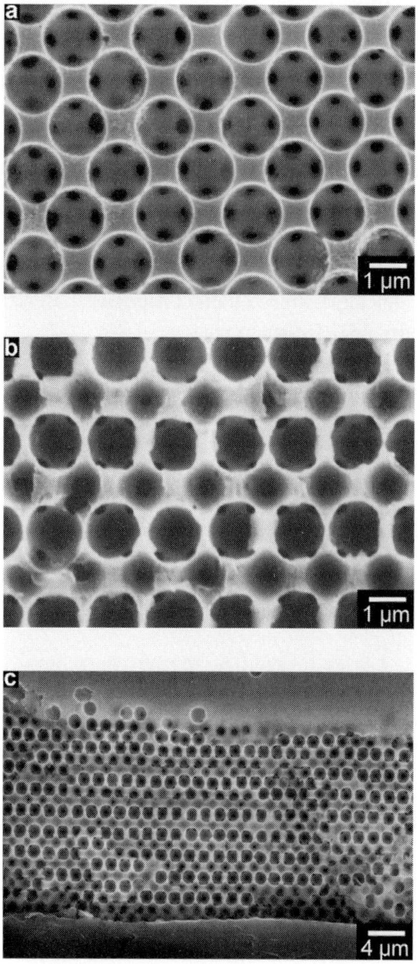

Fig. 3.41 Scanning electron micrographs of a selenium photonic crystal cut and polished to expose specific crystallographic faces. (a) Polished (001) facet, which is perpendicular to the settling direction. The contact points between the spheres on adjacent layers are clearly visible as pores, but within the (001) plane the voids do not appear to be interconnected. This pore morphology is the result of a 2% mismatch between the sphere diameter and the template size. (b) (110) face, proof that the selenium photonic crystal is indeed fcc. Again, the voids are generally not interconnected within the (001) plane, but are interconnected between (001) planes. (c) Low-magnification scanning electron micrograph of the (110) face, presenting a cross section of the entire thickness of the selenium photonic crystal. The sample was mounted in epoxy prior to polishing, which is seen below the photonic crystal; solid selenium is present above the photonic crystal. Importantly, the structure is highly ordered throughout the entire crystal, including the top layer. Because of an ~1.5° misorientation of the polished sample with respect to the (110) face, the size of the pores appears to change from left to right across the sample. Adapted from [138]

Infilling was accomplished by imbibing molten selenium under high pressure into the colloidal crystal, followed by quenching to vitrify the selenium. Subsequently, the colloidal crystal template was removed with HF, resulting in a macroporous selenium/air structure with a high contrast in refractive index.

3.5 Summary

The confluence of nanoscience, biotechnology, and materials chemistry offers great potential for discovery and fabrication of advanced composite materials. Great amounts of information still need to be gleaned from the study of biological systems, but we have now reached a point where the current body of knowledge on how biological systems can create highly functional nanocomposites is starting to enable the creation of advanced materials. For example, natural systems widely exploit self-assembly to create a great diversity of interesting and highly functional materials, and today we are beginning to also create synthetic systems by similar processes. We must always remember that biological organisms work with a limited subset of materials and take up to years to create nanocomposite structures, so, rather than attempting to create materials by direct mimicking of biology, it will likely be much more productive to create materials by exploiting the design rules expressed by biological systems and applying them to synthetic systems. The dividends for research into biologically inspired nanocomposite materials are great, and much progress is expected in the near future.

References

1 Corma A: From microporous to mesoporous molecular sieve materials and their use in catalysis. Chem. Rev. 1997, 97:2373–2419.

2 Brinker CJ: Oriented inorganic films. Curr. Opin. Colloid Interface Sci. 1998, 3:166–173.

3 Soten I, Ozin GA: New directions in self-assembly: materials synthesis over 'all' length scales. Curr. Opin. Colloid Interface Sci. 1999, 4:325–337.

4 Antonietti M: Surfactants for novel templating applications. Curr. Opin. Colloid Interface Sci. 2001, 6:244–248.

5 Sayari A, Hamoudi S: Periodic mesoporous silica-based organic: inorganic nanocomposite materials. Chem. Mater. 2001, 13:3151–3168.

6 Stroscio JA, Eigler DM: Science 1991, 254:1319–1326.

7 Whitesides GM, Mathias JP, Seto CT: Molecular self-assembly and nanochemistry: a chemical strategy for the synthesis of nanostructures. Science 1991, 254:1312–1319.

8 Mann S, Ozin GA: Synthesis of inorganic materials with complex form. Nature 1996, 382:313–318.

9 Nanostructures: Special Issue: Nanostructured Materials. Chem. Mater. 1996, 8:1569–1882.

10 Stupp SI, LeBonheur V, Walker K, Li LS, Huggins KE, Keser M, Amstutz A: Supramolecular Materials: Self-organized nanostructures. Science 1997, 276:384–389.

11 Stupp SI, Braun PV: Molecular manipulation of microstructures: biomaterials, ceramics, and semiconductors. Science 1997, 277:1242–1248.

12 Liu J, Kim AY, Wang LQ, Palmer BJ, Chen YL, Bruinsma P, Bunker BC, Exarhos GJ, Graff GL, Rieke PC, et al.: Self-assembly in the synthesis of ceramic materials and composites. Adv. Colloid Interface Sci. 1996, 69:131.

13 Shenton W, Pum D, Sleytr UB, Mann S: Synthesis of cadmium sulfide superlattices using self-assembled bacterial S-layers. Nature 1997, 389:585–587.

14 Neeraj, Rao CNR: Metal chalcogenide-organic nanostructured composites from self-assembled organic amine templates. J. Mater. Chem. 1998, 8:279–280.

15 Schnur JM: Lipid Tubules: A paradigm for molecularly engineered structures. Science 1993, 262:1669–1676.

16 Archibald DD, Mann S: Template mineralization of self-assembled anisotropic lipid microstructures. Nature 1993, 364:430.

17 Kresge CT, Leonowicz ME, Roth WJ, Vartuli JC, Beck JS: Ordered mesoporpous molecular sieves synthesized by a liquid-crystal template mechanism. Nature 1992, 359:710–712.

18 Beck JS, Vartuli JC, Kennedy GJ, Kresge CT, Roth WJ, Schramm SE: Chem. Mater. 1994, 6:1816–1821.

19 Monnier A, Schuth F, Huo Q, Kumar D, Margolese D, Maxwell RS, Stucky GD, Krishnamurty M, Petroff P, Firouzi A, et al.: Cooperative formation of inorganic-or-

ganic interfaces in the synthesis of silicate mesostructures. Science 1993, 261:1299–1303.
20 Huo Q, Margolese DI, Ciesla U, Feng P, Gier TE, Sieger P, Leon R, Petroff PM, Schuth F, Stucky GD: Generalized synthesis of periodic surfactant/inorganic composite materials. Nature 1994, 368:317–321.
21 Firouzi A, Kumar D, Bull LM, Besier T, Sieger P, Huo Q, Walker SA, Zasadzinski JA, Glinka C, Nicol J, et al.: Cooperative organization of inorganic-surfactant and biomimetic assemblies. Science 1995, 267:1138–1143.
22 Attard GS, Glyde JC, Goltner CR: Nature 1995, 378:366–368.
23 Beck JS, Vartuli JC: Curr. Opin. Solid State Mater. Sci. 1996, 1:76.
24 Hinman MB, Jones JA, Lewis RV: Trends Biotechnol. 2000, 18:374.
25 Hayashi CY, Shipley NH, Lewis RV: Int. J. Biol. Macromol. 1999, 24:271.
26 Gosline JM, Guerette PA, Ortlepp CS, Savage KN: J. Exp. Biol. 1999, 202:3294.
27 Hayashi CY, Lewis RV: Science 2000, 387.
28 Thiel BL, Viney C, Jelinski LW: β sheets and spider silk. Science 1996, 273:1477–1480.
29 Simmons AH, Michal CA, Jelinski LW: Molecular orientation and two-component nature of the crystalline fraction of spider dragline silk. Science 1996, 271:84–87.
30 Lazaris A, Arcidiacono S, Huang Y, Zhou J-F, Duguay F, Chretien N, Welsh EA, Soares JW, Karatzas CN: Spider silk fibers spun from soluble recombinant silk produced in mammalian cells. Science 2002, 295:472–476.
31 Harrison CC: Evidence for intramineral macromolecules containing protein from plant silicas. Phytochem. 1996, 41:37–41.
32 Mann S: Biomineralization: Principles and Concepts in Bioinorganic Materials Chemistry. Oxford: Oxford University Press; 2002.
33 Schultze-Lam S, Harauz G, Beveridge TJ: Participation of a cyanobacterial S layer in fine-grain mineral formation. J. Bacteriol. 1992, 174:7971–7981.
34 Schultze-Lam S, Fortin D, Davis BS, Beveridge TJ: Mineralization of bacterial surfaces. Chem. Geol. 1996, 132:171–181.
35 Weiner S, Addadi L: Design strategies in mineralized biological materials. J. Mater. Chem. 1997, 7:689–702.
36 Weiner S, Traub W, Wagner HD: Lamellar bone: structure–function relations. J. Struct. Biol. 1999, 126:241–255.
37 Weiner S, Wagner HD: The material bone: structure mechanical function relations. Annu. Rev. Mater. Sci. 1998, 28:271–298.
38 Douglas K, Clark NA, Rothschild KJ: Nanometer molecular lithography. Appl. Phys. Lett. 1986, 48:676–678.
39 Douglas K, Clark NA, Rothschild KJ: Biomolecular/solid-state nanoheterostructures. Appl. Phys. Lett. 1990, 56:692–694.
40 Coffer JL, Bigham SR, Li X, Pinizzotto RF, Rho YG, Pirtle RM, Pirtle IL: Dictation of the shape of mesoscale semiconductor nanoparticle assemblies by plasmid DNA. Appl. Phys. Lett. 1996, 69:3851–3853.
41 Mirkin CA, Letsinger RL, Mucic RC, Storhoff JJ: A DNA-based method for rationally assembling nanoparticles into macroscopic materials. Nature 1996, 382:607–609.
42 Alivisatos AP, Johnsson KP, Peng XG, Wilson TE, Loweth CJ, Jr, Bruchez MP, Schultz PG: Organization of 'nanocrystal molecules' using DNA. Nature 1996, 382:609–611.
43 Alivisatos AP, Schultz PG, Peng XG, Loweth CJ, Caldwell WB: DNA-based assembly of gold nanocrystals. Angew. Chem., Int. Ed. Engl. 1999, 38:1808–1812.
44 Soto CM, Srinivasan A, Ratna BR: Controlled assembly of mesoscale structures using DNA as molecular bridges. J. Am. Chem. Soc. 2002, 124:8508–8509.
45 Brown S: Engineered iron oxide-adhesion mutants of the Escherichia coli phage-lambda receptor. Proc. Natl. Acad. Sci. U.S.A. 1992, 89:8651–8655.
46 Brown S: Metal recognition by repeating polypeptides. Nat. Biotechnol. 1997, 15:269–272.
47 Whaley SR, English DS, Hu EL, Barbara PF, Belcher AM: Selection of peptides with semiconductor binding specificity for directed nanocrystal assembly. Nature 2000, 405:665–668.
48 Lee SW, Mao CB, Flynn CE, Belcher AM: Ordering of quantum dots using geneti-

cally engineered viruses. Science 2002, 296:892–895.
49 Addadi L, Weiner S: Control and design principles in biological mineralization. Angew. Chem., Int. Ed. Engl. 1992, 31:153–169.
50 Weller H: Colloidal semiconductor Q-particles: chemistry in the transition region between solid state and molecules. Angew. Chem., Int. Ed. Engl. 1993, 32:41–53.
51 Weller H: Quantized semiconductor particles: a novel state of matter for materials science. Adv. Mater. 1993, 5:88–95.
52 Weller H: Optical properties of quantized semiconductor particles. Philos. Trans. R. Soc. London, Ser. A 1996, 354:757–766.
53 Murray CB, Norris DJ, Bawendi MG: J. Am. Chem. Soc. 1993, 115:8706.
54 Dabbousi BO, Murray CB, Rubner MF, Bawendi MG: Langmuir–Blodgett manipulation of size-selected cdse nanocrystallites. Chem. Mater. 1994, 6:216–219.
55 Murray CB, Kagan CR, Bawendi MG: Self-organization of CdSe nanocrystallites into three-dimensional quantum dot superlattices. Science 1995, 270:1335–1338.
56 Alivisatos AP: Semiconductor nanocrystals. MRS Bull. 1995, 20:23–32.
57 Peng X, Wilson TE, Alivisatos AP, Schultz PG: Synthesis and isolation of a homodimer of cadmium selenide nanocrystals. Angew. Chem., Int. Ed. Engl. 1997, 36:145–147.
58 Brus LE: Electron–electron and electron–hole interactions in small semiconductor crystallites: the size dependence of the lowest excited electronic state. J. Chem. Phys. 1994, 80:4403–4409.
59 Siegel RW: Creating nanophase materials. Sci. Am. 1996, 275(December):74–79.
60 Fendler JH, Meldrum FC: The colloid chemical approach to nanostructured materials. Adv. Mater. 1995, 7:607–632.
61 Geppert L: Semiconductor lithography for the next millennium, IEEE Spectrum 1996, 33: 33–38.
62 Levenson MD: Solid State Technol. 1995 (Sept) 81.
63 Chang THP, Thomson MGR, Yu ML, Kratschmer E, Kim HS, Lee KY, Rishton SA, Zolgharnain S: Microelectron. Eng. 1996, 32:113.
64 Matsui S, Ochiai Y: Nanotechnology 1996, 7:247.
65 Smith HI, Schattenburg ML, Hector SD, Ferrera J, Moon EE, Yang IY, Bukhardt M: Microelectron. Eng. 1996, 32:143.
66 Marrian CRK, Snow ES: Microelectron. Eng. 1996, 32:173.
67 Zhao X-M, Xia Y, Whitesides GM: J. Mater. Chem 1997, 7:1069.
68 Fendler JH: Membrane-Mimetic Approach to Advanced Materials. Berlin: Springer-Verlag; 1994.
69 Peng XG, Manna L, Yang WD, Wickham J, Scher E, Kadavanich A, Alivisatos AP: Shape control of CdSe nanocrystals. Nature 2000, 404:59–61.
70 Manna L, Scher EC, Alivisatos AP: Synthesis of soluble and processable rod-, arrow-, teardrop-, and tetrapod-shaped CdSe nanocrystals. J. Am. Chem. Soc. 2000, 122:12700–12706.
71 Berman A, Addadi L, Weiner S: Nature 1988, 331:546–548.
72 Berman A, Addadi L, Kvick A, Leiserowitz L, Nelson M, Weiner S: Science 1990, 250:664–667.
73 Berman A, Hanson J, Leiserowitz L, Koetzle TF, Weiner S, Addadi L: Science 1993, 259:776–779.
74 Messersmith PB, Stupp SI: High-temperature chemical and microstructural transformations of a nanocomposite organoceramic. Chem. Mater. 1995, 7:454–460.
75 Mitzi DB, Feild CA, Harrison WTA, Guloy AM: Conducting tin halides with a layered organic-based perovskite structure. Nature 1994, 369:467–469.
76 Mitzi DB, Wang S, Feild CA, Chess CA, Guloy AM: Conducting layered organic-inorganic halides containing <110>-oriented perovskite sheets. Science 1995, 267:1473–1476.
77 Feng X, Fryxell GE, Wang Q-L, Kim AY, Liu J, Kemner KM: Functionalized monolayers on ordered mesoporous supports. Science 1997, 276:923–926.
78 Weiss D, Roukes ML, Mensching A, Grambow P, Klitzing Kv, Weimann G: Electron pinball and commensurate orbits in a periodic array of scatters. Phys. Rev. Lett. 1991, 66:2790–2793.

79 Hansen W, Kotthaus JP, Merkt U: Electrons in Laterally Periodic Nanostructures, vol. 35. Ed. Hansen W, Kotthaus JP, Merkt U. San Diego:Academic Press; 1992:279–380.

80 Schechter RS, Bourrel M (Ed): Microemulsions and Related Systems: Formation, Solvency, and Physical Properties. New York: Marcel Dekker; 1988.

81 Tanford C: J. Phys. Chem. 1974, 78:2469.

82 Rosoff M (Ed): Vesicles. New York: Marcel Dekker; 1996.

83 Ringsdorf H, Schlarb B, Venzmer J: Angew. Chem., Int. Ed. Engl. 1988, 27:114–158.

84 Laughlin RG: The Aqueous Phase Behavior of Surfactants. San Diego: Academic Press; 1994.

85 Degiorgio V, Corti M (Ed): Physics of Amphiphiles: Micelles, Vesicles and Microemulsions. Amsterdam: Elsevier Science Pub. Co.; 1985.

86 Israelachvili J: Thermodynamic and geometrical aspects of amphiphile aggregation into micelles, vesicles, and bilayers, and the interactions between them. In Physics of Amphiphiles: Micelles, Vesicles and Microemulsions. Ed. Degiorgio V, Corti M; 1985:24.

87 Bates FS: Polymer–polymer phase behavior. Science 1991, 251:898–905.

88 Wennerstrom H: J. Colloid Interface Sci. 1979, 68:589–590.

89 Tiddy G: Phys. Rep. 1980, 57:1.

90 Fribreg SE, Wang J: J. Dispersion Sci. Technol. 1991, 12:387–402.

91 Walsh D, Hopwood JD, Mann S: Crystal tectonics: constructruction of reticulated calcium phosphate frameworks in bicontinuous reverse microemulsions. Science 1994, 264:1576–1578.

92 Yang JP, Qadri SB, Ratna BR: Structural and morphological characterization of pbs nanocrystallites synthesized in the bicontinuous cubic phase of a lipid. J. Phys. Chem. 1996, 100:17255–17259.

93 Braun PV, Osenar P, Stupp SI: Semiconducting superlattices templated by molecular assemblies. Nature 1996, 380:325–328.

94 Osenar P, Braun PV, Stupp SI: Adv. Mater. 1996, 8:1022.

95 Tohver V, Braun PV, Pralle MU, Stupp SI: Counterion effects in liquid crystal templating of nanostructured CdS. Chem. Mater. 1997, 9:1495–1499.

96 Braun PV, Osenar P, Tohver V, Kennedy SB, Stupp SI: Nanostructure templating in inorganic solids with organic lyotropic liquid crystals. J. Am. Chem. Soc. 1999, 121:7302–7309.

97 Attard GS, Bartlett PN, Coleman NRB, Elliott JM, Owen JR, Wang JH: Mesoporous platinum films from lyotropic liquid crystalline phases. Science 1997, 278:838–840.

98 Attard GS, Edgar M, Goltner CG: Inorganic nanostructures from lyotropic liquid crystal phases. Acta Materialia 1998, 46:751–758.

99 Elliott JM, Birkin PR, Bartlett PN, Attard GS: Platinum microelectrodes with unique high surface areas. Langmuir 1999, 15:7411–7415.

100 Attard GS, Leclerc SAA, Maniguet S, Russell AE, Nandhakumar I, Bartlett PN: Mesoporous Pt/Ru alloy from the hexagonal lyotropic liquid crystalline phase of a nonionic surfactant. Chem. Mater. 2001, 13:1444.

101 Attard GS, Leclerc SAA, Maniguet S, Russell AE, Nandhakumar I, Gollas BR, Bartlett PN: Liquid crystal phase templated mesoporous platinum alloy. Microporous Mesoporous Mater. 2001, 44:159–163.

102 Nelson PA, Elliott JM, Attard GS, Owen JR: Mesoporous nickel/nickel oxide: a nanoarchitectured electrode. Chem. Mater. 2002, 14:524–529.

103 Nandhakumar I, Elliott JM, Attard GS: Electrodeposition of nanostructured mesoporous selenium films (H-I-eSe). Chem. Mater. 2001, 13:3840.

104 Gabriel T, Nandhakumar IS, Attard GS: Electrochemical synthesis of nanostructured tellurium films. Electrochem. Commun. 2002, 4:610–612.

105 Schick MJ: Nonionic Surfactants, Physical Chemistry. New York: Marcel Dekker; 1987.

106 Wanka G, Hoffmann H, Ulbricht W: Phase diagrams and aggregation behavior of poly(oxyethylene)-poly(oxypropylene)-poly(oxyethylene) triblock copolymers in aqueous solutions. Macromolecules 1994, 27:4145–4159.

107 Fontell K: Cubic phases in surfactant and surfactant-like lipid systems. Colloid Polym. Sci. 1990, 268:264–285.

108 Lo I, Florence AT, Treguier J-P, Seiller M, Puisieux F: The influence of surfactant HLB and the nature of the oil phase on the phase diagrams of nonionic surfactant-oil-water systems. J. Colloid Interface Sci. 1977, 59:319–327.

109 Treguier JP, Seiller M, Puisieux F, Orecchioni AM, Florence AT: Effect of a hydrophilic surfactant and temperature on water-surfactant-oil diagrams. In First Expo. Cong. Int. Technol. Pharm. 1977: Assoc. Pharm.; 1977:75–87.

110 Braun PV, Stupp SI: CdS mineralization of hexagonal, lamellar, and cubic lyotropic liquid crystals. Mater. Res. Bull. 1999, 34:463–469.

111 Blackburn JC, Kilpatrick PK: Using deuterium NMR line shapes to analyze lyotropic liquid crystalline phase transitions. Langmuir 1992, 8:1679–1687.

112 Schnepp W, Disch S, Schmidt C: 2H NMR study on the lyomesophases of the system hexaethylene glycol dodecyl methyl ether/water: temperature dependence of quadrupole splittings. Liq. Cryst. 1993, 14:843–852.

113 Attard GS, Goltner CG, Corker JM, Henke S, Templer RH: Liquid-crystal templates for nanostructured metals. Angew. Chem., Int. Ed. Engl. 1997, 36:1315–1317.

114 Park M, Harrison C, Chaikin PM, Register RA, Adamson DH: Block copolymer lithography: periodic arrays of $\sim 10^{11}$ holes in one square centimeter. Science 1997, 276:1401–1404.

115 Park M, Chaikin PM, Register RA, Adamson DH: Large area dense nanoscale patterning of arbitrary surfaces. Appl. Phys. Lett. 2001, 79:257–259.

116 Li RR, Dapkus PD, Thompson ME, Jeong WG, Harrison C, Chaikin PM, Register RA, Adamson DH: Dense arrays of ordered GaAs nanostructures by selective area growth on substrates patterned by block copolymer lithography. Appl. Phys. Lett. 2000, 76:1689–1691.

117 Thurn-Albrecht T, Schotter J, Kastle CA, Emley N, Shibauchi T, Krusin-Elbaum L, Guarini K, Black CT, Tuominen MT, Russell TP: Ultrahigh-density nanowire arrays grown in self-assembled diblock copolymer templates. Science 2000, 290:2126–2129.

118 Kim HC, Jia XQ, Stafford CM, Kim DH, McCarthy TJ, Tuominen M, Hawker CJ, Russell TP: A route to nanoscopic SiO2 posts via block copolymer templates. Adv. Mater. 2001, 13:795.

119 Lin ZQ, Kim DH, Wu XD, Boosahda L, Stone D, LaRose L, Russell TP: A rapid route to arrays of nanostructures in thin films. Adv. Mater. 2002, 14:1373–1376.

120 Bal M, Ursache A, Touminen MT, Goldbach JT, Russell TP: Nanofabrication of integrated magnetoelectronic devices using patterned self-assembled copolymer templates. Appl. Phys. Lett. 2002, 81:3479–3481.

121 Shin K, Leach KA, Goldbach JT, Kim DH, Jho JY, Tuominen M, Hawker CJ, Russell TP: A simple route to metal nanodots and nanoporous metal films. Nano Lett. 2002, 2:933–936.

122 Joannopoulos JD, Meade RD, Winn JN: Photonic Crystals: Molding the Flow of Light. Princeton: Princeton University Press; 1995.

123 Busch K, John S: Photonic band gap formation in certain self-organizing systems. Phys. Rev. E 1998, 58:3896–3908.

124 Velev OD, Kaler EW: Structured porous materials via colloidal crystal templating: from inorganic oxides to metals. Adv. Mater. 2000, 12:531–534.

125 Velev OD, Jede TA, Lobo RF, Lenhoff AM: Porous silica via colloidal crystallization. Nature 1997, 389:447–448.

126 Holland BT, Blanford CF, Stein A: Synthesis of macroporous minerals with highly ordered three-dimensional arrays of spherical voids. Science 1998, 281:538–540.

127 Wijnhoven JEGJ, Vos WL: Preparation of photonic crystals made of air spheres in titania. Science 1998, 281:802–804.

128 Zakhidov AA, Baughman RH, Iqbal Z, Cui CX, Khayrullin I, Dantas SO, Marti I, Ralchenko VG: Carbon structures with three-dimensional periodicity at optical wavelengths. Science 1998, 282:897–901.

129 Mguez H, Blanco A, Meseguer F, López C, Yates HM, Pemble ME, Fornés V, Mifsud A: Brag diffraction from indium phosphide

infilled fcc silica colloidal crystals. Phys. Rev. B 1999, 59:1563–1566.
130 Blanco A, Chomski E, Grabtchak S, Ibisate M, John S, Leonard SW, Lopez C, Meseguer F, Miguez H, Mondia JP, et al.: Large-scale synthesis of a silicon photonic crystal with a complete three-dimensional bandgap near 1.5 micrometers. Nature 2000, 405:437–440.
131 Vlasov YA, Bo XZ, Sturm JC, Norris DJ: On-chip natural assembly of silicon photonic bandgap crystals. Nature 2001, 414:289–293.
132 Vlasov YA, Yao N, Norris DJ: Synthesis of photonic crystals for optical wavelengths from semiconductor quantum dots. Adv. Mater. 1999, 11:165–169.
133 Subramania G, Constant K, Biswas R, Sigalas MM, Ho K-M: Optical photonic crystals fabricated from colloidal systems. Appl. Phys. Lett 1999, 74:3933–3935.
134 Subramanian G, Manoharan VN, Thorne JD, Pine DJ: Ordered macroporous materials by colloidal assembly: a possible route to photonic bandgap materials. Adv. Mater. 1999, 11:1261–1265.
135 Mguez H, Meseguer F, López C, Holgado M, Andreasen G, Mifsud A, Fornés V: Germanium FCC structure from a colloidal crystal template. Langmuir 2000, 16:4405–4408.
136 Jiang P, Cizeron J, Bertone JF, Colvin VL: Preparation of macroporous metal films from colloidal crystals. J. Am. Chem. Soc. 1999, 121:7957–7958.
137 Braun PV, Wiltzius P: Electrochemically grown photonic crystals. Nature 1999, 402:603–604.
138 Braun PV, Zehner RW, White CA, Weldon MK, Kloc C, Patel SS, Wiltzius P: Epitaxial growth of high dielectric contrast three-dimensional photonic crystals. Adv. Mater. 2001, 13:721–724.
139 Johnson SA, Ollivier PJ, Mallouk TE: Ordered mesoporous polymers of tunable pore size from colloidal silica templates. Science 1999, 283:963–965.
140 Jeon S, Braun PV: Hydrothermal synthesis of Er-doped luminescent TiO_2 nanoparticles, Chem. Mater. 2003, 15: 1256–1263.
141 Braun PV, Wiltzius P: Electrochemical fabrication of 3-D microperiodic porous materials. Adv. Mater. 2001, 13:482–485.
142 Sumida T, Wada Y, Kitamura T, Yanagida S: Macroporous ZnO films electrochemically prepared by templating of opal films. Chem. Lett 2001, 1:38–39.
143 Pandey RK, Sahu SN, Chandra S: Handbook of Semiconductor Electrodeposition. New York: Marcel Dekker; 1996.
144 Gurunathan K, Vadivel MA, Marimuthu R, Mulik UP, Amalnerkar DP: Electrochemically synthesised conducting polymeric materials for applications towards technology in electronics, optoelectronics and energy storage devices. Mater. Chem. Phys. 1999, 61:173–191.
145 Martin CR: Nanomaterials: a membrane based synthetic approach. Science 1994, 266:1961–1966.
146 Koopal CGJ, Feiters MC, Nolte RJM, Deruiter B, Schasfoort RBM: Third generation amperometric biosensor for glucose-polypyrrole deposted within a matrix of uniform latex particles as mediator. Bioelectrochem. Bioenerg. 1992, 29:159–175.
147 Sumida T, Wada Y, Kitamura T, Yanagida S: Electrochemical preparation of macroporous polypyrrole films with regular arrays of interconnected spherical voids. Chem. Commun. 2000:1613–1614.
148 Bartlett PN, Birkin PR, Ghanem MA, Toh CS: Electrochemical syntheses of highly ordered macroporous conducting polymers grown around self-assembled colloidal templates. 2001; J. Mater. Chem. 2001, 11:849–853.
149 Cassagneau T, Caruso F: Semiconducting polymer inverse opals prepared by electropolymerization. Adv. Mater. 2002, 14:34–38.
150 Ebbesen TW, Lezec HJ, Ghaemi HF, Thio T, Wolff PA: Extraordinary optical transmission through sub-wavelength hole arrays. Nature 1998, 391:667–669.
151 Moroz A: Three-dimensional complete photonic-band-gap structures in the visible. Phys. Rev. Lett. 1999, 83:5274–5277.
152 Moroz A: Photonic crystals of coated metallic spheres. Europhys. Lett. 2000, 50:466–472.
153 Zhang WY, Lei XY, Wang ZL, Zheng DG, Tam WY, Chan CT, Sheng P: Robust photonic band gap from tunable scatterers. Phys. Rev. Lett. 2000, 84:2853–2856.

154 Oldenburg SJ, Averitt RD, Westcott SL, Halas NJ: Nanoengineering of optical resonances. Chem. Phys. Lett. 1998, 288:243–247.

155 Graf C, van Blaaderen A: Metallodielectric colloidal core-shell particles for photonic applications. Langmuir 2002, 18:524–534.

156 Wijnhoven J, Zevenhuizen SJM, Hendriks MA, Vanmaekelbergh D, Kelly JJ, Vos WL: Electrochemical assembly of ordered macropores in gold. Adv. Mater. 2000, 12:888–890.

157 Xu L, Zhou WL, Frommen C, Baughman RH, Zakhidov AA, Malkinski L, Wang JQ, Wiley JB: Electrodeposited nickel and gold nanoscale meshes with potentially interesting photonic properties. Chem. Commun. 2000:997–998.

158 Luo Q, Liu Z, Li L, Xie S, Kong J, Zhao D: Creating highly ordered metal, alloy, and semiconductor macrostructures by electrodeposition, ion spraying, and laser spraying. Adv. Mater. 2001, 13:286–289.

159 Bartlett PN, Birkin PR, Ghanem MA: Electrochemical deposition of macroporous platinum, palladium and cobalt films using polystyrene latex sphere templates. Chem. Commun. 2000:1671–1672.

160 Kokorina VF: Glasses for Infrared Optics. Boca Raton, FL: CRC Press; 1996.

161 Dunin-Borkowski RE, McCartney MR, Frankel RB, Bazylinski DA, Pósfai M, Buseck PR: Magnetic microstructure of magnetotactic bacteria by electron holography. Science 1998, 282:1868–1870.

162 Zaremba CM, Morse DE, Mann S, Hansma PK, Stucky GD: Aragonite–hydroxyapatite conversion in gastropod (abalone) nacre. Chem. Mater. 1998, 10:3813–3824.

163 Ziv V, Sabanay I, Arad T, Traub W, Weiner S: Transitional structures in lamellar bone. Microsc. Res. Technique 1996, 33:203–213.

164 Wong KKW, Douglas T, Gider S, Awschalom DD, Mann S: Biomimetic synthesis and characterization of magnetic proteins (magnetoferritin). Chem. Mater. 1998, 10:279–285.

165 Yang J, Meldrum FC, Fendler JH: Epitaxial growth of size-quantized cadmium sulfide crystals under arachidic acid monolayers. J. Phys. Chem. 1995, 99:5500–5504.

166 Rosevear, FB: Liquid Crystals: the mesomorphic phases of surfactant compositions. J. Soc. Cosmetic Chemists 1968, 19:581–594.

4
Modeling of Nanocomposites

Catalin Picu and Pawel Keblinski

4.1
Introduction – The Need For Modeling

Polymer- as well as ceramic-based composites filled with nanofillers are a novel class of materials. These systems are compositionally similar to conventional composites. In both nanocomposites and regular composites (referred to here as microcomposites), the matrix is the same. For example, polymer forming a matrix may be essentially any thermoplastic or thermosetting polymer. The feature that distinguishes the two classes of composites is the size and nature of the filler. Nanocomposites are filled with nanoparticles or with carbon nanotubes. The characteristic dimensions of the filler are <100 nm and often in the range of 100 nm. The fillers may be spheroid, such as ceramic nanoparticles or carbon black; plate-like, such as layered silicates; or fiber-like, such as nanotubes.

Nanocomposite materials exhibit properties that are significantly different from those of the corresponding microcomposite. Examples are discussed at length in other chapters of this book. Polymers filled with nanoparticles have significantly enhanced stiffness and strength compared to similar materials filled with regular micrometer-size ceramic particles. Most importantly, a gain in these properties does not compromise ductility, as typically occurs with conventional materials. In other systems a significant ductility enhancement was measured with no decrease in strength [1, 2]. Also, ceramic nanocomposites exhibit increased fracture toughness. Other examples involve fluids such as water or glycol which, when filled with metal nanoparticles, exhibit thermal conductivity enhancement [3] that cannot be explained by classical approaches valid for standard composites [4]. This clearly indicates that, whereas a body of knowledge derived from standard continuum-level modeling approaches provides a good description of microcomposites, it is inadequate in its current form to describe properties and behaviors of nanocomposites.

The development of these new materials, and particularly an understanding of their properties, is still in its infancy. Consequently, to date, proofs of concept have been repeatedly and consistently made, but material optimization is still an elusive goal. For example, good filler dispersion has been demonstrated to be critical for achieving properties superior to those of microcomposites, but it is still unclear how other parameters such as filler distribution and filler surface functionalization influence the overall properties. Optimization of these materials will be possible only when the physics controlling the enhanced properties is properly understood, and when the processing routes leading to the desired nano- and microstructures are established. The current approach to investigating these issues is experimental and based on trial and error. Hence, modeling the material response would provide valuable information on the underlying physics and the desired structure–property linkages. Furthermore, once the desired structure is established, modeling has the capability to assist the development of material processing.

4.2
Current Conceptual Frameworks

The properties of the matrix, the distribution and properties of the filler, and the nature of their interface control the behavior of a typical microcomposite. The interface plays an important role in controlling several microcomposite properties such as its thermal conductivity, load-carrying capacity, toughness, and transport properties. However, for a characteristic filler particle sizes on the micrometer scale and larger, the main parameter controlling the macroscopic properties is still the filler volume fraction. In these materials, the interface may be viewed in the standard way, as a 2D surface separating the constituents. By contrast, when the matrix is filled with nanosized particles, a significant fraction of the material is either at the interface or within its immediate vicinity. For ceramic composites, one can argue that the interface width is still small relative to the filler size or the filler-to-filler distance; thus extension or modification of standard composite theory or modeling approaches may be sufficient. For polymer nanocomposites, the modification of the matrix may extend over several radii of gyration from the particle surface. In turn, this situation results in novel macroscopic properties that are dictated qualitatively, as a trend, and quantitatively, by the behavior of the confined polymer that forms the modified matrix.

Several attempts have been made to provide a conceptual framework for justifying the observed enhanced properties of polymer nanocomposites. One such theory, alluded to above, the 'bonded polymer layer' [5], is based on the observation that, at the same volume fraction, the total interfacial area in a nanocomposite is much larger than in a microcomposite. The total volume of 'bonded polymer' (confined polymer next to the filler wall) is conjectured to represent a large fraction of the volume of the nanocomposite, and therefore, the properties of this confined material determine the macroscopic properties of the system. However, the available experimental results on the existence of a bonded polymer layer are contradictory [6, 7]. It is also unclear how thick the layer surrounding each particle is and, most importantly, what the properties of the polymer in the interfacial region are.

More recently, the 'double network' theory was proposed. This idea is based on the observation that the reduction in particle size entails a similar reduction in the interparticle spacing and that special macroscopic properties are obtained when the average distance between particles becomes comparable to the bulk radius of gyration of the polymer chains. These events create conditions for a chain to connect two or more particles, hence forming a network that provides additional strength by increasing the number of effective entanglements in the system [8]. The intuitive appeal of this theory is high, but quantitative predictions that may be compared with experimental results require a thorough understanding of the structure of chains confined between spherical fillers.

4.3 Multiscale Modeling

Considering the multiple time and length scales characterizing nanocomposites, and polymer nanocomposites in particular, a comprehensive approach to modeling these systems requires multiscale modeling. Whereas multiscale modeling across all length scales, from electronic level to bulk-matter sizes, and all time scales is a future challenge, the various components of composites have however undergone extensive independent study. The dynamics of polymers, which control the mechanical and dielectric responses of these materials, have been studied for many years. Much literature exists on both molecular and continuum aspects of this problem. Several classical models of stress production and chain dynamics dominate current thinking. Following an idea introduced by Guth and Mark [9], the chains are regarded as entropic springs in tension. This concept is rooted in the observation that, as the material deforms, the chains are stretched and their entropy decreases. The change in entropy entails a free energy variation, leading to a retractive force. The force required to stretch the chain proves proportional to the change in the end-to-end distance, which suggests that the chain may be compared with a linear spring. The force is always retractive, and in vacuum, the chain should collapse into itself. The corollary is that all chains are in tension. Furthermore, the force is proportional to the absolute temperature, which is the signature of its entropic nature. The energetic component of the free energy is usually neglected in this theory. This view gained support from early experimental observations that established unequivocally that the nature of stress in rubbers is indeed entropic.

Two main models describe chain dynamics, and hence stress and dielectric relaxation. The Rouse model represents the dynamics of chains shorter than the average distance between entanglements, but long enough to behave as entropic springs [10]. This model is in agreement with the entropic spring description of a chain. The principal interactions accounted for are those along the chain, and the surrounding medium is represented in a homogenized way, by means of a frictional force acting on the representative chain. Chains longer than the entanglement length exhibit slower dynamics, which is described by the reptation model [11]. In this model, the effect of nonbonded interactions between monomers not directly connected by backbone covalent bonds is conjectured to lead to constraints that force the chain to diffuse primarily along its contour. Several other representations have been proposed [12–14].

A rich literature exists on the development of continuum constitutive models of polymer behavior. These models are mostly phenomenological and are developed based on macroscopic mechanical testing of both neat and filled polymeric materials. Several excellent treatises exist on this topic; [15] is a general overview.

Polymer interfaces and, in particular, the polymer chain behavior close to flat interfaces, have been studied by simulations and have a long history of theoretical treatment. The study of the conformations and dynamics of linear polymeric chains in the vicinity of flat walls or free surfaces [16–29] is motivated by the many applications in which the behavior of confined chains is important. Such applications include, but are not limited to, coatings, flow of molecular solutions in capillaries, and their interaction with filters and membranes.

In nanocomposites, the chain–filler interactions control to a large extent the matrix and macroscopic properties. Therefore, understanding chain conformations and dynamics of the confined chains next to filler surfaces is crucial. The physical picture that emerges from the cited studies is summarized below.

The conformations of a macromolecule located close to an impenetrable interface are restricted. They are determined by the configurational entropy of the chain, by the energetic interactions between polymers (cohesive interactions in bulk), by the nature and strength of the interaction between polymers and the confining wall, and by global system parameters such as temperature and density. When the chain approaches the wall, the configurational entropy decreases due to the reduced number of accessible chain configurations. This generates an entropic driving force that retracts the chain from the interface, with the force increasing linearly with temperature. This effect generates a low-density polymer layer in the immediate vicinity of the interface [30–32]. The variation in density has a significant effect on chain mobility as well as on the diffusion of small molecules. The retraction is balanced by packing in the bulk polymer. An attractive energetic interaction between polymers and the wall also promotes chain adsorption. The interplay of these forces determines the actual configuration.

Another entropic force promotes movement of chain-end monomers to the interface [21, 32]. This force is due to the fact that the reduction in the number of accessible chain configurations induced by confinement is less dramatic if the chain comes in contact with the wall with only one of its ends. Hence, chains having their center of mass within two radii of gyration of the wall adjust so as to replace chain-inner-monomer contacts with chain-end contacts.

In the immediate vicinity of a flat impenetrable wall, the polymer chains are preferentially aligned on multiple scales in a direction parallel to the interface. This leads to changes in the glass transition temperature and to a broader relaxation spectrum. The degree of alignment depends on the details of the energetic interaction with the wall. It is generally considered that a chain may adopt a 'docking' type configuration at high temperatures and relatively weak attractive interactions and may completely adsorb if the attraction is strong [33]. Chain collapse in the interface leads to pronounced alignment on the bond and larger scales.

These details are important for understanding the physics of polymer-filler interactions in systems filled with layered silicates, in which interfaces are mostly planar [34]. Another issue in these composites regards the dynamics of polymer intercalation between silicate layers. In polymers filled with nanoparticles and nanotubes, the interface is curved, and it is necessary to ask to what extent the curvature modifies the physical picture described above. This issue was studied [35, 36], resulting in the conclusion that, for realistic curvatures, the polymer structure next to a wall is close to that next to a flat surface in the presence of similar energetic interactions. Interestingly, these models showed that, even in the presence of strong hydrogen bonding of polymers to the curved wall, the shape and size of polymer coils having their center of mass close to the wall remains essentially similar to that in the bulk. However, the ellipsoidal coils undergo a 'docking transition', rotating with their long semiaxis in the direction tangential to the curved wall.

The connection between the physics on the atomic and molecular scales and the meso and macroscopic material behavior requires the development of continuum models. These models must incorporate the smaller-scale physics. Steps along this line have been taken for neat polymer and are reviewed in [37]. In nanocomposites, additional difficulty resides in the fact that the mesoscale continuum model needs to describe deformation on the scale of the inter-filler spacing. Because the fillers have dimensions comparable to those of the polymeric chains, and because the fields vary on the same scale, the homogeneity assumption for the representative volume element is no longer mandated. Under these conditions of high structural and field gradients, nonlocal constitutive laws are required. The development of such laws that incorporate molecular and atomistic information is an emerging field.

On a larger scale, nanoparticles tend to cluster before and during processing of the composite. Currently, it is generally accepted that good dispersion is required for enhanced macroscopic properties. Obtaining the appropriate dispersion in practice is largely a matter of experimental ability. Modeling may be used in this connection to elucidate two issues. On the one hand it is necessary to understand the lower limit of dispersion required for imposed macroscopic properties, and on the other, process modeling that would guide the routes for achieving the desired dispersion is highly desirable. Developing these types of models requires reliable constitutive equations.

On the scale of filler agglomerates and larger, the material may be described in the continuum sense by using the methodologies developed for classical microcomposites. Homogenization theory and asymptotic expansion techniques are useful tools in this endeavor. The standard homogenization theory leads to poor approximations of local fields in the presence of high-gradient regions. In fact, the use of this technique is possible only within the assumption of homogeneous fields over the representative volume considered. In practice, the adequacy (or lack of it) of the homogenization theory can be assessed on the basis of the uniform validity of the double-scale asymptotic expansion [38].

These techniques may be used to study the effect of filler distribution. Good dispersion of fillers, but nonuniform distribution, improved wear properties and scratch resistance of the composite. Most properties depend on filler distribution, and material optimization would require the development of models that may provide the gross behavior while accounting for nonuniform distributions of fillers.

4.4 Multiphysics Aspects

One of the principal advantages of polymer nanocomposites is multifunctionality. Through multifunctionality, the material exhibits enhanced properties with regard to several physical phenomena. The composite may be transparent to the visible spectrum and opaque to ultraviolet while being wear- and scratch-resistant; it may simultaneously provide good strength and diffusion barrier properties. All models predicting the macrosopic behavior should capture the relevant multiple physical processes with respect to which the material is being optimized.

Various physical phenomena are intimately coupled at the atomic and molecular scales and, as such, are captured by discrete models. In continuum representations, the physics is traditionally decoupled. Nanocomposite models should overcome this limitation. Hence, when going from discrete to continuum models, one has to ensure that discretely represented processes can be described by conventional partial differential equations, that the constitutive law selected for the 'smallest-scale continuum' is able to represent the (mean) 'fields', and that the time scale cutoff, which is associated with this transition, does not alter the essential physics studied. Accurate modeling of discrete-continuum transition hinges on an affirmative answer to all these questions.

Multiphysics coupling in the continuum sense may be performed either by adding additional terms in the continuum balance equations ('equation coupling') or by requiring the phenomenological coefficients entering the constitutive laws to depend on the fields to be coupled ('constitutive law coupling'). The coupling of fields (stress/strain, temperature, solute/reactants concentration) occurs through diverse phenomena, some of which are shown in the interaction matrix of Table 4.1. The interaction matrix is 'nonsymmetric', with cell (i, j) representing the phenomenon induced by the

Tab. 4.1 Coupling of physical phenomena in continuum models.

Deformation (displacements)	Deformation-induced heat production	Strain-controlled diffusion	Strain-controlled chemical activity
Thermal expansion	**Heat transfer (temperature)**	Temperature-controlled diffusion	Temperature-controlled chemical activity
Clustering, precipitation (eigenstrains)	Diffusant concentration-controlled heat transfer	**Diffusion (diffusant concentration)**	Transport of reactants
Reaction products accommodation (eigenstrains)	Heat production during chemical reactions	Transport of reaction products	**Chemical reactions**

process corresponding to field i and which influences field j. For instance, cell (1, 2) represents heating due to deformation, and cell (2, 1) corresponds to thermal expansion and thermal stresses. A fully coupled continuum analysis would consider all processes shown in the matrix, but an uncoupled approach would only consider the diagonal entries. As mentioned, the coupling of various physical processes in the mathematical model can be carried out either by considering additional terms in the field equations or by allowing the constitutive law to depend on the interacting field. An example from the first category is the effect of the temperature gradient on the stress field, which can be captured by adding a thermal stress term to equilibrium equations. A less familiar example within the same class of problems is the effect of diffusion (clustering of diffusants) and chemical reactions (reaction products) on the stress field, which can be accounted for by the eigenstrain formulation [39]. An example from the second category is the influence of temperature upon diffusion, which can be captured by simply considering the diffusion coefficients to be temperature dependent.

4.5 Validation

Development of faithful predictive multiscale models of nanocomposites will require a systematic validation effort. The multiscale multiphysics models must be calibrated and validated *at multiple scales* against experiments that resolve the finer scales. For polymer nanocomposites, the experimental data obtained by nuclear magnetic resonance (NMR), dielectric spectroscopy (DS) and quantitative dynamic mechanical analysis (QDMA) can be used to verify the ability of the developed models to predict the conformation and kinetics of polymer chains. NMR measurements probe segmental motions of the main chain and side chain movement over a frequency domain from several Hz to several MHz. Longer-time and larger-scale dynamic effects, which can be modeled by coarse-grained discrete and continuum approaches, would be validated against DS and QDMA measurements. By these techniques, short- and long-time relaxation modes are probed by mechanical or electric excitations.

The challenge to the experimental work will be to perform measurements on well characterized systems, such as those with well defined particle size and shape distribution and particle dispersion. Ideally, one would like to control and vary individual parameters so as to asses their roles. Some successes in this area involve the ability to change surface chemistry, thus controlling the strength of the particle-matrix interactions. However, in experiment, it is often very difficult or even impossible to control various parameters and characteristics independently of one another. The hope is that the combined experimental–modeling effort will allow the acquisition of better and deeper understanding of the structure–property–processing relationships in the quest for guided design of nanocomposites with desired, optimized properties.

References

1 C. B. Ng, B. J. Ash, L. S. Schadler, R. W. Siegel, Adv. Polym. Composites, 10, 101–111 (2001).
2 K. Yano, A. Usuki, A. Okada, T. Kurauchi, O. Kamigaito, Synthesis and properties of polyimide–clay hybrid, J. Polym. Sci., Part A: Polym. Chem. 31, 2493–2498 (1993).
3 S. Lee, S.U.-S. Choi, S. Li, J.A. Eastman, Measuring thermal conductivity of fluids containing oxides nanoparticles, Trans. ASME J. Heat Transfer, Heat Transfer, 121, 280–289 (1999).
4 R. L. Hamilton, O. K. Crosser, Thermal conductivity of heterogenous two-component systems, Ind. Eng. Chem. Fundam., 1, 187–191 (1962).
5 P.H.T. Vollenberg, D. Heikens, Polymer 30, 1656 (1989).
6 P.H.T. Vollenberg, J.W. deHaan, L.J.M. van de Ven, D. Heikens, Polymer 30, 1663 (1989).
7 A.J. Zhu, S.S Sternstein, S.S., MRS Symp. Proc., 661, KK4.3.1–4.3.7 (2001).
8 W.F. Reichert, D. Goritz, E.J. Duschl, Polymer 34, 1216 (1993).
9 E. Guth, H. Mark, Monatsh. Chem. 65, 93 (1934).
10 P.E. Rouse, J. Chem. Phys. 21, 1272 (1953).
11 M. Doi, S.F. Edwards, The Theory of Polymer Dynamics, Clarendon, Oxford, 1986.
12 M.F. Herman, B. Panajotova, K.T.J. Lorenz, J. Chem. Phys., 105, 1153 (1996).
13 R.C. Picu, G. Loriot, J.H. Weiner, J. Chem. Phys., 108, 4984 (1998).
14 R.C. Picu, Atomic level stress in polymeric systems: a review of principal results, In Recent Research Developments in Macromolecules, Ed. S.G. Pandalai, Research Signpost, 2003.
15 L. E. Nielsen, R. F. Landel, Mechanical Properties of Polymers and Composites, Marcel Dekker, New York, 1994.
16 M. Stamm, Adv. Poly. Sci. 100,357 (1992).
17 B.J. Factor, T.P. Russel, M.F. Toney, Macromolecules 26, 2847 (1993).
18 J.A. Forrest, K. Dalkoni-Veress, J.R. Stevens, J.R. Dutcher Phys. Rev. Lett. 77, 2002 (1996).
19 S. Sen, J.M. Cohen, J.D. McCoy, J.G. Curro, J. Chem. Phys. 101, 9010 (1994).
20 D.N. Theodorou, Macromolecules 21,1391 (1988).
21 P.G. de Gennes, Adv. Colloid Interface Sci. 27, 189 (1987).
22 J.M.H.M. Scheutjens, G.J. Fleer, J. Phys. Chem. 83, 1619 (1979).
23 M.H.M. Scheutjens, G.J. Fleer, Macromolecules 18,1882 (1985).
24 Y.Zhan, W.L. Mattice, Macromolecules 27, 7056 (1994).
25 K.F. Mansfield, D.N. Theodorou, Macromolecules 24, 6283 (1991).
26 S.K. Kumar, M. Vacatello, D.Y. Yoon, Macromolecules 23, 2189 (1990).
27 A.J. Yethiraj, Chem. Phys. 101, 2489 (1994).
28 E. Eisenriegler, K. Kremer, K. Binder, J. Chem. Phys. 77, 6296 (1982).
29 I.A. Bitanis, G. ten Brinke, J. Chem. Phys. 99, 3100 (1993).
30 J.H. Jang, W.L. Mattice, Polymer 40, 4685 (1999).
31 J. Baschnagel, K. Binder, Macromolecules 28, 6808 (1995).
32 P. Cifra, E. Nies, F.E. Karasz, Macromolecules 27, 1166 (1994).
33 A. Bellemans, J. Orban, J. Chem. Phys. 75, 2454 (1981).
34 E. Manias, H. Chen, R. Krishnamoorti, J. Genzer, E.J. Kramer, E.P. Ginnelis, Macromolecules 33, 7955, 2000.
35 M.S. Ozmusul, R.C. Picu, Polymer 43, 4657 (2002).
36 R.C. Picu, M.S. Ozmusul, M.S., submitted to Macromolecules, 2002.
37 R.B. Bird, C.F. Curtis, R.C. Armstrong O. Hassager, Dynamics of Polymeric Liquids: Kinetic Theory, Wiley-Interscience, New York, 1987.
38 J.Fish, P.Nayak, M.H.Holmes, Comp. Mech. 14, 323–338 (1994).
39 J. Fish, Q. Yu, K.L. Shek, Int. J. Numer. Methods Eng. 45(11), 1657–1679 (1999).

Index

q

r